Handbook of Pulsar Astronomy

Radio pulsars are rapidly rotating highly magnetised neutron stars. Studies of these fascinating objects have provided applications in solid-state physics, general relativity, galactic astronomy, astrometry, planetary physics and even cosmology. Most of these applications and much of what we know about neutron stars are derived from single-dish radio observations using state-of-the-art receivers and data acquisition systems. This comprehensive book is a unique resource that brings together the key observational techniques, background information and a review of the latest results, including the recent discovery of a double pulsar system. Useful software tools are provided which can be used to analyse example data, made available on a related website. This volume will be of great value not only to graduate students but also to researchers wishing to carry out and interpret a wide variety of radio pulsar observations.

DUNCAN LORIMER received his Ph.D. in 1994 from the University of Manchester for his work on the Galactic population of millisecond and normal pulsars. Since then he has been involved in numerous surveys for radio pulsars. From 1995 to 1998 he was a post-doctoral fellow at the Max-Planck-Institute for Radioastronomy (MPIfR) where, working with Michael Kramer, he discovered the first pulsars with the Effelsberg telescope. From 1998 to 2001 he was a staff scientist at the 305 m Arecibo telescope in Puerto Rico, USA. Since September 2001, he has held a Royal Society University Research Fellowship at Jodrell Bank Observatory. His main research interests are the origin and evolution of pulsars.

MICHAEL KRAMER received his Ph.D. from the University of Bonn, Germany, in 1995 for his work on pulsar observations with the 100 m Effelsberg radio telescope operated by MPIfR. In 1997, he received the Otto Hahn medal from the Max Planck Society, followed by a year as a visiting scholar at the University of California at Berkeley, USA. In September 1999 he took up a lectureship position at the University of Manchester where he is now Reader in Physics and Astronomy. He was the main editor of *Pulsar Astronomy: 2000 and Beyond* published by PASP. His main research interests cover pulsar timing, gravitational physics and the observations of pulsars at the highest radio frequencies ever used.

Both authors have extensive experience in using the largest radio telescopes for pulsar observations, including the telescopes at Jodrell Bank, Arecibo, Effelsberg, Green Bank, Parkes, GMRT and Westerbork. As a recent highlight they have both been involved in the discovery of the first double pulsar system.

Cambridge Observing Handbooks for Research Astronomers

Today's professional astronomers must be able to adapt to use telescopes and interpret data at all wavelengths. This series is designed to provide them with a collection of concise, self-contained handbooks which covers the basic principles peculiar to observing in a particular spectral region, or to using a special technique or type of instrument. The books can be used as an introduction to the subject and as a handy reference for use at the telescope, or in the office. They also promote an understanding of other disciplines in astronomy and a modern, multiwavelength, multi-technique approach to research. Although aimed primarily at graduate students and researchers, many titles in the series are of interest to keen amateurs and undergraduate students.

Series editors

Professor Richard Ellis, Institute of Astronomy, *University of Cambridge*

Professor John Huchra, Center for Astrophysics, *Smithsonian Astrophysical Observatory*

Professor Steve Kahn, Department of Physics, *Columbia University, New York*

Professor George Rieke, Steward Observatory, *University of Arizona, Tucson*

Dr Peter B. Stetson, Herzberg Institute of Astrophysics, *Dominion Astrophysical Observatory, Victoria, British Columbia*

HANDBOOK OF
PULSAR ASTRONOMY

D. R. LORIMER
JODRELL BANK OBSERVATORY

M. KRAMER
UNIVERSITY OF MANCHESTER

CAMBRIDGE
UNIVERSITY PRESS

CAMBRIDGE UNIVERSITY PRESS
Cambridge, New York, Melbourne, Madrid, Cape Town,
Singapore, São Paulo, Delhi, Tokyo, Mexico City

Cambridge University Press
The Edinburgh Building, Cambridge CB2 8RU, UK

Published in the United States of America by Cambridge University Press, New York

www.cambridge.org
Information on this title: www.cambridge.org/9780521828239

First published 2005

A catalogue record for this publication is available from the British Library

Library of Congress Cataloguing in Publication Data
Lorimer, D. R. (Duncan Ross), 1969–
    Handbook of pulsar astronomy / D. R. Lorimer & M. Kramer.
    p.   cm. – (Cambridge observing handbooks for research astronomers ; v. 4)
    Includes bibliographical references and index.
    ISBN 0 521 82823 6 (hardback : alk. paper)
    1. Pulsars – Handbooks, manuals, etc.   I. Kramer, M. (Michael), 1967–
II. Title.   III. Cambridge observing handbooks for research astronomers ; 4.

    QB843.P8L67 2005
    523.8′874 – dc22   2004057044

ISBN 978-0-521-82823-9 Hardback

Contents

Introduction

Radio pulsars – rapidly rotating highly magnetised neutron stars – are fascinating objects to study. Weighing more than our Sun, yet only 20 km in diameter, these incredibly dense objects produce radio beams that sweep the sky like a lighthouse. Since their discovery by Jocelyn Bell-Burnell and Antony Hewish at Cambridge in 1967 (Hewish *et al.* 1968), over 1600 have been found. Pulsars provide a wealth of information about neutron star physics, general relativity, the Galactic gravitational potential and magnetic field, the interstellar medium, celestial mechanics, planetary physics and even cosmology.

Milestones of radio pulsar astronomy

Pulsar research has been driven by numerous surveys with large radio telescopes over the years. As well as improving the overall census of neutron stars, these searches have discovered exciting new objects, e.g. pulsars in binary systems. Often, the new discoveries have driven designs for further surveys and detection techniques to maximise the use of available resources. The landmark discoveries so far are:

- The Cambridge discovery of pulsars (Hewish *et al.* 1968). Hewish's contributions to radio astronomy, including this discovery, were recognised later with his co-receipt of the 1974 Nobel Prize for Physics with Martin Ryle.
- The first binary pulsar B1913+16[1] by Russell Hulse and Joseph Taylor at Arecibo in 1974 (Hulse & Taylor 1975). This pair of neutron stars

---

1 Pulsars are named with a PSR prefix followed by a 'B' or a 'J' and their celestial coordinates. Pulsars discovered before 1990 usually are referred to by their 'B' names (Besselian system, epoch 1950). Later discoveries are only referred to by their 'J' names (Julian system, epoch 2000).

orbit each other every 7.75 h and, in about 200 Myr, will coalesce due to the emission of gravitational radiation at the expense of orbital energy. The measurement of orbital shrinkage (1 cm day^{-1}) due to this effect was the first experimental demonstration of the existence of gravitational waves. Hulse and Taylor received the 1993 Nobel Prize for Physics in recognition of their achievement.

- The first millisecond pulsar B1937+21 by Shrinivas Kulkarni, Donald Backer and collaborators at Arecibo (Backer *et al.* 1982). This remarkable neutron star spins at 642 Hz, very close to the theoretical rotation limit (see Chapter 3), and turns out to be a highly stable clock on short timescales; PSR B1937+21 remains the most rapidly spinning neutron star known, despite the subsequent discovery of over a hundred millisecond pulsars in sensitive searches. High-precision timing of millisecond pulsars has provided a wealth of applications in astronomy and physics (see Chapter 2).

- The first pulsar in a globular cluster M21 by Andrew Lyne and collaborators at Jodrell Bank (Lyne *et al.* 1987). Since this discovery, eighty pulsars have been found in twenty-four globular clusters and are being used as powerful probes of the physical properties of clusters and the pulsars therein. One recent application is the first detection of ionised gas in the cluster 47 Tucanae (Freire *et al.* 2001; see also Chapter 2).

- The first pulsar planetary system B1257+12 by Alexander Wolszczan and Dale Frail at Arecibo in 1990 (Wolszczan & Frail 1992). This remarkable system contains two Earth-mass planets and one of lunar mass (Wolszczan 1994) and was the first extra-solar planetary system to be discovered (see Chapter 2).

- The first triple system B1620–26: a pulsar, a white dwarf and a Jupiter-mass planet by Stephen Thorsett and collaborators in the globular cluster M4 (Backer *et al.* 1993a; Thorsett *et al.* 1993). This system highlights the rich variety of evolutionary scenarios possible in globular clusters (Sigurdsson *et al.* 2003; see also Chapter 2).

- The discovery of J0737–3039, the first 'double pulsar' system by Marta Burgay and collaborators in 2003 (Burgay *et al.* 2003, Lyne *et al.* 2004). This fascinating system, that consists of a 22.7 ms pulsar in orbit around a 2.77 s pulsar, is the first double neutron star binary in which both components have been observed as radio pulsars. As described in Chapter 2, it promises to place even more stringent constraints on strong-field gravitational theories than PSR B1913+16.

Present-day progress

Many of the most exciting pulsar discoveries in the past five years, including J0737–3039, are being made using the Parkes telescope. The scientific output of this relatively modest 64-m telescope currently dominates the flood of exciting pulsars being found, using a state-of-the-art 'multibeam' receiver system to collect data from thirteen independent points in the sky simultaneously. Indeed, over half of all known pulsars have been discovered by this system. Current highlights include three double-neutron-star binaries, a number of pulsars with ultra-high (10^{14} G) magnetic fields and over two dozen millisecond pulsars.

Progress elsewhere makes use of new and/or recently refurbished instruments, in particular the Green Bank Telescope, the Giant Metre Wave Radio Telescope and the upgraded Arecibo telescope. These sensitive instruments are being used to perform deep targeted searches for pulsars in supernova remnants and globular clusters. One exciting recent find is PSR J0514–4002, a 4.99 ms pulsar in a highly eccentric ($e = 0.89$) binary system in the globular cluster NGC 1851 (Freire *et al.* 2004).

Pulsar astronomy is a truly multi-wavelength field. Not only are neutron stars the only astronomical sources that are observed across the electromagnetic spectrum, they are one of the prime sources for gravitational wave observatories now coming online. Although this book deals mainly with radio observations, new high-energy instruments, e.g. *Chandra* and *XMM*, provide complementary windows in the electromagnetic spectrum to study pulsars and their environments (see Chapter 9).

Open questions

Some of the many open questions[2] in pulsar astronomy are:

- How many pulsars are in the Galaxy and what is their birth rate?
- How are isolated millisecond pulsars produced?
- How many pulsars are in Globular clusters?
- How many pulsar planetary systems exist?
- Do the magnetic fields of isolated neutron stars decay?
- What are the minimum and maximum spin periods for radio pulsars?
- What is the relationship between core collapse in supernovae and neutron star birth properties?

2 This list reflects our personal bias somewhat. Our apologies to those who have answered these questions to their satisfaction already, or have other unanswered questions which we have not included here!

- How many pulsar–black hole binaries exist?
- Are all magnetised neutron stars radio pulsars?
- How and where is the radio and high-energy emission produced?
- What is shape and structure of the radio beam?
- What is the role of propagation effects in pulsar magnetospheres?
- What is the composition of neutron star atmospheres, and how do they interact with the strong magnetic fields?

While we do not provide the answers in this book, it is hoped that the techniques described and the ever-increasing advances in observational sensitivity will go some way to expanding our knowledge in the future.

The future

Pulsar astronomy currently is enjoying one of the most productive stages in its relatively short (35 year) existence. The future of the field looks even brighter. As pulsar surveys continue to become more and more sensitive, the true character of the Galactic neutron star population is being revealed. The 305 m Arecibo radio telescope is being fitted with a multibeam receiver that will permit extremely sensitive large-scale surveys over the next 5–10 years that are expected to yield up to 1000 pulsars. Looking further ahead, the next-generation radio telescope – the Square Kilometre Array (SKA) – should be close to operational by 2015. This telescope, with a sensitivity of more than 100 times that of Parkes, is expected to detect essentially all of the $\sim 20,000$ active radio pulsars in the Galaxy, the emission beams of which intersect our line of sight. A full Galactic census should reveal a number of exciting systems for future study, including one of the 'holy grails' of pulsar astronomy: a pulsar orbiting a stellar-mass black hole (see Chapter 2).

Aims and layout of this book

There is now a growing need for astronomers to become proficient with a variety of techniques and instruments across the electromagnetic spectrum. Building on the excellent existing monographs describing the theory and observations of pulsars (Manchester & Taylor 1977; Michel 1992; Beskin, Gurevich & Istomin 1993; Lyne & Smith 2005) we describe the techniques of radio pulsar astronomy. Our aim is to make this field more accessible to a wider audience of astronomers.

In addition to describing the latest results and observational techniques and summarising relevant mathematical formulae, the book web site (see below and Appendix 3) with links to software packages and sample data sets. We aim to provide the reader with the necessary techniques to analyse radio pulsar observations. It is our hope that this combination will appeal as an introduction to graduate students and astronomers from other fields, as well as being a handy reference for seasoned pulsar observers. Although we strive to use the SI system of units wherever possible, we adopt other units (e.g. Gauss, parsec, Jansky) where in standard use by pulsar astronomers. In our compendium of useful formulae (see Appendix 2) we list the appropriate conversion factors for those who wish to convert these quantities into SI.

The first three chapters of the book discuss basic pulsar properties (see Chapter 1), uses of pulsars as physical tools (see Chapter 2) and some relevant theoretical background for interpreting observations (see Chapter 3). In Chapter 4, we discuss the effects of the interstellar medium on pulsar observations in some detail with particular attention to observable quantities. The most commonly used devices and techniques for pulsar data acquisition are then covered in Chapter 5. Our discussion of the fundamental observing techniques for single-dish pulsar astronomy is divided into pulsar searching (see Chapter 6), routine observations of known pulsars (see Chapter 7) and timing observations (see Chapter 8). Many of the techniques described in these chapters are brought into a single reference for the first time. Observations with multiple radio telescopes and instruments outside of the radio spectrum are reviewed briefly in Chapter 9. Appendices cover basic radio astronomy concepts (see Appendix 1), a collection of the most useful equations mentioned throughout the book (see Appendix 2) and a list of web-based resources which are available on the book web site (see Appendix 3). The index is designed to provide speedy access to key parts of the book and also serves as a glossary for some of the more commonly used abbreviations.

While this book can be read from cover to cover, it is perhaps best 'dipped into' for specific information as required. Each chapter contains a reading list for further references and details of relevant software freely available via the book web site

http://www.jb.man.ac.uk/~pulsar/handbook

We welcome all comments and feedback on this site and all contents of the book via email at the following address: handbook@jb.man.ac.uk.

Acknowledgements

We are deeply indebted to our colleagues for providing material used in some of the figures, making their data and software available or for invaluable feedback on earlier drafts of the chapters. In particular, our thanks go to Dimitris Athanasiadis, Don Backer, Ramesh Bhat, Dipankar Bhattacharya, Walter Brisken, Adam Chandler, Jim Cordes, Ali Esamdin, Andy Faulkner, Paulo Freire, Tim Hankins, Jon Hagen, George Hobbs, Aidan Hotan, Bryan Jacoby, Axel Jessner, Simon Johnston, Aris Karastergiou, Bernd Klein, Oliver Löhmer, Andrew Lyne, Olaf Maron, Maura McLaughlin, Peter Müller, Dipanjan Mitra, Dave Moffett, Steve Ord, Scott Ransom, Graham Smith, Ingrid Stairs, Steve Thorsett, Willem van Straten, and Norbert Wex. Our thanks also go to Jacqueline Garget at Cambridge University Press for her patience as we blatantly missed several 'deadlines' during the completion of this work. Frequent use was made of NASA's Astrophysical Data Service and the Los Alamos preprint server in the literature searches.

DRL would like to express his thanks to the Royal Society for financial support and for providing the intellectual freedom to pursue lengthy and, at times, seemingly never-ending projects such as this book. His wife Maura and their parrot Noel gave essential support and encouragement during the long months of late nights and early mornings of the winter of 2003/4 spent preparing the manuscript and in helping to edit in all the corrections from the proof reading stage. Without their love and patience, his contribution to this book would not have been possible.

MK would like to thank his colleagues and students for being patient at times when the work on the book seemed to drag on forever. Norbert Wex is thanked in particular for offering a seemingly endless stream of information about literature, profound background knowledge and countless enjoyable discussions. None of MK's contributions would have been remotely possible without the constant support from his wife Busaba and family. Without Busaba's love, reassurance and comfort but also her wake-up calls, many more nights of work would have been necessary.

1

The pulsar phenomenon

Observations of radio pulsars carried out over the past four decades have provided a wealth of information about the properties of neutron stars. In this chapter, we review briefly the current state of our knowledge based on the roughly 1600 pulsars now known. For the purposes of this discussion, we consider three main areas: (a) pulsar emission properties used to help understand and constrain mechanisms for the radio emission; (b) interstellar propagation effects on the pulsar signal; (c) pulsar population properties and their implications for the Galactic population of neutron stars.

1.1 Emission properties

The various properties of the observed emission from pulsars are explained most naturally by a simple picture in which the observed radiation is produced by the acceleration of charged particles along the field lines of highly magnetised rotating neutron stars (Gold 1968; Pacini 1968). In the following, we concentrate on the main results in support of this idea. The model will be explored further in Chapter 3.

1.1.1 The lighthouse effect

Figure 1.1 demonstrates the basic pulsar signal in the form of a recording of intensity received by a radio telescope as a function of time. This 22 s time series from PSR B0301+19 shows regularly spaced pulses with a repetition period $P = 1.38$ s. The now standard interpretation is that the pulses are produced by the *lighthouse effect* of an emission beam of a neutron star as it sweeps past our line of sight once per rotation. In this case, the rotation frequency $\nu = 1/P = 0.725$ Hz.

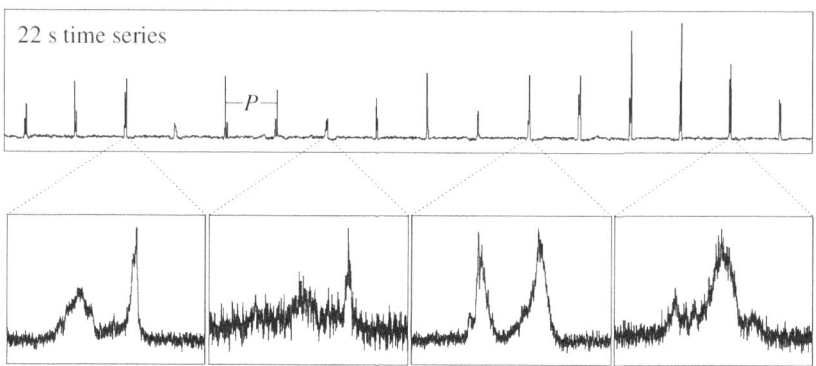

140 ms zoom in on individual pulses

Fig. 1.1. A 22 s time series from the Arecibo radio telescope showing single pulses from PSR B0301+19. Insets show expanded views of selected pulses.

When placed in context of the range of spin periods observed, PSR B0301+19 is a fairly 'common-or-garden' pulsar. The most rapidly rotating neutron star currently known is PSR B1937+21 (Backer *et al.* 1982) with a period of only 1.56 ms. The corresponding rotation frequency of 642 Hz for this star rules out some models of the equation of state for neutron stars (see Chapter 3). In contrast, the longest period observed for any radio pulsar so far is 8.5 s (or a sedate 0.12 Hz) for PSR J2144–3933 (Young, Manchester & Johnston 1999). The distribution in spin period is distinctly bimodal. As we discuss further in Section 1.3, the short-period 'millisecond pulsars' form a separate (older) population with different evolutionary histories from the long-period 'normal pulsars'. Throughout our tour of the emission properties, we shall compare and contrast the millisecond and normal pulsars where appropriate.

1.1.2 Integrated profiles

Pulsars are very weak radio sources. Even with the sensitivity of current radio telescopes, individual pulses such as those shown in Figure 1.1 are observable from only the strongest sources. Most pulsars require the coherent addition of many hundreds or even thousands of pulses together, a process known as *folding*, in order to produce an *integrated pulse profile* that is discernible above the background noise of the receiver.

Remarkably, for a given pulsar, even though the individual pulses have different shapes (see Figure 1.1), the integrated pulse profile is usually very stable for any observation at the same radio frequency. As such it can be thought of as a 'fingerprint' showing a cross-sectional cut through each neutron star's emission beam[1]. A selection of integrated pulse profiles is shown in Figure 1.2. The time scale required to achieve a stable integrated profile varies between a few hundred to a few thousand pulse periods (Helfand; Manchester & Taylor 1975; Rankin & Rathnasree 1995). This turns out to be a key property for timing observations which we discuss further in Chapter 8.

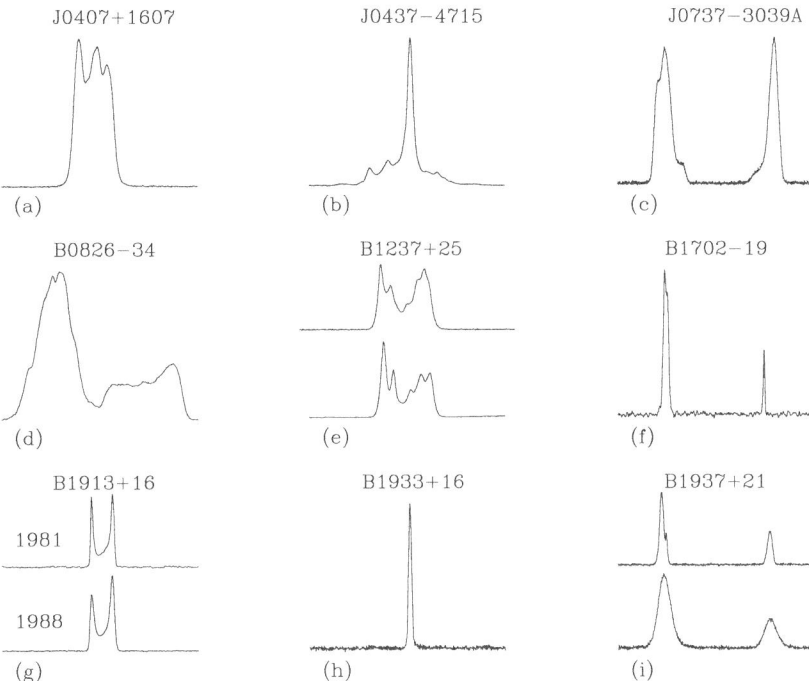

Fig. 1.2. Integrated pulse profiles for a sample of nine pulsars. With the exception of PSR B1237+25, each profile shows 360° of rotational phase. An expanded view of the profile of this pulsar is shown for two different modes. For PSR B1913+16 we show two 430 MHz profiles taken at separate epochs to highlight the profile evolution due to geodetic precession. These profiles are freely available on-line as part of the EPN database (see Appendix 3). For PSR B1937+21, we show two profiles observed with the Effelsberg radio telescope at 1.4 GHz. The upper profile was coherently de-dispersed and shows the true pulse shape. The lower profile was obtained with an incoherent filterbank system which results in much poorer time resolution (see Chapter 5).

---

1 As we explain in Chapter 3, the observed pulse shape depends critically on the size and structure of the emission beam, as well as the angle between our line of sight and the beam centre.

1.1.2.1 Pulse shapes

From the examples shown in Figure 1.2, it can be seen that the pulse shapes vary considerably in complexity. This sample has been chosen for illustrative purposes, often showing extreme cases. In the simplest case, as shown for PSR B1933+16, the pulse consists of a single component that is essentially Gaussian in form. A number of pulsars exhibit a characteristic double-peaked structure (e.g. PSR B1913+16) and even more complicated shapes with multiple components, e.g. PSR B1237+25.

While information about the structure of the emission beam can be inferred from an analysis of these pulse shapes (see Chapter 3), we note here simply that the pulse shapes fall into a number of different morphological categories depending on the number and placement of pulse components (Backer 1976; Rankin 1983a,b). Although it was thought initially that millisecond pulsar profiles were more complex than normal pulsars, studies identifying the number of components in profiles show these to be comparable for the two classes. On average, we observe 3 ± 1 components for normal pulsars and 4 ± 1 components for millisecond pulsars (Kramer *et al.* 1998).

In some cases, as shown dramatically for PSR B0826–34, pulsars show emission over the whole pulse period. This suggests strongly that the magnetic and rotation axes of such pulsars are aligned essentially and our line of sight remains inside the emission beam at all times.

1.1.2.2 Interpulses

The astute reader will notice three examples of *interpulses* in Figure 1.2. These are secondary pulses separated by about 180° from the main pulse. The simplest interpretation for this phenomenon is that the main pulse and interpulse originate from opposite magnetic poles of the neutron star. We see both poles if the magnetic and spin axes are almost orthogonally aligned. As shown in Chapter 3, polarisation measurements can be used to identify such cases. An alternative hypothesis, first presented by Manchester & Lyne (1977), is that interpulses represent the emission from extreme edges of a single wide beam. One example of this is the 22 ms relativistic binary pulsar J0737–3039A shown in Figure 1.2 that has two symmetric pulses (Burgay *et al.* 2003).

1.1.2.3 Profile evolution with time

Although almost all profiles are stable from epoch to epoch, there are a number of important exceptions. Temporal changes in pulse profiles have so far been observed to be caused by either a change in the orientation of the radio beam with respect to our line of sight due to some form of precession (see, for example, Kramer (1998), Stairs, Lyne and Shemar (2000)), or a transition between two or more competing pulse shapes. We shall return to the latter *mode-changing* phenomenon when discussing single pulse properties in Section 1.1.4.5

Evidence for geodetic precession in binary pulsars was first seen as a secular change in the pulse shape of the double neutron star binary B1913+16 (Weisberg, Romani & Taylor 1989). More recent observations show that both the pulse amplitude and separation are changing as the emission beam precesses out of our line of sight (Kramer 1998). Current estimates are that this pulsar will no longer be visible by the year 2025 (Istomin 1991; Kramer 1998). Geodetic procession has also been observed for another double neutron star binary B1534+12 (Arzoumanian *et al.* 1999; Stairs *et al.* 2000b) and is expected to be detectable soon for PSR J0737–3039A (Burgay *et al.* 2003). These binary systems are discussed in Section 1.3.3 and Chapter 2.

Evidence for free precession in isolated pulsars has been observed for PSR B1828–11 (Stairs, Lyne & Shemar 2000). This unique pulsar shows highly periodic and correlated variations in its pulse shape and rotation rate. The simplest explanation for this behaviour is that we observe a 0.3° change in the line of sight through the emission beam as the neutron star wobbles about its spin axis. Why no other radio pulsar displays this behaviour is currently unclear.

1.1.2.4 Profile evolution with frequency

As shown in Figure 1.3 for PSRs B1133+16 and J2145–0750, there is often a significant evolution in pulse shape as a function of observing frequency. Most normal pulsars, like PSR B1133+16, show a systematic *increase* in pulse width and separation of profile components when observed at *lower frequencies*. This effect was first proposed by Komesaroff (1970) to be a direct consequence of emission at higher frequencies being produced closer to the neutron star surface than at lower frequencies. This model is now known as *radius-to-frequency mapping* (Cordes 1978). As shown for the millisecond pulsar J2145–0750, this effect is far less pronounced and, indeed, millisecond pulsars in general show very little evolution of pulse width and component separation with frequency (Kramer *et al.* 1999b). As we shall see in Chapter 3, this is consistent with the idea that the size of the emission region of millisecond pulsars is much smaller than for normal pulsars.

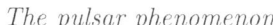

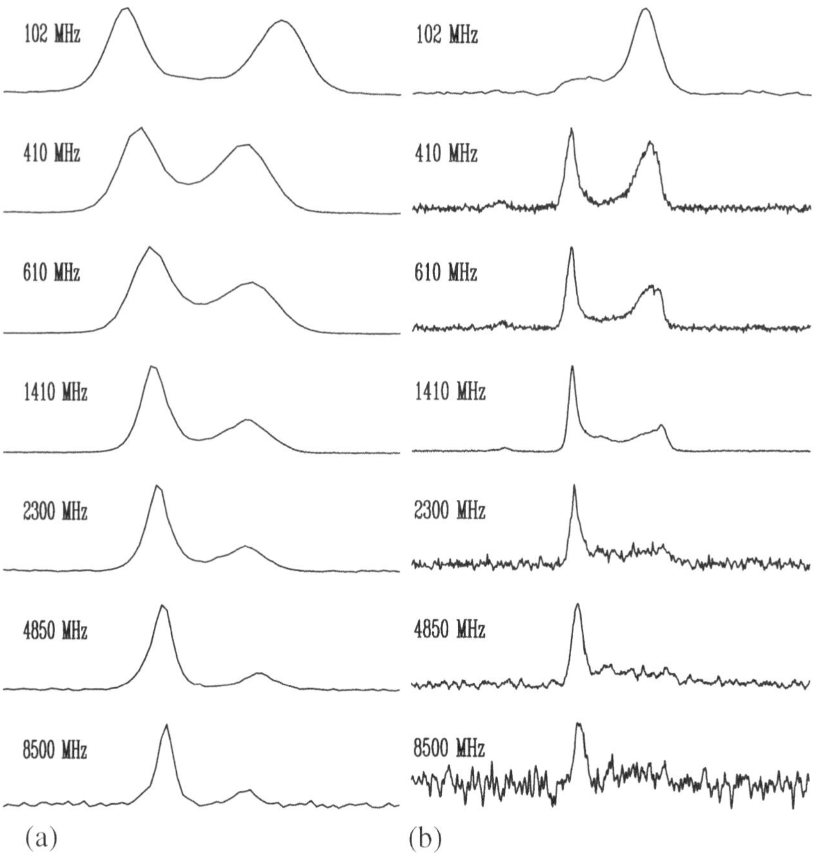

Fig. 1.3. Multi-frequency pulse profiles for two pulsars: (a) The 1.16 s pulsar B1133+16; (b) The 16 ms pulsar J2145–0750. In contrast for the profiles for J2145–0750 which show the full pulse period, the profiles for B1133+16 zoom in on the on-pulse region for clarity. These profiles are freely available on-line as part of the EPN database (see Appendix 3).

Another phenomenon often observed is the change in the number and/or relative intensity of the various pulse components as a function of frequency. This is clearly the case for both pulsars shown in Figure 1.3 and usually is related to geometrical factors and a different spectral index of emitting regions.

1.1.3 Flux density spectra

For properly calibrated data (see Section 7.3.3), the unit of *flux density* measured by the telescope is the Jansky, Jy (where $1 \ \text{Jy} \equiv 10^{-26} \ \text{W m}^{-2} \ \text{Hz}^{-1}$; see Appendix 1 for further discussion). Two useful measures of flux density are the peak flux density (the maximum intensity of a pulse profile) and the mean flux density (the integrated intensity of the pulse profile averaged over the pulse period). Generally speaking, pulsars are weak radio sources. For the current sample of 908 pulsars in the public domain catalogue[2] with flux densities measured at 1.4 GHz, the median value is 0.8 mJy with a range between 20 μJy and 5 Jy.

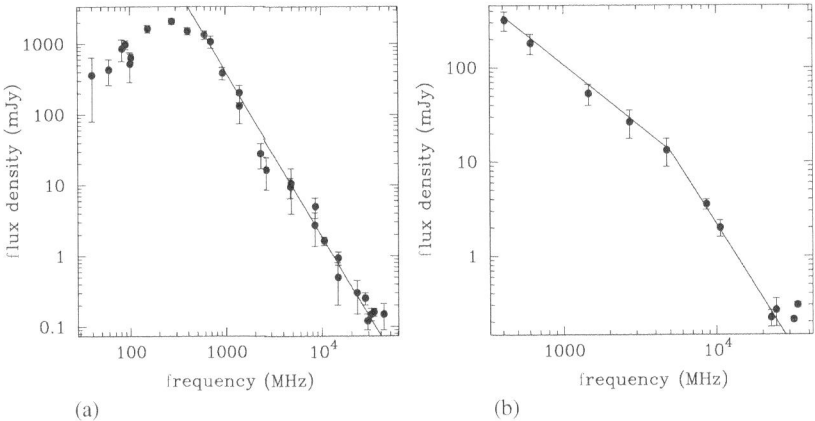

Fig. 1.4. Sample flux density spectra for two pulsars showing different types of spectral behaviour. (a) A low-frequency turnover in PSR B0329+54. (b) A broken power law fit and possible high-frequency turn-up in PSR B1929+10.

As shown in Figure 1.4, the mean flux densities of pulsars have a strong inverse dependence with observing frequency (see, for example, Sieber (1973) for an early study). For most pulsars observed above 100 MHz, the dependence can be approximated by a simple power law: $S_{\text{mean}}(f) \propto f^{\xi}$, where f is the observing frequency and ξ is the 'spectral index'. In recent work by Maron *et al.* (2000), about 5 per cent of the 281 pulsars studied showed a more complex spectral behaviour that required a two-component power law model. Other intriguing deviations from a single power-law behaviour are the roll over in the spectra often seen at low frequencies (see, for example, Malofeev (1996)) and the tantalising hints of a spectral increase or turn-up seen for a few pulsars at millimetre

2 http://www.atnf.csiro.au/research/pulsar/psrcat.

wavelengths (Kramer *et al.* 1996b). The highest radio frequency at which pulsars have been observed successfully at is 87 GHz (Morris *et al.* 1997).

For pulsar spectra that can satisfactorily fit a single power law, the range of observed spectral indices is broad: $0 \gtrsim \xi \gtrsim -4$, with a mean value of -1.8 ± 0.2 (Maron *et al.* 2000). The spectra of millisecond pulsars do not appear to be significantly steeper or more complex than normal pulsars. Independent studies by Kramer *et al.* (1998) and Toscano *et al.* (1998) find the mean spectral indices for millisecond pulsars to be consistent with the normal pulsar population.

1.1.4 Individual pulses

A closer inspection of the individual pulses in Figure 1.1 shows a rich variety of structure on a range of timescales. When sufficient signal to noise (S/N) is available, high time resolution observations of individual pulses provide a wealth of information about the emission process.

1.1.4.1 Microstructure

Early observations with high time resolution initially revealed structure on timescales below 200μs in PSR B0950+08 (Craft *et al.* 1968). Later observations with higher time resolution (see, for example, Bartel and Sieber (1978)) revealed features with a duration of a few microseconds. This phenomenon, known as *microstructure*, has been shown to be both broad-band (Rickett *et al.* 1975) and often quasi-periodic (Boriakoff *et al.* 1981). As shown by Lange *et al.* (1998), for observations with sufficient sensitivity, microstructure is detected commonly in normal pulsars, in which the fraction of individual pulses that exhibit microstructure is 30–70 per cent.

The smallest features resolved to date occur on nanosecond timescales in pulses from the 33 ms pulsar B0531+21 that powers the Crab nebula[3] (Hankins *et al.* 2003), shown in Figure 1.5. Simple light travel time arguments can be made to show that, in the absence of relativistic beaming effects, these incredibly bright bursts originate from regions less than 1 m in size. The demands on instrumentation required for such exquisite time resolution will be discussed in Chapter 5.

---

3 The Crab nebula is the remains of the supernova explosion observed by Chinese astronomers in 1054 AD (see, for example, Stephenson & Green (2002)). The discovery of PSR B0531+21 – usually known as 'the Crab pulsar' – provided one of the first observational links between pulsar birth and supernovae (Staelin & Reifenstein 1968).

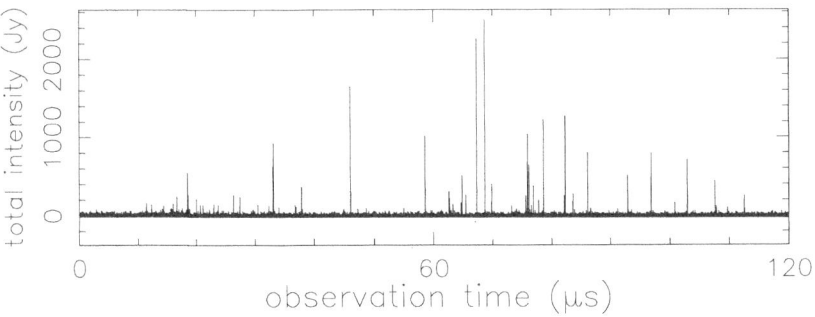

Fig. 1.5. A 120 μs window centred on a giant pulse from the Crab pulsar showing high-intensity nanosecond bursts. Figure provided by Tim Hankins.

1.1.4.2 Giant pulses

The nanosecond bursts observed in the Crab pulsar are the latest in a long series of observations aimed at understanding the so-called 'giant pulses'. Unlike the steady pulses seen in Figure 1.1, the Crab emits occasional giant pulses with an intensity up to 1000 times that of an individual pulse (see Cordes *et al.* (2004) and references therein) of which the nanosecond structure is now being resolved. Indeed, the Crab pulsar was first discovered through its giant pulse emission (Staelin & Reifenstein 1968) and for a long time was the only pulsar known to emit giant pulses. Advances in instrumentation and sensitivity resulted in the detection of giant pulses from the millisecond pulsars B1937+21 (Wolszczan, Cordes & Stinebring 1984; Sallmen & Backer 1995; Cognard *et al.* 1996) and B1821–24 (Romani & Johnston 2001) as well as the young Crab-like pulsar B0540–69 in the large Magellanic cloud (Johnston & Romani 2003). A possibly related class are the giant 'micro-pulses' seen in the young pulsars Vela[4] (Johnston *et al.* 2001) and PSR B1706–44 (Johnston & Romani 2002). There seems to be a connection between the giant pulses and the high-energy emission of these pulsars, which we discuss in more detail in Chapter 9.

1.1.4.3 Pulse nulling

Backer (1970a) first noticed that the emission from some pulsars appears to abruptly 'switch off' for many pulse periods. This behaviour is known as *nulling* and is shown to good effect for a recent observation of

---

4 Like the Crab, the 89 ms pulsar B0833–45 (Large, Vaughan & Mills 1968) is often simply known by its associated remnant, i.e. 'Vela'.

PSR B1944+17 in Figure 1.6. The duration of the nulls is observed to vary greatly from pulsar to pulsar, with the fraction of pulses not seen due to nulling ranging between essentially zero and unity! In the most extreme cases, the pulsar may be visible only in short bursts in between nulls making it very difficult to detect. For PSR B1944+17, the nulling fraction is about 55 per cent (Deich *et al.* 1986).

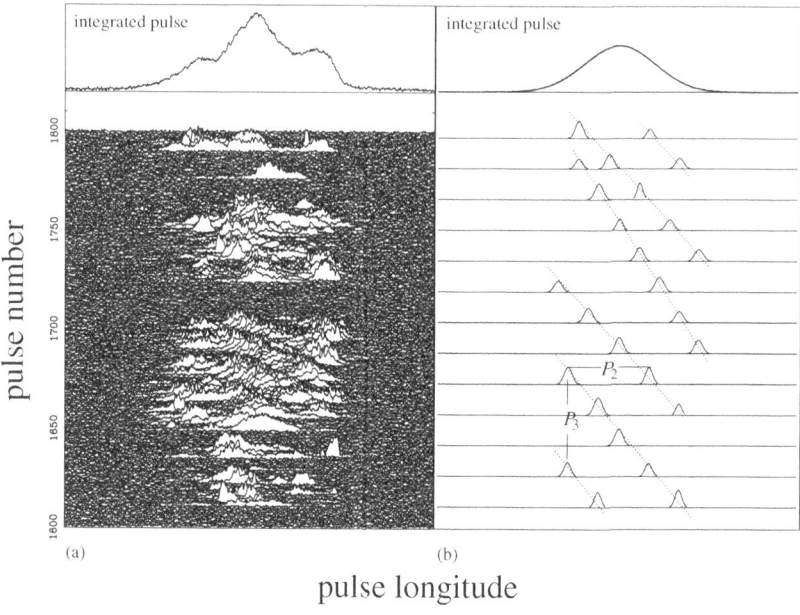

pulse longitude

Fig. 1.6. (a) A sequence of single pulses for PSR B1944+17 taken with the Arecibo telescope at 430 MHz showing both nulling and drifting sub-pulses. (b) Schematic view of the drifting sub-pulse phenomenon showing the periodicities P_2 and P_3 (see text). In some cases, a change in the drift rate is observed as indicated by the different slopes in the schematic. The top panel of each plot shows the integrated pulse profile.

An early analysis by Ritchings (1976) suggested that the null fraction was correlated inversely with the age of a pulsar. Biggs (1992) later found some evidence for an inverse correlation between the nulling fraction and pulse period. While these results are suggestive of a faltering of the emission process as the pulsar reaches the end of its active lifetime as a radio source, as pointed out by Rankin (1986), there is a large scatter in the observed nulling fractions so that a clear nulling-age effect is not obvious. In this regard, it is interesting to note that the only millisecond pulsar for which single pulses have been studied in detail, PSR J0437–4715, shows no evidence for any nulling (Vivekanand *et al.* 1998).

It is probably fair to say that a consensus on nulling is yet to be reached. Given the large increase in the number of nulling pulsars found in the recent Parkes multibeam survey (Manchester *et al.* 2001), further investigations are now being carried out.

1.1.4.4 Drifting sub-pulses

Also evident from Figure 1.6 are clearly ordered sub-pulses that appear to drift across the main pulse window at a fixed rate. These *drifting sub-pulses* were first noted by Drake and Craft (1968). Following Backer (1973) we represent this schematically in Figure 1.6 and, in addition to the pulse period $P = P_1$, identify the characteristic spacing between sub-pulses, P_2, and the period at which a pattern of pulses crosses the pulse window, P_3. Measurements of these periodicities will be discussed in Chapter 7.

An explanation was offered by Ruderman and Sutherland (1975) as being due to a rotating carousel of sub-beams within a hollow emission cone (see Chapter 3). This idea has gained significant observational credence with the cartographic mapping technique developed by Desphande and Rankin (1999) and the two-dimensional Fourier transform approach of Edwards and Stappers (2002) both of which allow the distribution and number of emitting regions to be determined (see Chapter 7 for further details).

1.1.4.5 Mode changing

If nulling can be viewed as the transition between an 'on' and 'off' state of the pulsar emission, it is likely to be related to a similarly abrupt phenomenon known as *mode changing*. First recognised by Backer (1970b) in PSR B1237+25, mode changing refers to the transition between two or more different forms of the integrated profile. The emission process usually favours one state (known as the *normal mode*) for most of the time and switches sporadically to the *abnormal mode(s)* at other times. The normal and abnormal modes of PSR B1237+25 are shown in Figure 1.2. For those pulsars exhibiting drifting sub-pulses, the drift rates are seen clearly to change when a mode switch occurs (see, for example, Taylor *et al.* (1975) and Fowler *et al.* (1981)). Although the mechanism is not understood, it seems likely that nulling, drifting and mode changing are related to highly ordered changes in the emission process.

1.1.5 Polarisation studies

The observations discussed so far have dealt with just the total intensity of the pulsar radiation. In fact, pulsars are among the most polarised of all known radio sources. Using a well-calibrated polarimeter (see Chapters 5 and 7), one obtains the four Stokes parameters, I, Q, U and V (see Appendix 1), where I is the total intensity, $L = \sqrt{Q^2 + U^2}$ is the linearly polarised intensity, and V is the circularly polarised intensity.

Based on a sample of 300 pulsars studied at 600 and 1400 MHz (Gould & Lyne 1998), the average degree of linear polarisation $\langle L/I \rangle$ is about 20 per cent. For circular polarisation, $\langle |V|/I \rangle$ is about 10 per cent. It should be noted that there is a great deal of variation about both these mean values, and individual pulsars can be up to 100 per cent polarised. Observations at higher frequencies generally show a decrease in the degree of linear polarisation (see, for example, Manchester (1971) and Xilouris *et al.* (1996)). An interesting result not seen at the lower frequencies is that the degree of circular polarisation observed at 5 GHz appears to be correlated strongly with the rate of loss of rotational kinetic energy (von Hoensbroech *et al.* 1998).

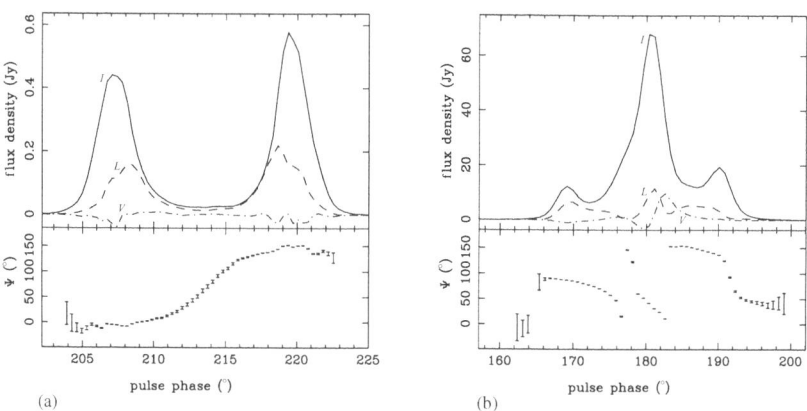

Fig. 1.7. Example 1.4 GHz polarisation profiles. (a) An Effelsberg observation of PSR B0525+21 (von Hoensbroech & Xilouris 1997). (b) A Jodrell Bank observation of PSR B0329+54 (Gould & Lyne 1998). In both cases, the total intensity I is the solid line, linearly polarised intensity L is the dashed line and circularly polarised intensity V is the dash-dotted line. The lower panels show the position angle of linear polarisation, Ψ. These profiles are freely available on-line as part of the EPN database (see Appendix 3).

Two examples of individual polarisation profiles are shown in Figure 1.7. In addition to the linearly and circularly polarised flux density, it is also of interest to study the position angle of linear polarisation $\Psi = \frac{1}{2}\tan^{-1}(U/Q)$ as a function of pulse phase. Normally the position angle of the linearly polarised flux varies throughout the pulse in a smooth and regular manner that is independent of observing frequency. The 'text-book' form of this curve is the characteristic S-shape as observed for PSR B0525+21. This behaviour was first noticed and interpreted by Radhakrishnan and Cooke (1969) in their *rotating vector model*. As discussed in Chapter 3, fits to this model allow the determination of the beam size and inclination angle with respect to the rotation axis.

As noted by several authors, not all pulsars show such smooth position angle behaviour. Millisecond pulsars in particular often exhibit very flat or 'disturbed' position-angle variations (see, for example, Xilouris *et al.* (1998)). From observations of single pulses in normal pulsars, Manchester *et al.* (1975) showed that several sources exhibit rapid jumps in their position angle curves, reaching nearly 90°, e.g. as shown for PSR B0329+54. Such discontinuities in an otherwise monotonic angle swing result from the presence of two *orthogonal polarisation modes* in the radiation. This represents a jump from one S-curve to another when the dominance of one mode over the other changes. Evidence for non-orthogonal jumps has also been observed (Backer & Rankin 1980). While this phenomenon is noticed mainly in the behaviour of the linear polarisation angle, Cordes *et al.* (1978) demonstrated that the circularly polarised radiation can undergo similar jumps between left-handed and right-handed senses of circular polarisation.

1.2 Propagation effects in the interstellar medium

As the signals from pulsars traverse the interstellar medium (ISM), three distinct propagation effects occur: (a) dispersion; (b) scintillation; (c) scattering. An additional effect is the Faraday rotation of the polarisation position angle by the Galactic magnetic field (see Chapter 2). Deferring a detailed discussion of the theory of all these effects until Chapter 4, we now summarise their observational consequences.

1.2.1 Pulse dispersion

One of the effects clearly noted in the discovery of pulsars (Hewish *et al.* 1968) was the effect of pulse dispersion. As shown in Figure 1.8, pulses observed at higher radio frequencies arrive earlier at the telescope than their lower frequency counterparts.

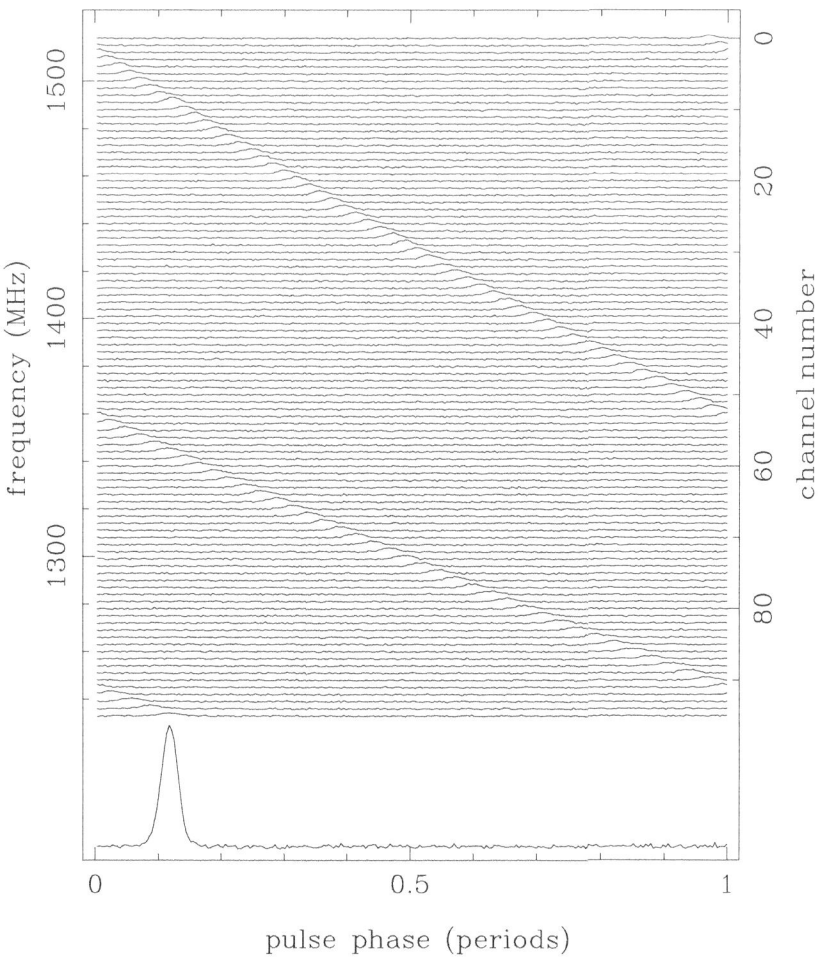

Fig. 1.8. Pulse dispersion shown in this Parkes observation of the 128 ms pulsar B1356–60. The dispersion measure is 295 cm^{-3} pc. The quadratic frequency dependence of the dispersion delay is clearly visible. Figure provided by Andrew Lyne.

Hewish *et al.* (1968) interpreted correctly the effect as the frequency dependence of the group velocity of radio waves as they propagate through the ionised component of the ISM. In this case, the delay in pulse arrival times is inversely proportional to observing frequency. The constant of proportionality is known as the dispersion measure (DM), the integrated column density of free electrons along the line of sight (see Chapter 4). As a *very crude* rule of thumb, if we assume a constant electron density (e.g. $n_e = 0.03$ cm^{-3} (Ables & Manchester 1976)), we can get a rough estimate of the distance to the pulsar $d \sim \text{DM}/n_e$ from a measurement of DM.

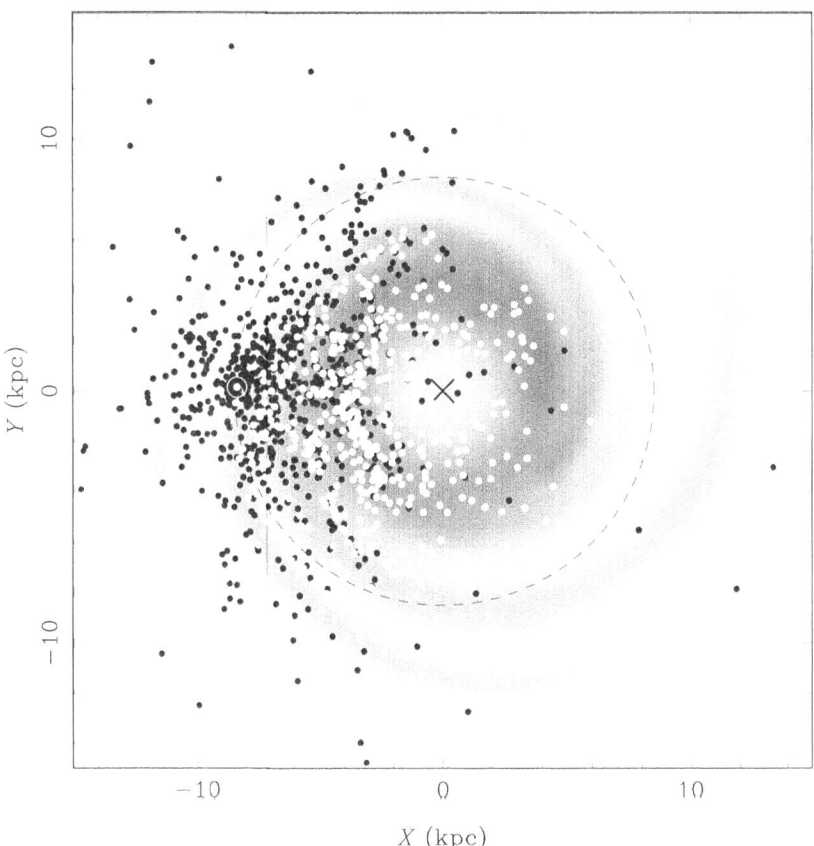

Fig. 1.9. The observed pulsar distribution (points) and model electron density distribution (grey scale) projected onto the Galactic plane. In these coordinates, the Sun is at $(-8.5, 0.0)$ kpc and the Galactic centre is at the origin. The darker points are those pulsars discovered in low-frequency (predominantly 430 MHz) searches. The lighter points are pulsars discovered in high-frequency (1.4 GHz) surveys of the Galactic plane. The electron density distribution is the NE2001 model by Cordes and Lazio (2002a). Darker areas correspond to regions of enhanced electron density. Figure provided by Bernd Klein.

A more sophisticated, and ultimately preferable, approach is to use pulsars with independently measured distances (e.g. via HI absorption described in Chapter 7 or parallax measurements described in Chapters 8 and 9) to calibrate a model for the distribution of n_e in the Galaxy (see, for example, Cordes and Lazio (2002a,b)). This distribution then can be integrated hopefully to obtain a more accurate distance estimate. Figure 1.9 shows the application of this model to the current sample of pulsars to project their distances onto the Galactic plane.

1.2.2 Scintillation

In addition to being magnetised and ionised, the ISM is highly turbulent and inhomogeneous. These irregularities produce phase modulations on the propagating pulsar signal that cause the observed intensity to fluctuate on a variety of bandwidths and timescales. This effect, shown in Figure 1.10 and known as interstellar scintillation, is similar to the familiar optical 'twinkling' of stars caused by the atmosphere of the Earth. It was recognised first as strong-intensity modulations in pulsar observations by Lyne and Rickett (1968). The basic theory of scintillation was developed by Scheuer (1968) who modelled the turbulent ISM as a *thin screen* of irregularities midway between the Earth and the pulsar. By considering the phase perturbations produced by such a screen, Scheuer demonstrated (see Chapter 4) that the intensity fluctuations should be correlated over a characteristic *scintillation bandwidth* $\Delta f \propto f^4$, where f is the observing frequency. This basic scaling agreed well with early observations (Rickett 1969).

1.2.3 Pulse scattering

Closely linked to scintillation are the characteristic 'exponential tails' in profiles of distant pulsars as shown in Figure 1.11. In the thin-screen model, this effect can be related directly to the variable path lengths and, hence, arrival times of the rays as they are scattered by irregularities in the ISM. As a result, there is a delayed arrival of the scattered rays that combine to broaden an otherwise intrinsically sharp pulse. In Chapter 4 we show that a simple model predicts the broadening as the convolution of the true pulse shape with a one-sided exponential with a $1/e$ time constant known as the *scattering time*, τ_s. Exponential fits to the scattered profiles shown in Figure 1.11 demonstrate that this simple model provides a good description of the phenomenon.

Measurements of τ_s for a large sample of pulsars show that it is correlated strongly with DM (see, for example, Bhat *et al.* (2004)). More

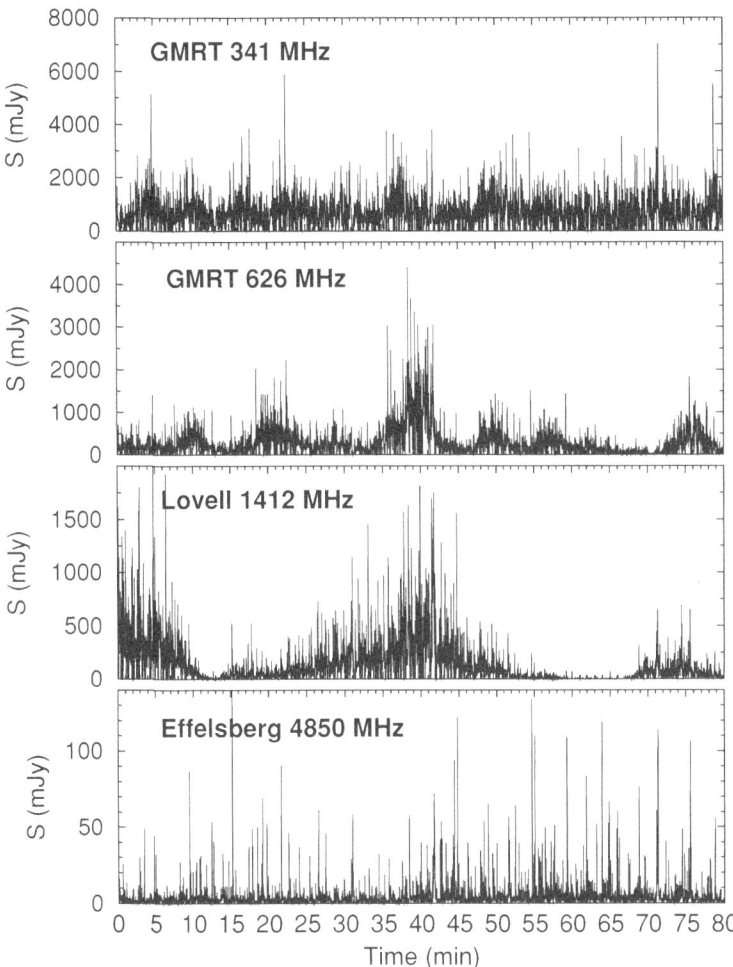

Fig. 1.10. Flux density of PSR B1133+16 simultaneously observed at four different frequencies with the Giant Metre-wave Radio Telescope (GMRT) at 341 MHz and 626 MHz, the Lovell telescope at 1412 MHz and the Effelsberg telescope at 4850 MHz. Strong intensity modulations due to interstellar scintillation are observed at 626 MHz and 1412 MHz. At the other frequencies the effect is less pronounced due to a relatively large observing bandwidth at 341 MHz and weak scintillation at 4850 MHz (see Chapter 4 for further details).

distant pulsars with larger DMs are, in general, more likely to be scattered. When searching for pulsars (see Chapter 6) this effect is highly undesirable since the scattering effectively stretches the true pulse shape which results in a reduction in the S/N ratio. However, as can be seen in Figure 1.11 the scattering time decreases when the pulsar is observed at higher frequencies. This is to be expected in the thin-screen model since $\tau_s \propto 1/\Delta f \propto f^{-4}$ (see Chapter 4), although recent work (see, for example, Löhmer *et al.* (2001)) suggests that the scaling index with frequency deviates from this simple model and can be as low as –2.8.

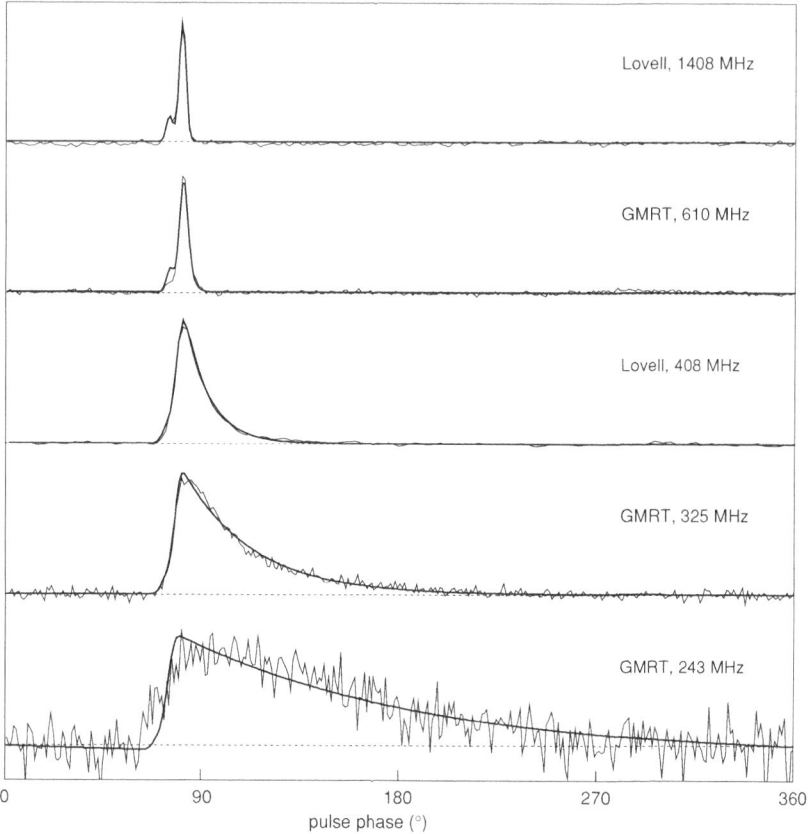

Fig. 1.11. Pulse profiles for PSR B1831–03 observed at five different frequencies with the Lovell telescope and the GMRT. These data show clearly the increasing effects of scatter broadening at lower frequencies. The solid lines show exponential model fits to the data (see Chapter 4 for further details). Figure provided by Oliver Löhmer.

The strong inverse frequency dependence of scattering clearly favours searches carried out at high frequencies. This is one of the main reasons why the Parkes multibeam Galactic plane survey, carried out at 1.4 GHz (Manchester *et al.* 2001), was so much more effective at finding highly dispersed and scattered pulsars in the inner Galaxy than previous lower-frequency searches.

1.3 Population properties

Aside from eight relatively bright pulsars in the large and small Magellanic clouds (Crawford *et al.* 2001) and eighty pulsars in Galactic globular clusters, the remainder of the ~ 1600 known radio pulsars largely populate the disk of our Galaxy. Since pulsars are intrinsically weak sources, and their pulses are distorted heavily by the propagation effects discussed above, the current sample of objects represents just the 'tip of the iceberg' of a much larger population of perhaps 10^6 active pulsars in the Galaxy. The observational bias is apparent in the projection of the known population onto the Galactic plane (see Figure 1.9) which shows the pulsars to be clustered around the Sun. Although the clustering of sources around the Sun seen in the left panel of Figure 1.9 would be consistent with Ptolemy's geocentric picture of the heavens, it is clearly at variance with what we now know about the Galaxy, where the massive stars show a radial distribution about the Galactic centre. Mindful of these selection effects, we summarise what is currently known about neutron star demography.

1.3.1 Galactic distribution and space velocities

From the sky distribution in Galactic coordinates in Figure 1.12, it is immediately apparent that pulsars are, like the massive O and B stars, concentrated strongly along the Galactic plane. This observation is consistent with the standard picture of the birth of neutron stars in the core collapse supernova explosions of massive stars. Currently, about twenty associations between radio pulsars and supernova remnants have been proposed (see Kaspi and Helfand (2002) for a recent review). In about a dozen of these cases, the supporting evidence in favour of a genuine association (e.g. proper motion and independent age/distance estimates) is very convincing. The remainder await further observational data to confirm or refute the associations.

From such a violent birth in supernovae, it is perhaps not surprising to learn that pulsars are high-velocity objects. This was first recognised by Gunn and Ostriker (1970) from the statistics of their height above and below the Galactic plane, z. The distribution in z is approximately exponential in form with $1/e$ scale height of about 300 pc, significantly larger than that for the massive progenitor stars. This implies that pulsars receive an impulsive 'kick' of roughly several hundred km s^{-1} at birth. The origin of these kicks is not yet clear, but may lie in small asymmetries in the supernova explosions (Lyne & Lorimer 1994). As we shall see in Chapter 2, proper motion measurements confirm these high inferred space velocities and show that a significant fraction of pulsars will escape the Galactic gravitational potential.

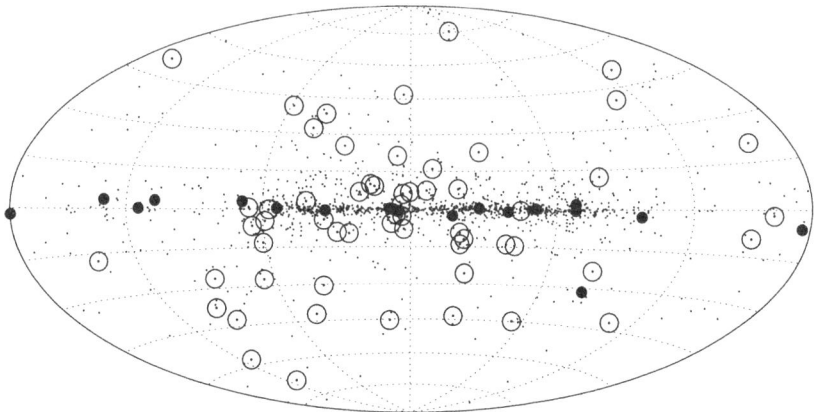

Fig. 1.12. Hammer–Aitoff projection showing the distribution of pulsars in Galactic coordinates. Pulsar–supernova remnant associations and millisecond pulsars are shown by the filled and open circles respectively.

Due largely to the scattering effects discussed above, millisecond pulsars are more strongly biased by observational selection effects than normal pulsars. As a result, we detect mostly local ($d < 3$ kpc) millisecond pulsars that therefore appear more isotropically distributed on the sky (Figure 1.12). While statistical studies suggest that the underlying population of millisecond pulsars is similar in size to the normal pulsars (see, for example, Lyne *et al.* (1998)), measurements of their proper motions show that their average space velocities are ~ 100 km s^{-1} (see, for example, Toscano *et al.* (1999)), significantly lower than those of the normal pulsars.

1.3.2 Spin evolution and the 'P–$\dot{P}$' diagram

Like many things in life, the observed emission from radio pulsars does not come for free. It takes place at the expense of the rotational kinetic energy of the neutron star. As a result, in addition to observing the pulsar's spin period, P, we also observe the corresponding rate of spin-down, $\dot{P}$. As we will describe in Chapter 8, both P and $\dot{P}$ can be obtained to very high levels of precision through timing measurements. These measurements give us unique insights into the spin evolution of neutron stars and are summarised on the 'P–$\dot{P}$ diagram' shown in Figure 1.13. Among other things, the diagram demonstrates clearly the distinction between the 'normal pulsars' ($P \sim 0.5$ s and $\dot{P} \sim 10^{-15}$ s s^{-1} and populating the 'island' of points) and the 'millisecond pulsars' ($P \sim 3$ ms and $\dot{P} \sim 10^{-20}$ s s^{-1} and occupying the lower left part of the diagram).

The differences in P and $\dot{P}$ imply fundamentally different ages and magnetic field strengths for the two populations. Deferring the details until Chapter 3, for a model[5] in which the spin-down is assumed to be due to magnetic dipole radiation, the inferred age $\tau \propto P/\dot{P}$ and magnetic field strength $B \propto \sqrt{P\dot{P}}$. Lines of constant B and τ are drawn on Figure 1.13 from which we infer typical magnetic fields and ages of 10^{12} G and 10^7 yr for the normal pulsars, and 10^8 G and 10^9 yr for the millisecond pulsars. The rate of loss of rotational kinetic energy $\dot{E} \propto \dot{P}/P^3$ (also known as the 'spin-down luminosity') is also indicated. As expected, these are highest for the young and millisecond pulsars.

A plausible 'evolutionary track' for a normal pulsar would be birth at short spin periods (in the upper left-hand region of Figure 1.13) followed by rapid spin down into the pulsar island on a timescale of 10^{5-6} yr, eventually becoming too faint to be detectable after 10^7 yr.

1.3.3 Binary systems

In addition to spin behaviour, a very important additional difference between normal and millisecond pulsars is binarity. Orbiting companions are observed around about 80 per cent of all millisecond pulsars but less than 1 per cent of all normal pulsars. As we detail in Chapter 8, timing measurements constrain the masses of orbiting companions that, often supplemented by observations at other wavelengths, tell us a great deal about their nature.

The orbiting companions are either white dwarfs, main sequence stars, or other neutron stars. Binary pulsars with low-mass companions ($\lesssim 0.5$ M$_\odot$ – predominantly white dwarfs) usually have millisecond spin periods and essentially circular orbits with orbital eccentricities in the range of

---

5 We shall also discuss the validity of this model in Chapter 3.

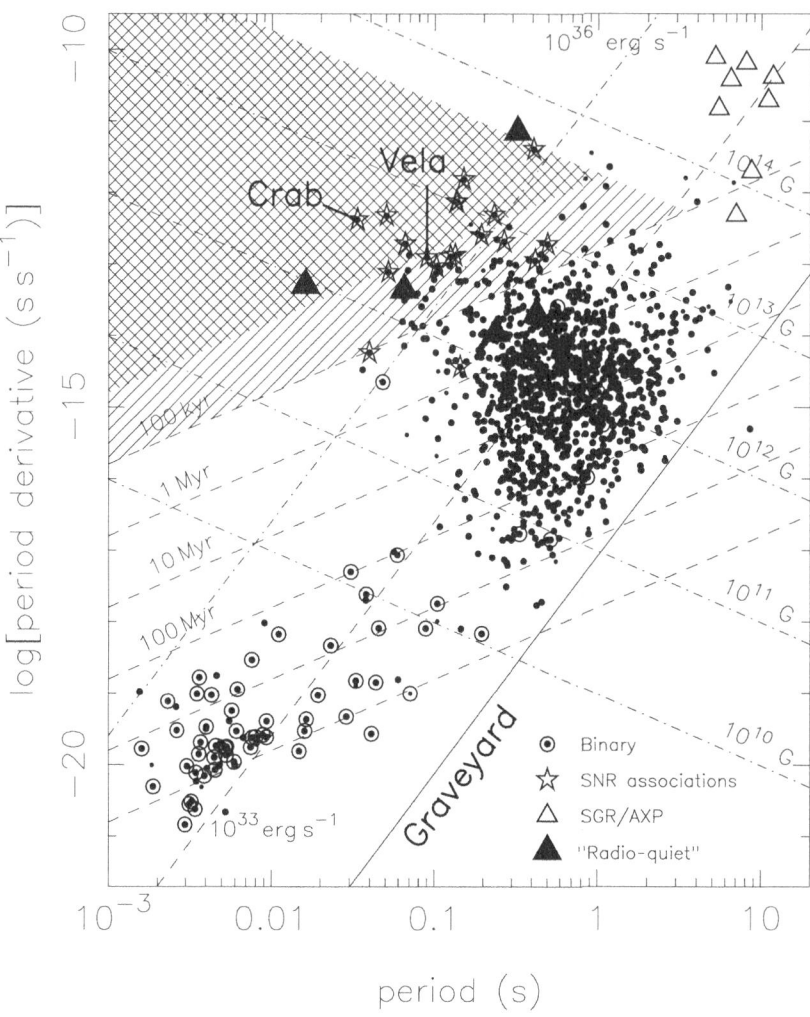

Fig. 1.13. The P–$\dot{P}$ diagram shown for a sample consisting of radio pulsars, 'radio-quiet' pulsars, soft-gamma repeaters (SGRs) and anomalous X-ray pulsars (AXPs). We discuss just the radio pulsars in this chapter (see Chapter 9 for details of the other objects). Pulsars known to be members of binary systems are highlighted by a circle. Pulsar–supernova remnant associations are shown by the stars. Lines of constant characteristic age τ, magnetic field B and spin-down luminosity $\dot{E}$ are also shown. The single hashed region shows 'Vela-like' pulsars with ages in the range 10–100 kyr, while the double-hashed region shows 'Crab-like' pulsars with ages below 10 kyr. The grey regions are areas where radio pulsars are not predicted to exist by theoretical models (see Chapter 3).

$10^{-5} \lesssim e \lesssim 10^{-1}$. Measurements of white-dwarf 'cooling ages' (see, for example, van Kerkwijk (1996)) agree generally with millisecond pulsar characteristic ages and support the idea that these binary systems have typical ages of a few Gyr. Binary pulsars with high-mass companions ($\gtrsim 1 \, M_\odot$ – neutron stars or main sequence stars) have larger spin periods ($\gtrsim 20$ ms) and are in more eccentric orbits: $0.1 \lesssim e \lesssim 0.9$.

The existence of binary pulsars can be understood by a simple evolutionary scenario which starts with two main-sequence stars (see, for example, Bhattacharya & van den Heuvel (1991)). The initially more massive (primary) star evolves first and eventually explodes in a supernova to form a neutron star. The high velocity imparted to the neutron star at birth and dramatic mass loss during the supernova usually is sufficient to disrupt most (90 per cent or more) binary systems (see, for example, Radhakrishnan and Shukre (1985)). Those neutron stars remaining bound to their companions spin down as normal pulsars for the next 10^6–10^7 yr. At some later time, the remaining (secondary) star comes to the end of its main sequence lifetime and begins a red giant phase. Depending on the orbital parameters of the system, the strong gravitational field of the neutron star attracts matter from the red giant, forming an accretion disk and making the system visible as an X-ray binary. A possible precursor to these systems is the pulsar-Be star binary B1259–63 (Johnston *et al.* 1992).

The accretion of matter transfers orbital angular momentum to the neutron star, spinning it up to short periods and dramatically reducing its magnetic field (Bisnovatyi-Kogan & Kronberg 1974; Shibazaki *et al.* 1989). A limiting spin period is reached due to equilibrium between the magnetic pressure of the accreting neutron star and the ram pressure of the infalling matter (Bhattacharya & van den Heuvel 1991; Arzoumanian, Cordes & Wasserman 1999). Such 'spun-up' neutron stars are often referred to in the literature as *recycled pulsars*. Unlike the young pulsars with high spin-down rates, the now weakly-magnetised recycled pulsars appear in the lower-left hand part of the P–$\dot{P}$ diagram and spin down much more gradually and over a longer timescale.

The ultimate fate of the binary system depends on the mass of the secondary star. We identify two main binary outcomes:

- *Double neutron star binaries*: A sufficiently massive secondary star will eventually undergo a supernova explosion forming a young, second neutron star. If the stars remain bound following this explosion, the resulting system is a pair of neutron stars in an eccentric orbit

with very different magnetic field strengths and spin-down properties. Until recently, only the first-born neutron star was observed in
these systems, e.g. B1913+16 and B1534+12. The recent discovery
of the double-pulsar binary J0737–3039   (Burgay *et al.* 2003; Lyne
et al. 2004), in which the second-born neutron star is observed as a
normal long-period pulsar is in excellent agreement with this scenario.

- *Millisecond pulsar-white dwarf binaries*: If the secondary star is not
  sufficiently massive to undergo core collapse, the mass transfer stage
  will last considerably longer (perhaps up to 10^8 yr). As a result, the
  neutron star will be spun up to millisecond spin periods. When the
  companion star eventually sheds its outer layers, the resulting system is a millisecond pulsar–white dwarf binary. Although consistent
  with the existence of the millisecond radio pulsars, strong independent
  support for this scenario was provided by the discovery of an X-ray
  millisecond pulsar in a low-mass binary system (Wijnands & van der
  Klis 1998; Chakrabarty & Morgan 1998).

The origin of isolated millisecond pulsars like PSR B1937+21 is still
not well understood. One possibility is that they ablated their companions by their strong relativistic particle winds during the X-ray phase
(Ruderman *et al.* 1989). Likewise, the end-products of more massive
binary systems are unclear. An outstanding problem is the paucity of
isolated recycled pulsars like PSR J0609+2130 which appears to be the
descendant of a high-mass binary that disrupted at the second supernova. According to the standard picture outlined above, such pulsars
should be much more common (Lorimer *et al.* 2004a).

1.4 Further reading

In this brief review, we have only highlighted some of the main results
from over 35 years of radio pulsar observations. A number of excellent
books and review articles provide further insights. Although now somewhat dated, Manchester and Taylor (1977) still provide one of the best
summaries of the early results in the field. The seminal papers by Rankin
(1983a,b; 1986) and Lyne and Manchester (1988) are essential reading
for those wishing to learn more about the basic emission properties and
profile classification schemes. An up-to-date review of the observations
of pulsar emission can be found in Smith (2003) and the most recently
revised edition of *Pulsar Astronomy* (Lyne & Smith 2005). Propagation
effects will be discussed further in Chapter 4. Excellent reviews of pulsar

demography and evolutionary scenarios can be found in Bhattacharya and van den Heuvel (1991) and Phinney and Kulkarni (1994).

2
Pulsars as physical tools

Pulsars are unique and versatile objects that can be used to study an extremely wide range of physical and astrophysical problems. In many respects, pulsars are a physicist's dream come true: besides testing theories of gravity one can study the gravitational potential and magnetic field of the Galaxy, the interstellar medium, stars, binary systems and their evolution, solid state physics and the interiors of neutron stars. Investigating the radio emission of pulsars provides insight into plasma physics under extreme conditions. This subject is particularly difficult, since a proper theoretical description of the observations is still lacking. Fortunately, for most applications, it is not necessary to understand how pulsars work; instead, it is sufficient usually to consider them as natural clocks emitting highly polarised coherent signals. To motivate the techniques described in subsequent chapters, we outline now a selection of the most wide-ranging applications of pulsars so far.

2.1 High-precision timing and time keeping

Most applications of pulsars involve a powerful technique known as *pulsar timing* - the measurement of the arrival time of photons emitted by the pulsar. Techniques to measure and model these arrival times will be described in detail in Chapter 8. The amount of useful information that can be extracted critically depends upon the measurement precision of the pulse arrival times which scales as the pulse width divided by the signal to noise (S/N) ratio. Bright millisecond pulsars with short periods and hence narrow pulse widths therefore are ideal tools for high-precision measurements. The current state of the art is the 5.75-ms pulsar J0437−4517, which yields a root mean square residual between measured and model arrival times of about 100 ns over an observing time of 1 yr or longer (van Straten *et al.* 2001). The corresponding rotational stability is similar to that achieved by the best atomic clocks. This amazing property is a direct result of the large rotational kinetic energy, E, of a millisecond pulsar and its relatively low spin-down energy loss rate, $\dot{E}$. For a typical neutron star moment of inertia of 10^{38} kg m^2 (see Chapter 3), the rotational kinetic energy of the millisecond pulsar B1937+21 is $E \sim 10^{45}$ J. The observed spin-down rate implies $\dot{E} \sim 10^{29}$ J s^{-1}.

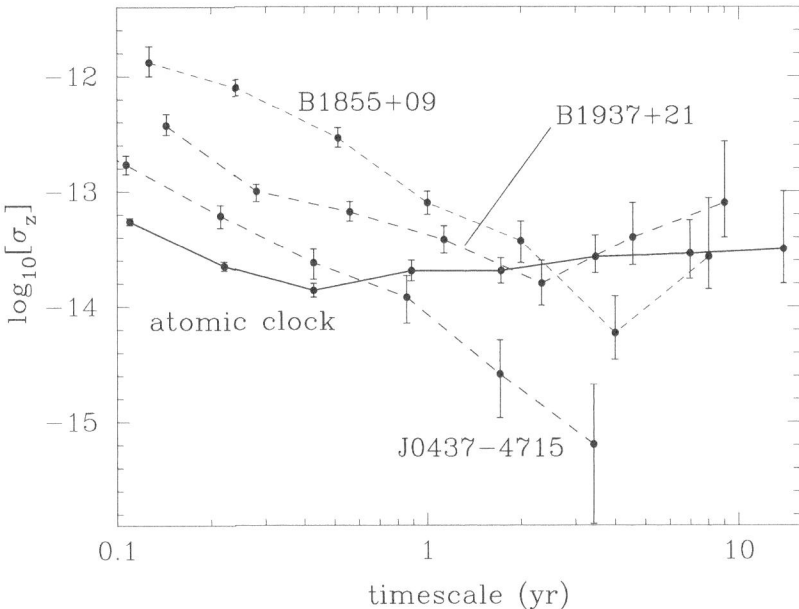

Fig. 2.1. The fractional stability of three millisecond pulsars compared to atomic clocks. Both PSRs B1855+09 and B1937+21 are comparable, or just slightly worse than, the atomic clock behaviour over timescales of a few years (data taken from Matsakis *et al.* 1997). More recent timing of the millisecond pulsar J0437–4715 indicates that it may be a more stable clock (unpublished data provided by Aidan Hotan).

The performance of different clocks can be compared using their fractional stabilities. After Matsakis *et al.* (1997) this fractional stability is expressed commonly by the parameter σ_z which resembles closely the *Allan variance* used by the clock community (Allan 1966). This quantity usually decreases over a period of time during which clock behaviour is predictable. Although millisecond pulsars are less stable than terrestrial atomic clocks on short timescales, their stabilities become similar over time spans of months and years (see Figure 2.1). Ultimately, rotational instabilities intrinsic to the pulsar known as *timing noise* can become important, and the internal stability of the pulsar clock is limited. The processes causing timing noise are not understood fully, but young pulsars show a much larger degree of timing noise than older ones. A trend seen among the sample of normal and millisecond pulsars indicates that the timing stability increases with smaller period derivatives and is, hence, greatest for millisecond pulsars (see Chapter 8).

If millisecond pulsars perform better than terrestrial clocks, it is conceivable that a time standard could be defined by timing several of the most stable millisecond pulsars against each other. However, each pulsar would have different spin and spin-down rates that are determined by rather poorly understood mechanisms (see Chapter 3). Additionally, while the current definition of the second is based on a hyperfine transition of Caesium 133, a pulsar time standard would not have such a link to a physical phenomenon that is reproducible in any laboratory. Unless pulsars can be monitored continuously with dedicated telescopes, terrestrial clocks would still be needed to interpolate between the observations and to tie pulsar time to some terrestrial endpoint times. While it may be possible that a smoothly running 'mean pulsar clock' could be used to compare different terrestrial time standards, the latest improvements in the development of fountain atomic clocks (see, for example, Bauch (2003)) suggests that physicists rather than astronomers will maintain the final word in time keeping.

2.2 Celestial mechanics and astrometry

As we shall discuss in Chapter 8, pulsar timing observations require transformation of the observed arrival times to the Solar System barycentre (SSB) using a planetary ephemeris. This transformation is amazingly sensitive to small changes in the assumed position and, when carried out correctly, allows the determination of the astrometric parameters of pulsars to phenomenal precision. For example, the positions of normal pulsars usually can be determined with uncertainties of a fraction of an arcsecond or less. Millisecond pulsars often permit sub-milliarcsecond precision; the best example so far is PSR J0437–4715, in which the positional uncertainty from timing observations is less than 10 μas (van Straten *et al.* 2001). For pulsars in globular clusters, the determination of accurate positions can be extremely useful for determining the physical properties of the cluster (see, for example, Phinney (1992)).

Comparing these so-called *timing positions*, which are determined in an ecliptic reference frame, with measurements using other techniques (see Chapter 8), is of interest for studying the relationship between different astrometric reference frames. For example, observing the pulsar with a radio interferometer (see Chapter 9), the pulsar's *interferometric position* can be determined in an equatorial reference frame that is tied to the rotation of the Earth and the position of quasars. The relative orientation of both coordinate systems depends on the position and inclination of the spin axis of the Earth, and a comparison of the timing and interferometric positions of a sample of pulsars can be used to tie these systems together (see, for example, Bartel *et al.* (1996)). In addition, a number of millisecond pulsars have white dwarf companions that can be detected optically. Measuring the precise position of the white dwarfs allows an additional tie to an equatorial reference frame based on optical observations of stars.

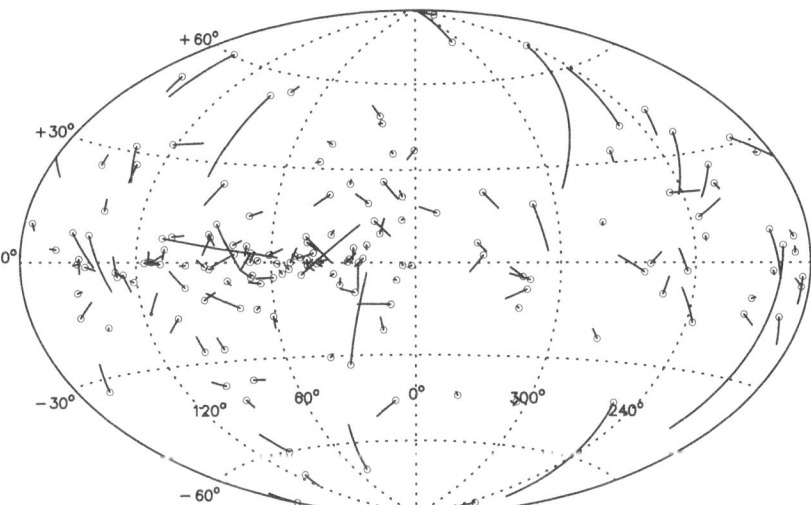

Fig. 2.2. Hammer-Aitoff projection showing the proper motion vectors of pulsars in Galactic coordinates. The current position of each pulsar is shown by the open circle. The solid lines are tracks showing the proper motion over the last million years. Figure provided by George Hobbs.

As shown in Figure 2.2, proper motions currently are measured for about 300 pulsars using either interferometric methods (see Chapter 9) or timing techniques (see Chapter 8). The measurement precision ranges between a few to tens of milliarcseconds per year for normal pulsars and often < 1 mas yr^{-1} for millisecond pulsars. Studies of these proper motions are extremely valuable for constraining the origin of pulsar velocities that are thought to have a broad spectrum extending up to and beyond 1000 km s^{-1} due to pre-supernova orbital velocities of binary progenitors and asymmetries in the supernova explosion (see Chapter 1). For the twenty-two pulsars known in the globular cluster 47 Tucanae, proper motion measurements from timing are now beginning to reveal the motion of these objects relative to the cluster centre of mass (Freire *et al.* 2001).

Parallax measurements, via timing and interferometric observations, are currently only available for a selection of bright nearby objects. They are vital, together with other independent distance estimates, for calibrating the pulsar distance scale (for a review, see Weisberg (1996)).

2.3 Gravitational physics in the strong-field regime

In astrophysical experiments – unlike terrestrial laboratories where we can modify the experimental set-up and control the environment – we remain observers, deriving all our information simply from the photons received. As a result, terrestrial experiments are typically more precise and, most importantly, reproducible in any other laboratory on Earth. However, when probing the limits of our understanding of gravitational physics, we are interested in extreme conditions that are not encountered on Earth. In some cases, it is possible to perform the experiment from a space-based satellite observatory. Indeed, Solar System tests provide a number of very stringent tests of general relativity (see, for example, Will (2001)). However, no such experiments made or proposed for the future will ever be able to test the strong-field limit. For such studies, binary pulsars are, and will remain, the only way to test the predictions made by general relativity or competing theories of gravity.

2.3.1 Tests using double neutron star binaries

Since neutron stars are very compact massive objects, double neutron star (DNS) binaries can be considered as almost ideal point sources for testing theories of gravity in the strong-gravitational field limit. With the recent discovery of three DNS binaries (Burgay *et al.* 2003; Champion *et al.* 2004; Faulkner *et al.* 2004) eight such systems are now known (see Table 2.1) with orbital periods ranging from 2.4 h to 18.8 days.

For all these DNS systems, the orbital separation is much larger than

Table 2.1. *Table summarising the properties of the eight DNS binaries currently known. From left to right we list the pulsar name, spin period (P), binary period (P_b), projected semi-major axis ($a_p \sin i$), orbital eccentricity (e), rate of advance of periastron ($\dot{\omega}$) and number of possible tests of general relativity using post-Keplerian (PK) parameters currently measured (see text).*

PSR	P (ms)	P_b (d)	$a_p \sin i$ (light-s)	e	$\dot{\omega}$ (deg yr^{-1})	N_{test}
J0737–3039	22.7/2770	0.102	1.42/1.51	0.09	16.9	4
B1534+12	37.9	0.421	3.73	0.27	1.76	2
J1518+4904	40.9	8.63	20.0	0.25	0.01	–
J1756–2251	28.5	0.320	2.76	0.18	2.60	–
J1811–1736	104.2	18.8	34.8	0.83	0.01	–
J1829+2456	41.0	1.18	7.24	0.14	0.30	–
B1913+16	59.0	0.323	2.34	0.62	4.23	1
B2127+11C	30.5	0.335	2.52	0.68	4.46	1

the size of the compact companion. As a result, neither mass transfer nor tidal effects are present and the system can be treated as a simple pair of point masses that can be compared readily to predictions by theories of gravity. An excellent example is the original binary pulsar B1913+16 in which the orbital period is observed to decrease at the rate predicted by general relativity due to gravitational-wave damping (Taylor and Weisberg 1989; Weisberg and Taylor 2003, 2004). This phenomenal measurement, shown in Figure 2.3, corresponds to an orbital shrinkage of 1 cm day^{-1} and provided the first evidence for the existence of gravitational radiation as predicted by general relativity.

The tests of general relativity for B1913+16 and other DNS binaries are made possible by the measurement of 'post-Keplerian' (PK) parameters – relativistic additions to the standard Keplerian orbital parameters. As we discuss in detail in Chapter 8 (see also Damour and Taylor (1992)), for point masses with negligible spin contributions, the PK parameters can be written as functions of only the pulsar and companion masses, m_p and m_c, and the Keplerian parameters. If two PK parameters are measured, a given theory of gravity will produce values for the pulsar and companion mass. More generally, a measurement of n PK parameters describes n curves in the m_p–m_c plane, the shapes and positions of which depend on the theory of gravity. For any viable

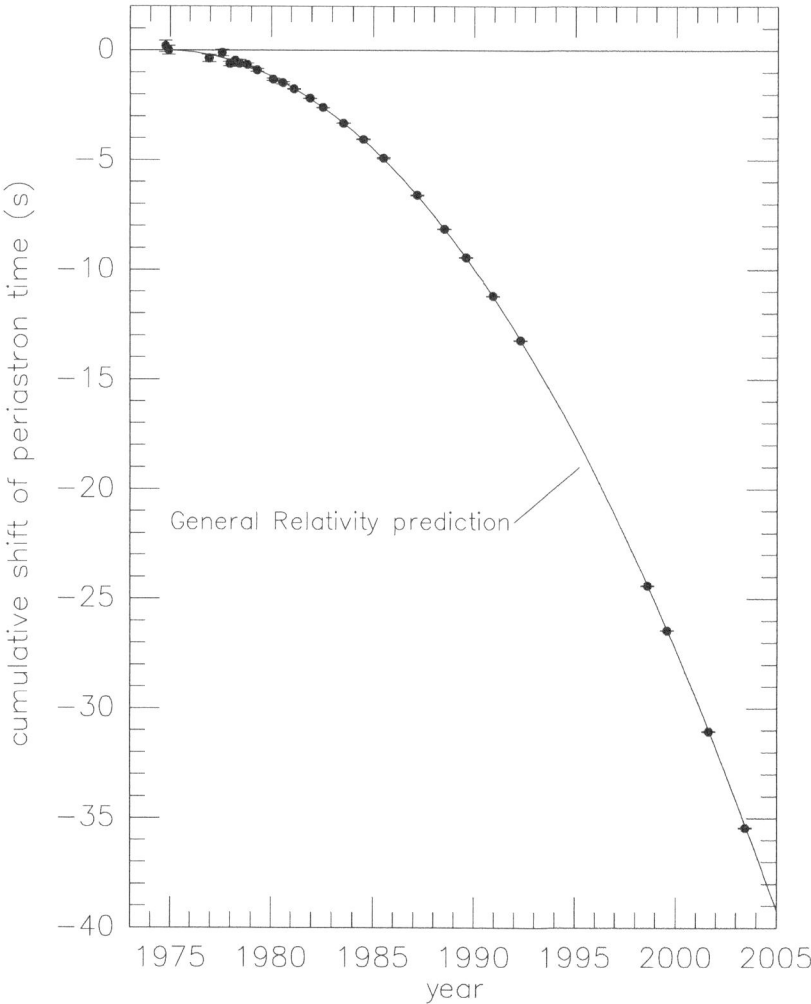

Fig. 2.3. The shift in the periastron passage of the binary pulsar B1913+16 plotted as a function of time. The non-zero shift results from orbital energy loss due to the emission of gravitational radiation. The agreement between the data, now spanning almost 30 yr, and the curve predicted by general relativity is better than 0.5 per cent. Figure provided by Joel Weisberg and Joe Taylor.

theory of gravity, all n curves should meet in a single point, allowing us to perform $N_{\text{test}} = (n - 2)$ possible tests of general relativity.

The most beautiful example of this approach is the recently discovered double pulsar system J0737–3039, a 22.7 ms pulsar (hereafter referred to as 'A' (Burgay *et al.* 2003)) in orbit around a 2.8 s pulsar (hereafter, 'B' (Lyne *et al.* 2004)). After only 12 months of observations, timing measurements for A already provide five PK parameters. The resulting $m_A - m_B$ diagram is shown in Figure 2.4. However, the detection of the

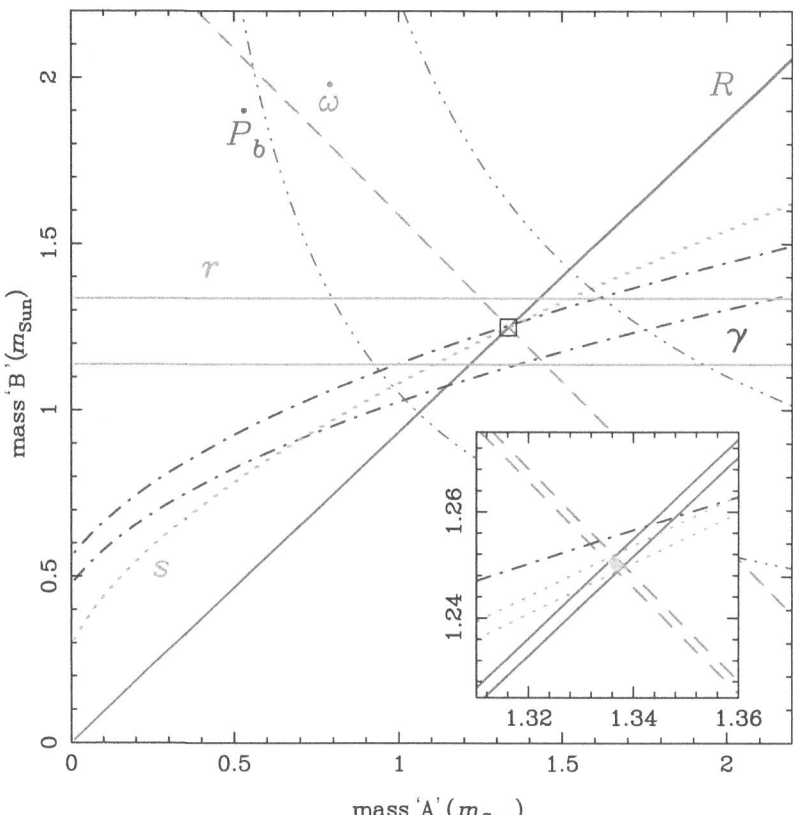

Fig. 2.4. 'Mass–mass' diagram showing the observational constraints on the masses of the neutron stars in the double pulsar system J0737 3039. The shaded regions are those that are excluded by the Keplerian mass functions of the two pulsars. Further constraints are shown as pairs of lines enclosing permitted regions as predicted by general relativity: (a) the measurement of $\dot{\omega}$ gives the total system mass $m_A + m_B = 2.59$ M$_\odot$; (b) the measurement of the mass ratio $R = m_A/m_B = 1.07$; (c) the measurement of the gravitational redshift/time dilation parameter γ; (d) the measurement of the two Shapiro delay parameters r and s; (e) the detection of an orbital decay, $\dot{P}_b$, due to gravitational wave emission. Inset is an enlarged view of the small square encompassing the intersection of the three tightest constraints (see text).

B pulsar opens up opportunities to go well beyond what is possible with previously known DNSs. A large part of the $m_A - m_B$ plane is forbidden by the individual Keplerian mass functions (see Section 8.3.1.4) of A and B due to the requirement that the angle between the plane of the orbit and the plane of the sky $i < 90°$. In fact, with a measurement of the projected semi-major axes of the orbits of both A and B, we obtain a precise measurement of the mass ratio $R = m_A/m_B$ of the two stars which provides a further constraint displayed in Figure 2.4.

For every realistic theory of gravity, we can expect the mass ratio, R, to follow this simple relation (Damour and Taylor 1992). Most importantly, the R-line is not only theory-independent, but also independent of strong-field (self-field) effects, which is not the case for PK parameters. This provides a stringent and new constraint for tests of gravitational theories as any intersection of the PK parameters *must* be located on the R-line in Figure 2.4. With five PK parameters already available, this additional constraint makes J0737–3039 the most overdetermined DNS system to date in which the most relativistic effects can be studied in the strong-field limit. With a more precise measurement of orbital period decay in the future, more tests can be performed than with any other system, assuming that certain kinematic effects (see Section 8.3.4.1) can be determined or at least controlled.

The position of the allowed region in the mass–mass diagram also determines the inclination of the plane of the orbit and the plane of the sky, $i = 88 \pm 1°$, and shows that the system is observed very nearly edge on. An independent, although less precise, measurement of i has recently been made using scintillation speed measurements (Ransom *et al.* 2004); (see Chapters 4 and 7 for a discussion of this technique).

Due to space-time curvature near massive objects, the spin axes of both pulsars will precess about the total angular momentum vector, changing the orientation of the pulsar as seen from Earth (Damour and Ruffini 1974). For the orbital parameters of J0737–3039, general relativity predicts periods of this *geodetic precession* of only 75 yr for A and 71 yr for B. Hence, the relative orientation of the spin axes of the pulsars within the system geometry is expected to change on short timescales. This should lead to measurable changes in the profiles of A and B (as has already been observed for B1913+16 (Weisberg *et al.* 1989; Kramer 1998), and perhaps also to measurable changes in the aberration effects due to the rotation of A and B (see Section 8.3.2.6).

The relativistic effects observed in the double pulsar system are so large that corrections to higher post-Newtonian order may soon need to be considered. For example, the rate of periastron advance, $\dot{\omega}$, may be measured precisely enough to require terms of second post-Newtonian order to be included in the computations (Damour and Schäfer 1988). Moreover, in contrast to Newtonian physics, general relativity predicts that the spins of the neutron stars affect their orbital motion via spin–orbit coupling. This effect would be visible clearest as a contribution to the observed $\dot{\omega}$ in a secular (Barker and O'Connell 1975a), and periodic fashion (Wex 1995b). For the J0737−3039 system, the expected contribution is about an order of magnitude larger than for PSR B1913+16 (Lyne *et al.* 2004). As the exact value depends on the pulsars' moment of inertia, a potential measurement of this effect allows the moment of inertia of a neutron star to be determined for the first time (Damour and Schäfer 1988). Such a measurement would be invaluable for studies of the neutron star 'equation-of-state' and our understanding of matter at extreme pressure and densities (see Chapter 3).

2.3.2 Tests using neutron star–white dwarf systems

Like the DNS binaries, most neutron star–white dwarf (NS–WD) binaries can be considered as pairs of point masses. While significant measurements of PK parameters can be obtained in a few cases (e.g. Lyne and Bailes 1990; Kaspi *et al.* 2000b; van Straten *et al.* 2001; Bailes *et al.* 2003), relativistic effects for NS–WDs are generally much smaller than for DNS binaries. Nevertheless, NS–WD systems can also be used to test certain aspects of gravitational theories which we now outline.

Unlike general relativity, some alternative theories of gravity (e.g. tensor–scalar theories) predict effects that depend strongly on the difference between the gravitational self-energy of two orbiting bodies. While a number of these effects can be constrained significantly using Solar System tests (see, for example, Will (2001)), this applies only to the weak-field limit. Indeed, it has been shown that one can develop alternative tensor–multiscalar theories that would pass all solar system tests but only show deviations from general relativity in the strong-field limit (Damour and Esposito Farèse 1992; 1996). Binary pulsars sometimes can provide equally stringent limits to the weak-field solar system tests in the strong-field regime. This provides a further test of any plausible gravitational theory, including general relativity.

The gravitational self-energy of a body of mass M and radius R is normally expressed in units of its rest mass energy, i.e. $\epsilon = E_{\text{grav}}/Mc^2 \sim -GM/Rc^2$, where G is the gravitational constant and c is the speed of light. This difference is particularly large for a neutron star ($\epsilon \sim -0.15$) and a white dwarf ($\epsilon \sim -10^{-4}$). Similarly, while general relativity predicts a quadrupole moment as the lowest contribution to gravitational-wave emission, some alternative theories predict the existence of gravitational dipole radiation. Dipolar effects would be most noticeable in NS–WD binaries due to the large differences in their gravitational self-energies. The most stringent limit to date on the gravitational dipole radiation is placed by timing observations of the NS–WD binary J1012+5307 (Lange *et al.* 2001).

Many of the testable effects can be related to possible violations of the *strong equivalence principle* (SEP) which is satisfied in general relativity but not necessarily in alternative theories. Unlike the weak equivalence principle (which dates back to Galileo's demonstration that all matter free falls in the same way) and the Einstein equivalence principle from special relativity (which states that the result of a non-gravitational experiment is independent of rest frame velocity and location), the SEP states that free fall is completely independent of gravitational self energy, ϵ. If the SEP is violated, then the ratio of inertial mass to gravitational mass of a test particle differs from unity by an amount $\Delta = \eta\epsilon + \eta'\epsilon^2 + \cdots$, where η parameterises the violation in terms of ϵ.

A violation of the SEP would mean that two bodies of unequal mass would fall differently in an external gravitational field. This effect – now known as the *Nordvedt effect* – was first proposed for the Earth–Moon system in the gravitational field of the Sun where an SEP violation would result in a 'polarisation' of the orbit in the direction of the Sun (Nordvedt 1968a, b). While lunar laser ranging places fairly stringent constraints ($|\eta| < 0.0016$) (see Will (2001) for a review), since $\epsilon_{\text{Earth}} \sim -10^{-10}$ and $\epsilon_{\text{Moon}} \sim -10^{-11}$, this applies only to the weak-field regime. However, the analogous influence of the Galactic gravitational field on the orbits of NS–WD binaries, which have significantly different self energies, tests the SEP in the strong-field regime. Long-period circular orbit binaries, as observed for many of the NS–WD systems, are particularly sensitive probes (Damour and Schäfer 1991). Using an ensemble of NS–WD binaries, Wex (1997b; 2000) found that $\Delta < 0.009$. This limit has recently been improved following the discovery of two long-period circular orbit NS–WD systems J0407+1607 and J2016+1947 (Lorimer and Freire 2004).

2.3.3 A pulsar timing array

A different aspect of gravitational physics could be probed by monitoring a large sample of millisecond pulsars. This so-called *pulsar timing array* (PTA) offers the means to detect the stochastic gravitational wave background that is expected, e.g. from the merger of massive black holes in early galaxy formation or in various cosmological theories (Foster & Backer 1990; Jaffe & Backer 2003). The PTA would be used to search for (correlated) structures in the timing residuals of millisecond pulsars distributed across the sky. Since the PTA is sensitive to low-frequency gravitational waves, it is complementary therefore to current ground-based gravitational wave detectors such as GEO600 (Danzmann *et al.* 1995), VIRGO (Caron *et al.* 1998), LIGO (Abramovici *et al.* 1992) and, as shown in Figure 2.5, even future detectors such as Advanced LIGO (Barish 2000) and LISA (Danzmann 2000).

One of the limitations of the PTA is the distribution of known millisecond pulsars across the sky that can be timed with sufficient accuracy. Current estimates suggest that timing precisions of 100 ns need to be obtained routinely for a sample of millisecond pulsars located in various different directions on the sky. The arrival times of the millisecond pulsars ultimately should be analysed simultaneously in order to detect the global terms present as correlated spatial structures. An internationally co-ordinated effort is underway to develop the techniques and to provide the data for the PTA (see also Chapter 8).

2.4 Extrasolar planets

The smallest bodies found orbiting around pulsars so far are the three planets detected around the 6.2 ms pulsar B1257+12 (Wolszczan & Frail 1992). The analysis of the pulse arrival times reveals one planet of Lunar mass (planet A) and two Earth-mass planets (planets B and C) with orbital periods of 25, 66 and 98 days, respectively. Rasio *et al.* (1992) and Malhotra *et al.* (1992) predicted that the 3:2 commensurability of the orbital periods of planets B and C should lead to deviations from a simple Keplerian model. These resonances have indeed been detected in subsequent timing analyses (Wolszczan 1994; Konacki *et al.* 1999) and provide irrefutable evidence for the planetary interpretation. This unexpected discovery of the first extrasolar planetary system[1] and demonstrates the

1 Although the planets around PSR B1257+12 predate the subsequent discoveries of over 100 Jupiter-mass planets around main-sequence stars using optical techniques,

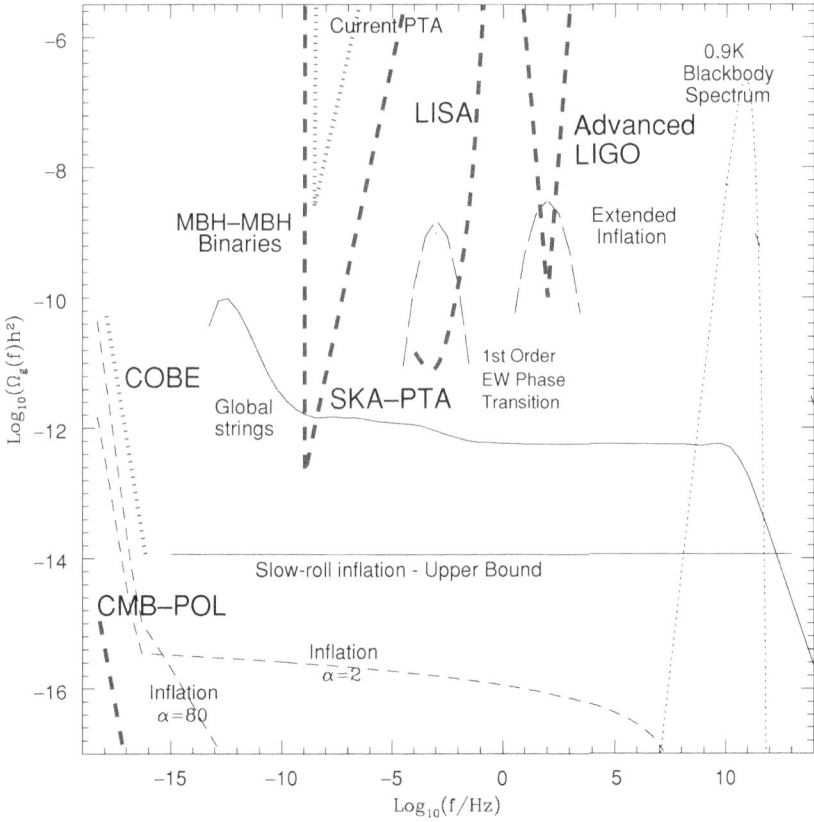

Fig. 2.5. Sensitivity curves for the detection of a stochastic gravitational wave background. The sensitivities of Advanced LIGO, LISA, current PTA and the future SKA PTA are shown. Estimated signal levels for various sources are indicated. Figure updated from an original design provided by Richard Battye.

versatility of pulsar timing to detect small bodies in orbit around neutron stars. As shown in Figure 2.6, pulsar timing is currently far more sensitive than optical searches for planets around main-sequence stars (see Marcy and Butler (2000) for a review).

Due to the violent conditions in which neutron stars are formed, planetary bodies orbiting around pulsars are not expected to be common. So far only one further 'pulsar planet' has been found around PSR B1620−26. This 11 ms pulsar in the globular cluster M4 was already

they are often unfortunately sidelined from extrasolar planet reviews due to the fact that neutron stars are not 'normal' (i.e. sun-like) host stars.

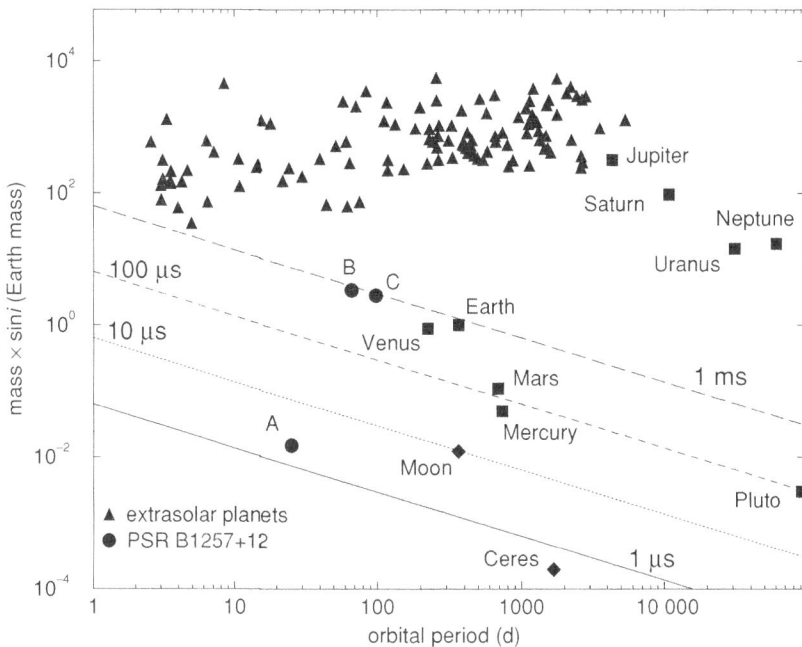

Fig. 2.6. Phase space for planetary companions that can be probed by pulsar timing assuming precisions of 1, 10, 100 and 1000μs. The parameters of the solar system planets and those of PSR B1257+12 are compared to those of the currently known extrasolar planets discovered by optical spectroscopy.

known to have a ~ 0.3 M$_\odot$ white dwarf companion in a 191 day orbit (Lyne *et al.* 1988). Further, longer-term timing measurements subsequently revealed the effects of a second companion in a much longer orbit (Backer *et al.* 1993a; Thorsett *et al.* 1993). More recent timing observations constrain the orbital parameters as a function of orbital eccentricity (Thorsett *et al.* 1999). The white-dwarf companion star has recently been detected in HST observations (Sigurdsson *et al.* 2003). From the mass measurements of the white dwarf, and the pulsar timing measurements, the likely planetary mass is in the range 1–2 Jupiter masses. The current estimates favour a roughly 45 yr low-eccentricity ($e \sim 0.16$) orbit with semi major axis ~ 25 astronomical units. The orbital parameters will be refined soon as the time span of the observations increases.

2.5 The Galaxy and the interstellar medium

The large observed proper motions of pulsars shown in Figure 2.2 imply velocities that are sufficiently large to carry pulsars far from their birth sites close to the Galactic plane. As a result, pulsars should exist in most regions of the Galaxy. The nature of pulsars as Galactic objects and the interaction of their polarised emission with the magnetised–ionised interstellar medium (ISM) makes them excellent probes to study the properties and structure of the Milky Way in a variety of ways.

2.5.1 Large-scale structure

As seen in Chapter 1, measurements of the delay of pulses as they propagate through the ISM provides the *dispersion measure* (DM) the integrated column density of free electrons along the line of sight to the pulsar. If the Galactic distribution of free electrons were known, a measurement of the DM would provide the distance to each pulsar. In practise, the electron distribution is uncertain but can be calibrated through pulsars with independent distance measurements, e.g. from parallax and neutral hydrogen absorption measurements. In this sense, pulsars are used as tools for mapping out the Galactic distribution of free electrons.

The most precise determinations of dispersion measures are possible for millisecond pulsars. Occasionally, the precision is sufficient to determine a variation of the dispersion measure over very small distances such as in the Globular cluster 47 Tucanae. Observing sixteen of the twenty-two millisecond pulsars in this cluster, Freire *et al.* (2001) determined a relative change of the dispersion measure of $\Delta DM/DM = 0.02$ over a distance of about 5 pc. This measurement implies the presence of ionised gas in this cluster with an electron density of 0.07 cm^{-3}.

In Chapter 1, we saw that pulsar signals are also highly polarised. As described in detail in Chapter 3, the magnetic field of the Galaxy acts as a Faraday screen rotating the position angle of linear polarisation by an amount proportional to the line of sight component of the Galactic field weighted by electron density. The *rotation measure* (RM) associated with this effect for pulsars over a wide range of locations in the Galaxy can, with care, be used to map out the large-scale structure of the Galactic magnetic field which shows field-line reversals between some neighbouring spiral arms. Currently, rotation measures for 535 pulsars are known (Han 2004). A subset of these are shown in Figure 2.7.

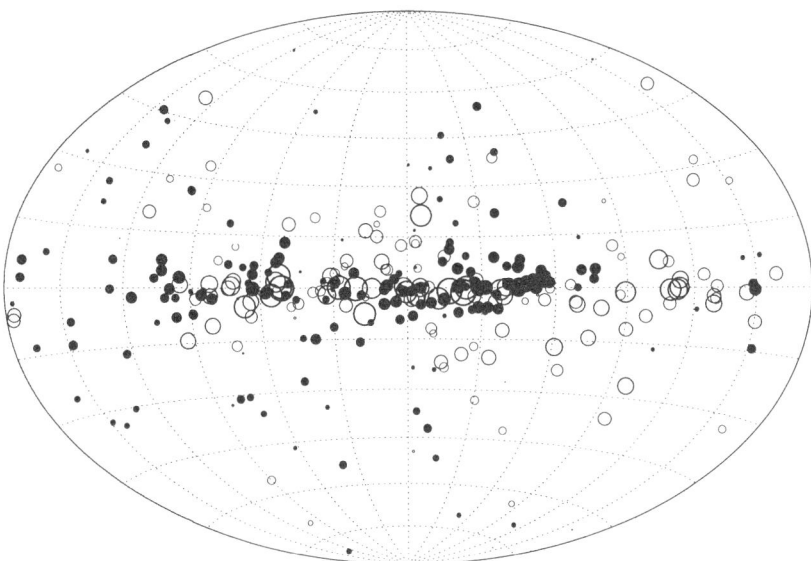

Fig. 2.7. Hammer–Aitoff projection showing the rotation measures (RMs) for 323 pulsars in Galactic coordinates. Open symbols denote negative RMs, while the filled points show positive values. The sizes of the points are proportional to the logarithm of the absolute value of each RM. Figure provided by Dimitris Athanasiadis.

2.5.2 Small-scale structure

It is often possible to study the dynamics of the ISM on small time and length scales through regular monitoring of pulsars. A prominent example is the 33 ms pulsar B0531+21 that powers the Crab supernova remnant. The observations of changing pulse profiles shown in Figure 9.3 are interpreted as being due to plasma clouds located in the filamentary interface between the synchrotron nebula of the supernova remnant and the supernova ejecta crossing the line of sight over a period of months (Backer *et al.* 2000; Lyne *et al.* 2001) (see Figure 2.8).

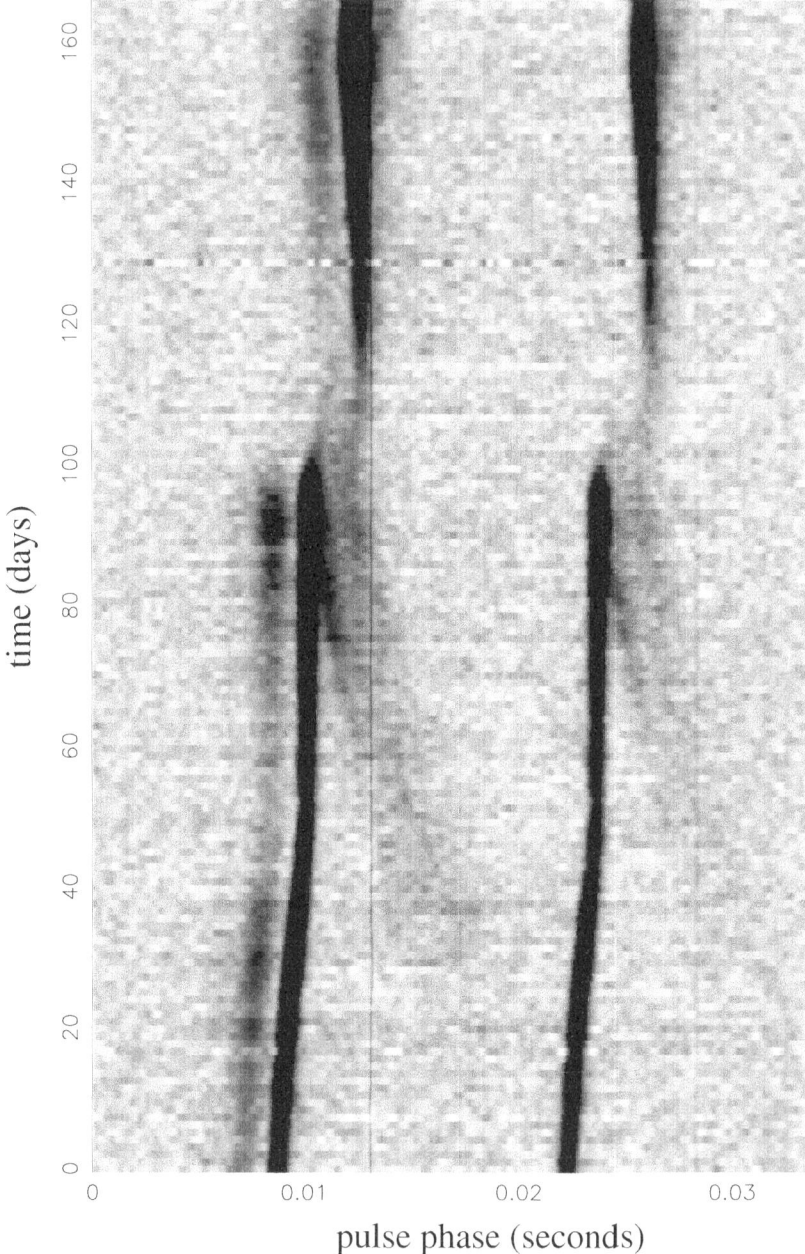

Fig. 2.8. Grey-scale showing observations of the Crab pulsar from 1997 (Lyne *et al.* 1999). One can see clearly an echo approaching the pulse, followed by a 'loss' of the pulsar signal at the point of intersection. The pulse reappears later, again with a visible echo but also shifted in pulse phase. The echo originates from a plasma cloud in the nebula approaching the line of sight and scattering the pulsar signal. Figure provided by Andrew Lyne and Graham Smith.

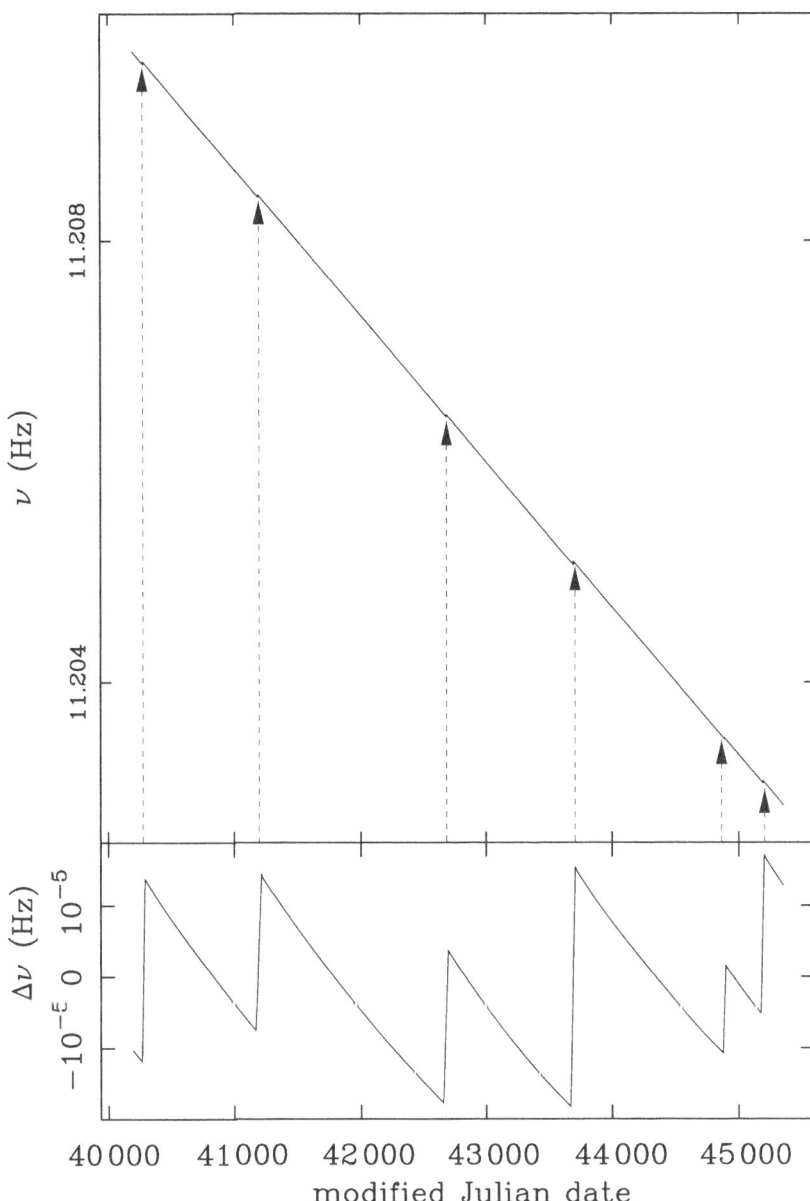

Fig. 2.9. Rotation frequency of the Vela pulsar as a function of time (Lyne 1999). The general slow down of the pulsar's rotation is clearly visible. The arrows indicate the occurrence of glitches when the pulsar spins up very abruptly. The glitches and the subsequent relaxation processes become more visible if the slope representing the general spin-down is subtracted (lower panel). The relaxation process provides information about the neutron star interior that is considered to be simultaneously super-fluid and super-conducting. Figure provided by Andrew Lyne.

2.6 Studying super-dense matter

While we cannot create matter of such extreme density that exists inside
neutron stars (see Chapter 3), observations of rotational instabilities in
pulsars, known as *glitches*, and the subsequent relaxation processes allow
us to probe neutron star material in a unique way that can be considered
as 'neutron star seismology'. Observations of glitches and their recovery
provide strong evidence for the existence of a fluid component inside the
solid outer crust of the neutron star (see Section 3.1.3).

Glitches are rare and mostly observed for young pulsars (see, for example, Shemar and Lyne (1996)). During a glitch, a small sudden increase
in rotation rate is observed in the sudden early arrival of the pulses (see
Figure 2.9). Often the glitch is followed by an exponential decay back
towards the pre-glitch rotation rate. The first model proposed by Baym
et al. (1969) explained glitches as the result of star quakes, where abrupt
changes in the oblateness of the neutron star crust cause a change in the
moment of inertia and, hence, in spin frequency as angular momentum
is conserved. While this model can explain the glitches observed for the
Crab pulsar, it cannot account for more frequent and larger glitches as
observed for the Vela pulsar (see, for example, Lyne (1992)). It is more
likely that the origin of glitches is related to the superfluid nature of the
liquid neutron star interior and how microscopic, quantised vortices in
the fluid develop and become pinned to the neutron star crust (Anderson
& Itoh 1975; Ruderman 1976). The discovery of free precession in PSR
B1828−11 by Stairs *et al.* (2000c) provides constraints on the strength
of this pinning.

2.7 Plasma physics under extreme conditions

The pulsar radio emission process and the associated plasma physics are
understood only poorly. This situation remains after almost 40 years
of research and in spite of the large number of studies and observations devoted to the identification of the relevant physical processes (see
Chapters 1 and 3). The gaps in our understanding are not caused by
a lack of available high-quality data; indeed, the opposite may be true
– the detail in which pulsars can be studied seems to complicate the
necessary separation of properties associated directly with the emission
process from geometrical or propagation effects.

It is important to recognise that the plasma processes to be understood
occur under extreme conditions of very high plasma densities, super-

strong magnetic fields and in a rather complex environment known as the *pulsar magnetosphere*. While being an extreme theoretical challenge, it seems difficult to identify new observational information that can help to finally solve the 'pulsar problem'. However, clues may come from unexpected discoveries such as the double pulsar system J0737–3039 which we highlight below.

The double pulsar binary system consisting of the 22.7 ms pulsar A (Burgay *et al.* 2003) and the 2.8 s pulsar B (Lyne *et al.* 2004), discussed above in the context of gravitational physics, also serves as a unique laboratory for plasma physics. What makes J0737–3039 particularly interesting is that the mean orbital separation of the two pulsars is only about 3 lt s (∼900 000 km). At this distance, the relativistic wind from the radiation beam of A is sufficiently energetic to penetrate into and probe the emission region of B, which is significantly distorted as a result. The two key observational facts relevant to such studies are: (a) the signal from A is not visible for about 30 s during each orbit, independent of observing frequency (Lyne *et al.* 2004; Kaspi *et al.* 2004); (b) the signal from B only occurs strongly in two distinct orbital phase ranges (Lyne *et al.* 2004) as shown in Figure 2.10.

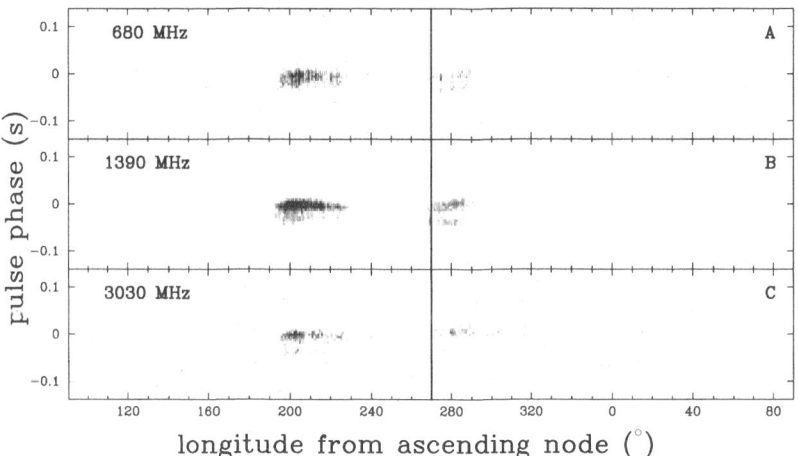

Fig. 2.10. Grey-scale images showing the pulse of PSR J0737–3039B as a function of orbital phase at three observing frequencies (Lyne *et al.* 2004).

The first point may be understood as an occultation of the radiation of A by the magnetosphere of B at conjunction. The fact that we view the orbit nearly edge-on means that, at conjunction, the line of sight to one pulsar passes within about 0.15 lt s of the other. Although much greater than that of the *light cylinder radius*[2] of A (0.004 lt s), it is substantially smaller than the radius of B's light cylinder (0.45 lt s). The size of the eclipsing region can be deduced from the duration of the eclipse and the relative transverse speeds of the two stars (660 km s^{-1}) and is found to be ∼0.044 lt s – only about one-tenth of the linear size of the light cylinder of B.

One reason for the size of the eclipsing region being so small may come from the fact that the total rate of spin-down energy loss from A is 3000 times greater than that from B. A simple calculation shows that the energy densities due to the relativistic wind of A and the magnetic field of B are in balance at a distance at about 40 per cent of the light cylinder radius of B. Since the point of pressure balance is far within the magnetosphere of B, the actual penetration of the wind of A into the magnetosphere of B will be a strong function of the orientation of the rotation and/or magnetic axes of B relative to the direction of the wind and, hence, will depend on the precessional and orbital phases of B. This is the most likely explanation for the large amplitude and pulse-shape changes of B shown in Figure 2.10.

As the pulsars move in their orbits, the line of sight from A passes through, and sweeps across, the magnetosphere of B, providing the opportunity to probe its physical conditions. The determination of changes in the radio transmission properties, including the dispersion and rotation measures, will potentially allow the plasma density and magnetic field structure to be probed. Additionally, the ∼ 70-year period geodetic precession discussed earlier will cause the line of sight to sample different trajectories through the magnetosphere of B over time.

2.8 Future applications

Many of the applications discussed above are undergoing incremental improvements as new pulsars are being discovered and more sensitive detectors being built. In the more distant future, these studies could well undergo a quantum leap as large numbers of pulsars are discovered and observed with the *Square Kilometre Array* (SKA). The sensitivity of the SKA means that it could find most pulsars in the Galaxy beaming towards the Earth! This clearly would allow the Galactic electron den-

---

2 The light cylinder radius is the point at which the co-rotation speed equals the speed of light and sets a natural limit on the size of the neutron star's magnetosphere (see Chapter 3).

sity distribution and magnetic field structure to be studied in far greater detail than presently is possible, particularly with the expected increase in parallax measurements made possible by the superior sensitivity of the SKA. In addition, studies of the significant numbers of extra-galactic pulsars expected to be detectable by the SKA would allow measurements of the inter-galactic, as opposed to the interstellar, medium.

The capability of the SKA to time the expected yield of $\sim$20 000 Galactic pulsars, including more than 1 000 millisecond pulsars, very precisely will clearly benefit the prospects of the PTA. The larger yield of pulsars effectively would sample every possible outcome of the evolution of massive binary stars, thereby guaranteeing the discovery of very exciting systems. Among the new SKA pulsar discoveries would almost certainly be a pulsar–black hole system. For tests of general relativity such a system would surpass all present and foreseeable competitors (Damour & Esposito-Farèse 1998). It may be possible, through pulsar timing, to determine the mass, spin and quadrupole moment of the black hole precisely (Wex & Kopeikin 1999). Such measurements would go beyond what currently is possible with binary pulsar tests of general relativity, perhaps even confronting the 'Cosmic censorship conjecture' and the famous 'no-hair' theorem for black holes (see, for example, Shapiro and Teukolsky (1983)).

2.9 Further reading

Further accounts of the many applications that pulsars have in other areas of astronomy and physics can be found in many of the contributions to the proceedings of the meeting *Pulsars as Physics Laboratories* (Blandford *et al.* 1992) as well as in *Pulsar Astronomy* (Lyne & Smith 2005). The excellent review by Stairs (2003) provides a recent account of the use of binary pulsars in the study of gravitational physics. Further discussions on the use of pulsars as gravitational-wave detectors can be found in the reviews by Romani (1992) and Backer (1996). Further details of the pulsar planetary system can be found in the proceedings of the conference *Planets around Pulsars* (Phillips *et al.* 1993). Two reviews dealing with pulsars as probes of the interstellar medium and the Galactic gravitational field that provide good starting-points for further reading are respectively Weisberg (1996) and Han (2004).

3

Theoretical background

In spite of almost four decades of intensive research on pulsars, it is a fair statement that pulsars are understood only poorly. The 'pulsar problem' is considered as one of the most difficult ones in modern astrophysics (see, for example, Bahcall and Ostriker (1997)) as it consists in fact of a range of unsolved problems. The questions to be answered address: (a) pulsars as neutron stars (e.g. *What is the structure of neutron stars?* or *What is the composition of neutron star atmospheres?*); (b) the surroundings of pulsars (e.g. *What is the structure of the magnetic and electric fields?*, *What are the plasma densities?* or *Are there regions of additional pair creation and how does that work?*); (c) the radiation processes themselves (e.g. *What is the radio emission mechanism?* or *How and where is the high-energy emission created?*).

Although much progress has been made, we do not know the definitive answer to any of the above questions[1]. Indeed, some of the problems are understood better than others, but theory is far from being able to provide 'the pulsar model' that explains some, or even most, of the observed pulsar properties (see Chapter 1) in a self-consistent way. Given these difficulties, this chapter cannot provide a detailed account of the rich literature that is available. Instead, we will take an (intentionally simplistic) observer's point of view and try to provide the basic theoretical background for interpreting the observed emission properties.

We discuss the relevant issues by following the generally assumed 'toy model' of a pulsar *magnetosphere* shown in Figure 3.1. After presenting the basic neutron star properties, we review neutron star spin-down, before introducing the concept of a pulsar magnetosphere. Then we discuss the radio beam geometry, emission altitudes, structure and inferred

---

1 Some colleagues may disagree with this statement!

radio luminosities. We conclude with considerations about acceleration gaps and suggested emission mechanisms for radio and high energies.

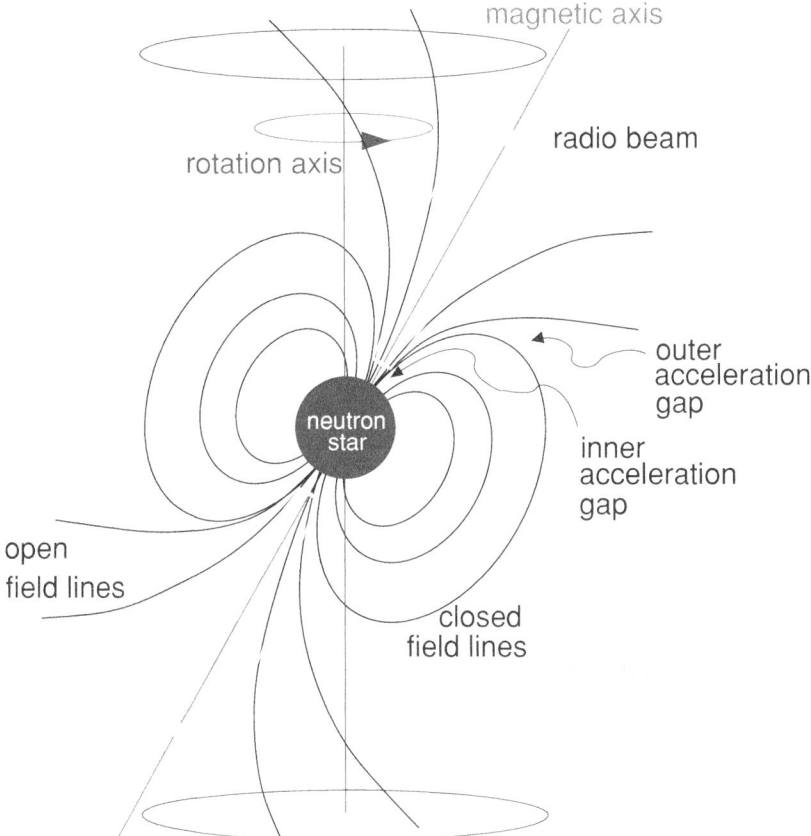

Fig. 3.1. Toy model for the rotating neutron star and its magnetosphere (not drawn to scale!). The various terms and regions are discussed in detail throughout the chapter.

3.1 Neutron stars

The internal structure of neutron stars is complex. Their composition, already addressed by Oppenheimer and Volkov (1939) long before the discovery of pulsars, depends sensitively on the *equation of state*, i.e. the relationship between density and pressure. Knowledge of the exact equation of state would allow us to deduce most of the neutron star's physical properties, most notably its radius for a given mass.

3.1.1 Mass and radius

Neutron stars consist of highly compressed matter that is very different from what can be studied in terrestrial laboratories. The theoretical calculations for neutron star equations of state are therefore necessarily uncertain. Typically, the models predict a maximum neutron star mass of about 2 $M_\odot$ (see Lattimer and Prakash (2001) for a comparison of various models) which can, however, increase due to effects of strong magnetic fields (Cardall *et al.* 2001). Direct and accurate mass measurements come from timing observations of binary pulsars (see Sections 2.3.1 and 8.3.3). Figure 3.2 summarises the available mass measurements as presented by Stairs (2004), following work by Thorsett and Chakrabarty (1999). The measurements are consistent with a typically assumed pulsar mass of 1.4 $M_\odot$, whereas a long recycling process appears to increase the mass as expected (see Section 1.3.3).

Reliable measurements of neutron star radii are difficult to obtain. One method is to observe thermal emission from the neutron star surface at optical and X-ray frequencies (see Chapter 9) where the observed luminosity can be used to infer the size of the emitting region. Although this method may yield the best estimates (see, for example, Lattimer and Prakash (2001) for a discussion) the strong gravitational fields and atmospheres of neutron stars complicate calculations somewhat. We now briefly discuss each of these issues in turn.

The neutron star's gravity causes redshifts in the observed thermal flux and temperature. As a result, a distant observer measures a temperature that is smaller than the real temperature at the surface. The corresponding radius $R_{\rm obs}$ inferred (see, for example, Lewin *et al.* (1993)) is larger than the intrinsic value R as follows:

$$R_{\rm obs} = \frac{R}{\sqrt{1 - 2GM/Rc^2}} = \frac{R}{\sqrt{1 - R_{\rm S}/R}}. \tag{3.1}$$

where M and R are the gravitational mass and radius of the star, c is the speed of light, G is Newton's gravitational constant and

$$R_{\rm S} = \frac{2GM}{c^2} \simeq 4.2 \text{ km} \left(\frac{M}{1.4 M_\odot}\right) \tag{3.2}$$

is the Schwarzschild radius.

Neutron stars are expected to have thin plasma layers as atmospheres at their surface (Pavlov & Shibanov 1978; Romani 1987; Shibanov *et al.* 1992). These atmospheres can modify the emitted luminosity considerably, resulting in a radiation spectrum that is very different from a

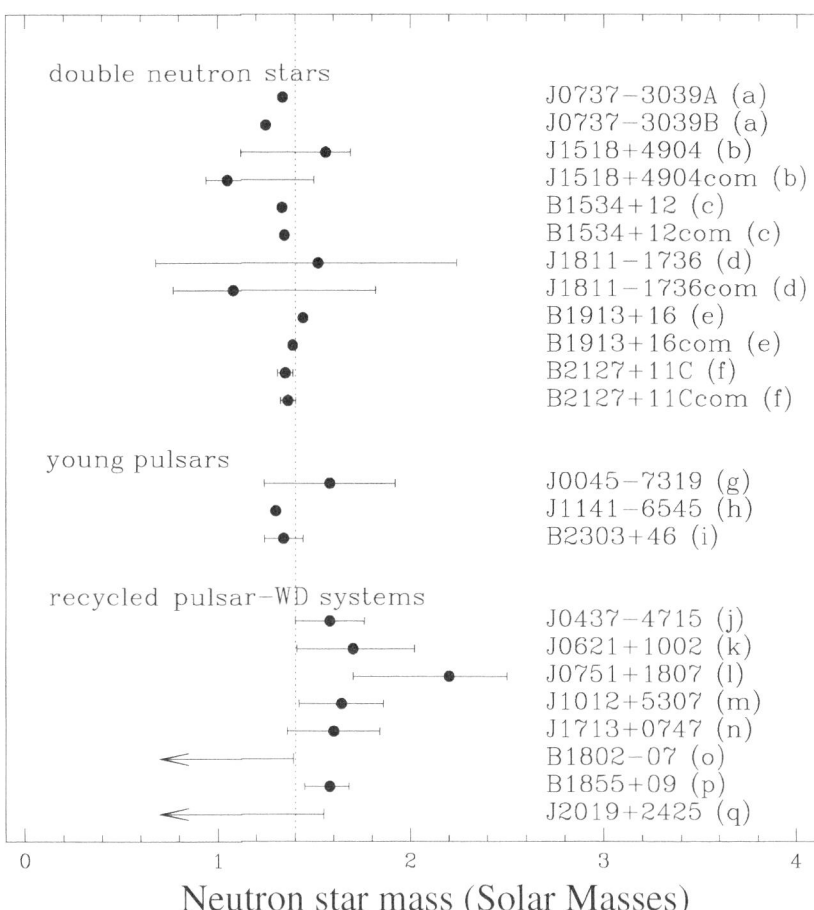

Fig. 3.2. Distribution of neutron star masses as inferred from timing observations of binary pulsars (Stairs 2004). The vertical dotted line shows the canonical neutron star mass of 1.4 $M_\odot$. Figure provided by Ingrid Stairs.

simple blackbody form. In particular, a strong magnetic field can modify the brightness distribution on the surface, creating hotter and cooler spots that may be misinterpreted. All these effects lead to wrongly inferred values for the neutron star radius if they are not taken into account correctly. Assumptions about the composition of the atmosphere and the effects of the strong magnetic field are particularly problematic (see, for example, Zavlin and Pavlov (2002)).

Based on arguments that the speed of sound should be smaller than the speed of light in a neutron star, and that the equation of state provides a smooth transition from low to high densities, Lattimer *et al.* (1990) and Glendenning (1992) derive a lower limit for the neutron star radius

$$R_{\min} \simeq 1.5 \ R_{\mathrm{S}} = \frac{3GM}{c^2} = 6.2 \ \mathrm{km} \ \left(\frac{M}{1.4\mathrm{M}_\odot} \right). \tag{3.3}$$

Conversely, an upper limit can be obtained by requiring that the neutron star is stable against break-up due to centrifugal forces. For a neutron star rotating with a period P, we find

$$R_{\max} \simeq \left(\frac{GMP^2}{4\pi^2} \right)^{1/3} = 16.8 \ \mathrm{km} \ \left(\frac{M}{1.4\mathrm{M}_\odot} \right)^{1/3} \left(\frac{P}{\mathrm{ms}} \right)^{2/3}. \tag{3.4}$$

For the fastest rotating pulsar known, PSR B1937+21 with a period $P = 1.56$ ms, this provides a limit of $R_{\max} = 22.6$ km for $M = 1.4$ M$_\odot$. Indeed, most models predict a radius of $R \sim 10 - 12$ km (Lattimer & Prakash 2001), consistent with the theoretical upper and lower limits. These values are only about 3 times larger than the Schwarzschild radius, demonstrating that pulsars are almost black holes, and as such are very dense objects and subject to large gravitational effects near the surface.

3.1.2 *Moment of inertia*

The moment of inertia $I = kMR^2$, where $k = 2/5 = 0.4$ for a sphere of uniform density. For a neutron star, the exact value of k depends on the density profile and, hence, on the equation of state. Most models predict a value of $k = 0.30 - 0.45$ for a mass–radius range of $M/R = 0.10 - 0.20$ M$_\odot$ km^{-1}. For most practical calculations, one adopts $I = 10^{38}$ kg m^2 = 10^{45} g cm^2, obtained for $M = 1.4$ M$_\odot$, $R = 10$ km and $k = 0.4$. Given the large uncertainties in the masses and radii of neutron stars, this canonical value for the moment of inertia may be uncertain by $\sim$70 per cent. As discussed in Section 2.3.1, it may be possible to measure the moment of inertia directly from future timing measurements of the double-pulsar binary J0737–3039.

3.1.3 *Structure*

Using the same canonical values as above, we find an average mass density $\langle \rho \rangle = 6.7 \times 10^{17}$ kg m^{-3} = 6.7×10^{14} g cm^{-3} which is even larger than the density of nuclear matter, 2.7×10^{14} g cm^{-3}. In reality, a neutron star is not a uniform sphere. Observations of glitches in the rotation of young pulsars (see Section 2.6) indicate a solid crust, containing $\gtrsim 1.4$ per cent of the moment of inertia (Link *et al.* 1999), that is tied to a liquid interior of increasing density which covers about 9 orders of magnitude between the crust and the central star. This is common to most neutron star models (see, for example, Shapiro and Teukolsky (1983) and Pines and Alpar (1985)). Near the surface, the solid crystalline crust consists mainly of iron nuclei and a sea of degenerate electrons, $\rho \simeq 10^6$ g cm^{-3}. Moving inward, the density increases to a point at which protons and electrons will combine to form neutrons, creating the neutron-rich nuclei of an inner crust. Passing the *neutron drip point* at $\rho \simeq 4 \times 10^{11}$ g cm^{-3}, several hundred metres below the surface, the number of neutrons released from the nuclei will increase rapidly. After the crust dissolves fully at $\rho \simeq 2 \times 10^{14}$ g cm^{-3}, the largest fraction of the neutron star is made up of a 'sea' of free, superfluid neutrons, mixed with about 5 per cent superconducting electrons and protons. The nature of the most inner core region differs for various theories. In the most extreme cases, the core consists of exotic matter like pions or quarks (see Shapiro and Teukolsky (1983)).

3.2 Spin evolution

Pulse periods are observed to increase with time. The rate of increase $\dot{P} = dP/dt$ can be related to a rate of loss of rotational kinetic energy

$$\dot{E} \equiv -\frac{dE_{\text{rot}}}{dt} = -\frac{d(I\Omega^2/2)}{dt} = -I\Omega\dot{\Omega} = 4\pi^2 I\dot{P}P^{-3}, \qquad (3.5)$$

where $\Omega = 2\pi/P$ is the rotational angular frequency and I the moment of inertia. This quantity $\dot{E}$ is called *spin-down luminosity* and it represents the total power output by the neutron star. For the canonical moment of inertia $I = 10^{45}$ g cm^2, we find

$$\dot{E} \simeq 3.95 \times 10^{31} \text{ erg s}^{-1} \left(\frac{\dot{P}}{10^{-15}}\right) \left(\frac{P}{\text{s}}\right)^{-3}. \qquad (3.6)$$

As we will discuss in Chapter 9, only a tiny, seemingly insignificant fraction of $\dot{E}$ is converted into radio emission. The bulk of the rotational energy is converted into high-energy radiation and, in particular, magnetic dipole radiation and a pulsar wind.

3.2.1 Braking index

Pulsars have strong dipole magnetic fields (see Section 3.2.4). According to classical electrodynamics (see, for example, Jackson (1962)), a rotating magnetic dipole with moment $|\mathbf{m}|$ radiates an electromagnetic wave at its rotation frequency. The radiation power

$$\dot{E}_{\text{dipole}} = \frac{2}{3c^3} |\mathbf{m}|^2 \, \Omega^4 \sin^2 \alpha, \qquad (3.7)$$

where α is the angle between the magnetic moment and the spin axis. By equating $\dot{E}_{\text{dipole}}$ with the spin-down luminosity (see Equation (3.5)), we derive the expected evolution of the rotation frequency,

$$\dot{\Omega} = - \left(\frac{2|\mathbf{m}|^2 \sin^2 \alpha}{3Ic^3} \right) \Omega^3. \qquad (3.8)$$

Expressing this equation more generally as a power law, and in terms of rotational frequency $\nu = 1/P$ rather than Ω, we have

$$\dot{\nu} = -K \, \nu^n, \qquad (3.9)$$

where n is the *braking index* and K is usually assumed to be constant. We expect, from Equation (3.8), $n = 3$ for pure magnetic dipole braking. However, other dissipation mechanisms may exist (e.g. a wind of outflowing particles) that also carry away some of the rotational kinetic energy. Therefore, while Equation (3.8) is valid for the vacuum case, the presence of a plasma leads to a spin-down even for aligned rotators. Both magnetic dipole braking and pulsar winds lead to the same scaling as given in Equation (3.8) (see, for example, Michel and Li (1999) and Spitkovsky (2004)).

The braking index can be determined if, in addition to $\dot{\nu}$, a second spin frequency derivative, $\ddot{\nu}$, can be measured. Differentiating Equation (3.9) and eliminating K, we find

$$n = \nu\ddot{\nu}/\dot{\nu}^2. \qquad (3.10)$$

Usually, the observed $\ddot{\nu}$ is contaminated by timing noise (see Section 8.4). However, for a few pulsars, a determination of the braking index that reflects the true spin-down behaviour has been determined. Values range from $n = 1.4$ to $n = 2.9$ (for a review see, for example, Kaspi and Helfand (2002)). These deviations indicate that the assumption of $n = 3$ is usually not correct! Nevertheless, this supposition is usually made to *define* a number of useful quantities that are used to characterise the basic properties of a radio pulsar. It should always be kept in mind that the resulting quantities are rough model-dependent estimates.

3.2.2 Age estimate

The spin-down model of Equation (3.9), expressed in terms of pulse period, becomes $\dot{P} = K\,P^{2-n}$. This is a simple first-order differential equation which may be integrated, assuming a constant K and braking index $n \neq 1$, to obtain the age of the pulsar

$$T = \frac{P}{(n-1)\dot{P}}\left[1 - \left(\frac{P_0}{P}\right)^{n-1}\right],\qquad(3.11)$$

where P_0 is the spin period at birth. Under the assumption that the spin period at birth is much shorter that the present value (i.e. $P_0 \ll P$) and that the spin-down is due to magnetic dipole radiation (i.e. $n = 3$) the above equation simplifies to the *characteristic age*

$$\tau_c \equiv \frac{P}{2\dot{P}} \simeq 15.8\;\mathrm{Myr}\left(\frac{P}{\mathrm{s}}\right)\left(\frac{\dot{P}}{10^{-15}}\right)^{-1}.\qquad(3.12)$$

Due to the assumption of a negligible initial spin period and $n = 3$, this quantity does not necessarily provide a reliable age estimate – for the Crab pulsar, one obtains a characteristic age of 1240 yr, which is comparable to the known age of the pulsar of about 950 years. However, discrepancies can be much larger. The youngest known radio pulsar, J0205+6449, was born in the historical supernova of AD 1181 (see Section 9.4.1.1), but the value of τ_c suggests an age of 5370 yr. An extreme case is PSR J0538+2817, where $\tau_c = 620$ kyr; however, proper motion measurements indicate a birth only 30 kyr ago (Kramer *et al.* 2003c). These cautionary cases again highlight that, while the characteristic age is computed readily from the observations, it should be interpreted with some care.

3.2.3 Birth period

By rearranging Equation (3.11) and using the definition of the characteristic age, τ_c (Equation (3.12)), we find

$$P_0 = P\left[1 - \frac{(n-1)}{2}\,\frac{T}{\tau_c}\right]^{\left(\frac{1}{n-1}\right)}.\qquad(3.13)$$

If the true age of the pulsar, T, is known independently (e.g. if the pulsar can be associated with a historical supernova), and ideally the braking index has been measured, then the birth period can be estimated. While it had been assumed in the past that pulsars are born with periods similar to that estimated for the Crab pulsar, $P_0 = 19$ ms (Lyne *et al.* 1993), recent estimates suggest a wide range of initial spin periods from 14 up to 140 ms (see, for example, Migliazzo *et al.* (2002) and Kramer *et al.* (2003c)). In most cases, the estimated birth periods are significantly larger than expected from core-collapse theories of massive stars that have difficulties explaining even spin periods as long as a few tens of milliseconds (Heger *et al.* 2003). It has also been proposed, on the basis of population analyses, that a significant fraction of pulsars are 'injected' into the population with longer spin periods (see, for example, Vivekanand and Narayan (1981)), perhaps up to 0.5 s. However, the injection hypothesis remains controversial (see, for example, Lorimer *et al.* (1993)) and further analyses are required in order for us to determine the distribution of neutron star birth spin periods.

3.2.4 Magnetic field strength

Observational estimates of the surface magnetic fields of neutron stars come from the detection of cyclotron radiation features in the spectra of X-ray binaries (see, for example, Trümper *et al.* (1978) and Wheaton *et al.* (1979)) and, more recently, for an isolated neutron star (Bignami *et al.* 2003). These estimates are in the range 10^{11-12} G. Although no direct measurements are possible for radio pulsars, we can obtain an estimate of the magnetic field strength by assuming that the spin-down process is dominated by dipole braking. The magnetic moment is related to the magnetic field strength according to $B \approx |\mathbf{m}|/r^3$. By rearranging Equation (3.8), we obtain an expression for the field strength at the surface:

$$B_{\mathrm{S}} \equiv B(r = R) = \sqrt{\frac{3c^3}{8\pi^2} \frac{I}{R^6 \sin^2 \alpha} P\dot{P}}. \qquad (3.14)$$

For the canonical neutron star with moment of inertia $I = 10^{45}$ g cm^2 and radius $R = 10$ km, assuming $\alpha = 90°$, we find

$$B_{\mathrm{S}} = 3.2 \times 10^{19} \mathrm{G} \ \sqrt{P\dot{P}} \simeq 10^{12} \ \mathrm{G} \left(\frac{\dot{P}}{10^{-15}}\right)^{1/2} \left(\frac{P}{\mathrm{s}}\right)^{1/2}. \qquad (3.15)$$

As for the characteristic age, the basic observables P and $\dot{P}$ provide a handy estimate for the field strength at the pulsar surface. We refer to it as the *characteristic magnetic field*. However, given the fact that α is usually unknown, that radius and moment of inertia are uncertain (see Section 3.1) and that other processes are contributing to the pulsar spin-down, Equation (3.15) is essentially an order of magnitude estimate. With this in mind, it seems hardly to matter that some authors (see, for example, Shapiro and Teukolsky (1983) and Usov and Melrose (1995)) have pointed out that Equation (3.15) gives the field strength at the magnetic *equator*, whereas the field strength at the magnetic *poles* should be a factor of two higher.

3.3 The pulsar magnetosphere

The previous discussion has assumed that the neutron star is rotating in vacuo – a situation that is not maintained in reality. Since the Lorentz forces on the charges in the neutron star interior are huge compared to the gravitational forces, the neutron star can be viewed as a highly magnetised, rotating superconducting sphere. As first shown for a neutron star with aligned rotation and magnetic axes by Goldreich and Julian (1969) based on the work by Deutsch (1955), such a case leads to external electric fields and, ultimately, to the extraction of plasma from the neutron star surface. The plasma-filled surrounding dominated by the magnetic field is known as the *pulsar magnetosphere*.

Even though the Goldreich–Julian model does not describe a realistic case (see, for example, criticism by Michel and Li (1999)), it illustrates some basic principles that are useful to understand the observations of pulsars. Attempts to extend the model to more realistic assumptions such as an inclined magnetic axis (see, for example, Mestel (1971) and Mestel and Pryce (1992)) have not been successful. It was realised that the structure of, and the processes in, a pulsar magnetosphere probably cannot be solved analytically in a closed form. Today, the best prospects for solving the 'pulsar problem' are provided by numerical simulations (see, for example, Krause-Polstorff and Michel (1985) and Contopoulos *et al.* (1999)) (see Spitkovsky (2004) for a recent review). Therefore, we make no attempt to discuss the severe problems of the Goldreich–Julian model in any detail, ignoring, for example the problem of the neglected central charge of the neutron star (see, for example, Michel (1991) and Michel and Li (1999)) or the general problem of a return current (see, for example, Beskin *et al.* (1993)). Instead, we use the following description by Goldreich and Julian (1969) to illustrate the basic concepts.

At any point within a rotating magnetised sphere there will be an induced electric field, $(\mathbf{\Omega} \times \mathbf{r}) \times \mathbf{B}$, due to the presence of the magnetic field, $\mathbf{B}$. For a perfectly conducting sphere this will be balanced by a distribution of charge giving an electric field, $\mathbf{E}$, so that at any point $\mathbf{r}$

inside the sphere a force-free state is obtained,

$$\mathbf{E} + \frac{1}{c}\left(\mathbf{\Omega} \times \mathbf{r}\right) \times \mathbf{B} = 0. \tag{3.16}$$

If there is a vacuum outside the sphere, the surface charges induce an external quadrupole field (unipolar induction) given by

$$\Phi(r, \theta) = \frac{B_{\mathrm{S}}\Omega R^5}{6cr^3}\left(3\cos^2\theta - 1\right), \tag{3.17}$$

where (r, θ) are polar coordinates of a star-centred coordinate system. The corresponding electric field at the stellar surface,

$$E_{\parallel} = \left.\frac{\mathbf{E} \cdot \mathbf{B}}{B}\right|_{r \approx R} = -\frac{\Omega B_{\mathrm{S}} R}{c}\cos^3\theta \tag{3.18}$$

causes an electric force $(F = qE_{\parallel})$ on the charged particles on the pulsar surface that exceeds gravity by 10 orders of magnitude or more for typical pulsar parameters! Unless particles are heavily bound, they are easily extracted from the surface and pulled into the surrounding vacuum (see Section 3.5). The starting condition of a vacuum state cannot be maintained, and the strong magnetic field of a neutron star leads to a dense surrounding plasma. The charge distribution is given by

$$\rho_{\mathrm{e}}(r, \theta) = \frac{1}{4\pi}\nabla\mathbf{E} = -\frac{\Omega\mathbf{B}}{2\pi c} = -\frac{B_{\mathrm{S}}\Omega R^3}{4\pi cr^3}\left(3\cos^2\theta - 1\right). \tag{3.19}$$

Charges above the equatorial region will be of opposite sign to charges above the magnetic poles, whereas ρ_{e} changes sign for $\cos\theta = \sqrt{1/3}$ (see Figure 3.3). Once this charge distribution is arranged outside the star, the electric fields, $E_{\parallel}$, are shielded and the force-free state of Equation (3.16) is also maintained outside the neutron star. Assuming complete charge separation, the number density, $n = \rho_{\mathrm{e}}/e$, at the magnetic pole, $(r = R, \theta = 0)$,

$$n_{\mathrm{GJ}} = \frac{\Omega B_{\mathrm{S}}}{2\pi ce} \simeq \frac{B_{\mathrm{S}}}{ceP} = 7 \times 10^{10}\mathrm{cm}^{-3}\left(\frac{P}{\mathrm{s}}\right)^{-1/2}\left(\frac{\dot{P}}{10^{-15}}\right)^{1/2} \tag{3.20}$$

is known as the *Goldreich–Julian density* and represents a maximum value unless further processes exist (see Section 3.5).

The plasma outside the neutron star experiences the same $\mathbf{E} \times \mathbf{B}$ field as the neutron star interior, forcing it to co-rotate rigidly with the star. However, co-rotation can only be maintained up to a (maximum) distance where the plasma speed reaches the speed of light. This limit

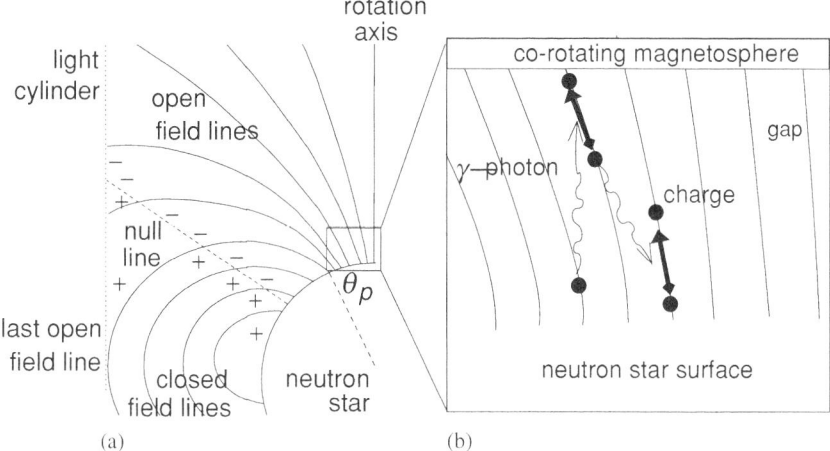

Fig. 3.3. Pulsar magnetosphere in the Goldreich–Julian model. The existence of a polar gap with pair creation cascades is indicated. See text for further details.

defines an imaginary surface known as the *light cylinder* with radius

$$R_{\rm LC} = \frac{c}{\Omega} = \frac{cP}{2\pi} \simeq 4.77 \times 10^4 \ {\rm km} \ \left(\frac{P}{\rm s}\right). \tag{3.21}$$

The magnetic field strength at the light cylinder radius

$$B_{\rm LC} = B_{\rm S}\left(\frac{\Omega R}{c}\right)^3 \simeq 9.2 \ {\rm G} \ \left(\frac{P}{\rm s}\right)^{-5/2} \left(\frac{\dot{P}}{10^{-15}}\right)^{1/2}, \tag{3.22}$$

where we have assumed the same values for radius and moment of inertia as in Equation (3.15).

As shown in Figure 3.1, the existence of the light cylinder divides the dipolar magnetic field lines into two groups: (a) field lines that close within the light cylinder radius (*closed field lines*); (b) field lines that do not close (*open field lines*). The open field line region defines the *polar cap* on the neutron star surface, centred on the magnetic pole. Its boundary, given by $(R, \theta_{\rm p})$, is defined by the 'last open field line' which is tangential to the light cylinder (see Figure 3.3). For a given dipolar field line, the expression $\sin^2\theta/r$ is a constant. For the last open field line, we find

$$\frac{\sin^2\theta}{r} = \frac{1}{R_{\rm LC}} = \frac{2\pi}{c\,P} = \frac{\sin^2\theta_{\rm p}}{R}. \tag{3.23}$$

If the radius of the polar cap measured on the surface, $R_{\rm p}$, is not too large, we can use Equation (3.23) and compute it as

$$R_{\rm p} \simeq R \sin \theta_{\rm p} = \sqrt{\frac{2\pi R^3}{cP}} = 150 \text{ m} \left(\frac{P}{\text{s}}\right)^{-1/2} \tag{3.24}$$

for $R = 10$ km. The polar cap parameters can be used to estimate the potential drop between the magnetic pole and the edge of the polar cap. From Equation (3.17),

$$\Delta\Phi = \frac{B_{\rm S}\Omega^2 R^3}{2c^2} = 2 \times 10^{13} \text{ V} \left(\frac{P}{\text{s}}\right)^{-3/2} \left(\frac{\dot{P}}{10^{-15}}\right)^{1/2}. \tag{3.25}$$

While we expect the overall magnetic field structure to be dipolar, very close to the pulsar surface higher-order multipole moments of the magnetic field may be important. Although the strength of the multipoles is small compared to the dipolar field further away from the surface, multipoles above the polar cap may be relevant for the creation of additional plasma that is eventually responsible for the observed radio emission (see, for example, Jones (1980) and Zhang and Qiao (1996) and Gil and Mitra (2001)).

In the presence of plasma, an azimuthal sweep-back of the magnetic field lines occurs at the light cylinder, which causes the pulsar to spin-down even in the case of aligned rotation and magnetic axes (see, for example, Spitkovsky (2004) and Section 3.2).

3.4 Simple models of the radio beam

The average pulse profiles shown in Section 1.1.2 can be considered as long-exposed images of the time-variable plasma processes that create the individual pulses. The properties of the average profiles should be determined mostly by time-stable factors, e.g. geometry or the dominance of the strong magnetic field. Most of the models proposed to explain the profile properties and discussed below lack a proper theoretical justification. They are motivated by the observations and as such are quite successful in describing the data and even making predictions, e.g. the evolution of the integrated pulse profiles with observing frequency.

3.4.1 Beam geometry

The open field line region provides a narrow and confined site for a stable radio beam. Although other possibilities had been proposed (see, for example, Smith (1969; 1970)), the generally accepted model of a cone-shaped beam centred on the magnetic axis (Radhakrishnan & Cooke 1969, Komesaroff 1970) can account for many of the observed profile properties.

In the simple picture, plasma flows from the surface along the open field lines, emitting photons in a direction that is tangential to the open field lines at the point of emission. Consequently, the opening angle of the resulting conical envelope – usually defined as the half-opening angle ρ – depends upon the width of the open field region at the emission height, r_{em}. Conal beams emitted at a lower height are expected to be narrower than those emitted at a higher altitude. The observed pulse width, however, depends on geometrical factors, i.e. how the observer's line of sight cuts the emission cone.

The geometrical factors are presented in Figure 3.4. The emission cone of angular radius, ρ, is centred on the magnetic axis, and is inclined to the rotation axis by an angle α. As the pulsar rotates, the observer's line of sight traces a curved path within the beam. The emission along this path is observed as the pulse profile and the path length as the pulse width. The observed pulse width, W, measured in longitude of rotation can be calculated by applying simple spherical geometry:

$$\sin^2\left(\frac{W}{4}\right) = \frac{\sin^2(\rho/2) - \sin^2(\beta/2)}{\sin\alpha \cdot \sin(\alpha + \beta)} \tag{3.26}$$

(Gil *et al.* 1984). Sometimes, the equivalent form

$$\cos\rho = \cos\alpha\cos(\alpha + \beta) + \sin\alpha\sin(\alpha + \beta)\cos\left(\frac{W}{2}\right) \tag{3.27}$$

is more practical. The impact angle or *impact parameter* β represents the closest approach of the line of sight to the magnetic axis. One distinguishes between *inner* and *outer* lines of sight. For an outer line of sight (as shown in Figure 3.4), β is positive for $\alpha < 90°$, and negative for $\alpha > 90°$. Correspondingly, an inner line of sight implies $\beta < 0$ for $\alpha < 90°$ and $\beta > 0$ for $\alpha > 90°$. In all cases, $|\beta| \leq \rho$, as the beam would otherwise be missed. Note that W can be larger than 2ρ due to the curved path of the line of sight through the beam.

The midpoint of the observed profile is expected to coincide with the plane that contains the rotation and magnetic axes and the vector pointing from the pulsar to the observer. This geometrically defined *fiducial plane* represents the proper reference point for timing measurements

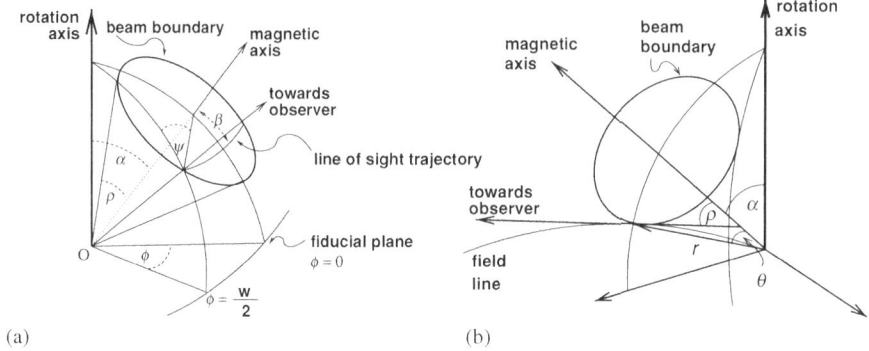

Fig. 3.4. Geometry of the pulsar emission beam. The rotation and magnetic axes are inclined by an angle α. (a) The emission cone with an opening angle ρ is cut by the observer's line of sight, the closest approach of which defines the impact parameter β, measured at the location of the fiducial plane, i.e. at longitude zero. The position angle of the linearly polarised emission, Ψ, is measured with respect to the projected direction of the magnetic axis. (b) The relationship between the polar coordinates of the emission point, (r, θ), and angular radius of the emission cone, ρ.

(see Section 8.1). One often defines the rotation phase as $\phi = 0$ for this *fiducial point*.

If the emission cone is confined by the last open magnetic field lines, we can relate the opening angle of the cone, ρ, to the polar coordinates (r, θ) of these field lines. This relationship is given (see, for example, Gangadhara and Gupta (2001)) by

$$\tan \theta = -\frac{3}{2 \tan \rho} \pm \sqrt{2 + \left(\frac{3}{2 \tan \rho}\right)^2} \qquad (3.28)$$

and is shown in Figure 3.4. For regions close to the magnetic axis (i.e. $\theta \lesssim 20°$ and $\rho \lesssim 30°$), this relationship simplifies to $\theta \approx 2\rho/3$. Assuming Equation (3.23) is valid for the last open dipolar field lines, the opening angle of the cone

$$\rho \approx \frac{3}{2}\theta_{\rm em} \approx \sqrt{\frac{9\pi r_{\rm em}}{2cP}} \text{ radians} = 1.24° \left(\frac{r_{\rm em}}{10 \text{ km}}\right)^{1/2} \left(\frac{P}{\rm s}\right)^{-1/2}, \qquad (3.29)$$

where $(r_{\rm em}, \theta_{\rm em})$ are the coordinates of the emission point. When $\theta \gtrsim 20°$ and/or $\rho \gtrsim 30°$ this approximation breaks down and the exact formulae (Equations (3.23) and (3.28)) must be used.

According to Equation (3.29), this simple model predicts a relationship $\rho \propto P^{-1/2}$ for a given emission height. This is indeed observed (Gil *et al.* 1993b; Rankin 1993 ; Gould 1994; Kramer *et al.* 1994). A conversion of an observed pulse width, W, into an opening angle ρ requires knowledge of the angles α and β, which can be estimated from polarisation measurements (see Section 3.4.4). In Figure 3.5 we show the widths and derived opening angles as a function of period. While the more slowly-rotating normal pulsars follow the expected $\propto P^{-1/2}$ dependence quite closely, this is not the case for the millisecond pulsars. Possible reasons are: (a) deviations from a dipolar field structure in the emission region; (b) an open field line region that is only partly active, producing 'unfilled' emission beams, as argued by Kramer *et al.* (1998; 1999a). The *beaming fraction* (i.e. the fraction of the celestial sphere illuminated by the pulsar beam) is, hence, somewhat smaller than expected for millisecond pulsars. However, it is generally the case that short-period pulsars have larger beams, so they are more likely to intersect with our line of sight than their long-period counterparts.

The above equations have been derived for the case of an aligned rotator. The corrections that need to be applied for an oblique dipole rotator have been discussed by Kapoor and Shukre (1998) who also investigated the possible changes due to space time curvature and special relativistic aberration. The combination of all these effects surprisingly leaves the Goldreich–Julian-type beam essentially unaltered, owing to the mutually opposing nature of these effects. General relativistic effects give, at most, a 4 per cent distortion of the beam.

3.4.2 Emission heights

The widths of pulse profiles for normal pulsars are known to *decrease* with increasing observing frequency (see Section 1.1.2.4). According to Equation (3.29) this is expected if high radio frequencies are emitted from closer to the surface rather than low radio frequencies. It is suggested in Figure 1.3 that the profiles trace the open field line region for different emission heights. Assuming this *radius-to-frequency mapping* (RFM) model, we can use Equation (3.29) to compute the corresponding emission heights from the derived opening angles (Cordes 1978). Expressing the emission height as a fraction of the light-cylinder radius

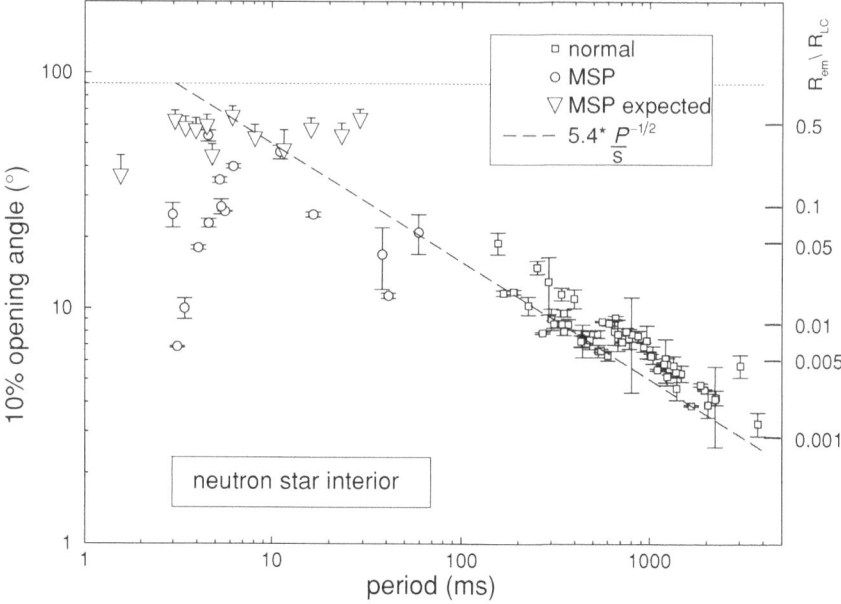

Fig. 3.5. Derived opening angles of the emission cone for normal pulsars (squares) and millisecond pulsars (circles). The opening angle scaling $\rho \propto P^{-0.5}$ is indicated by the dashed line. Upper limits for ρ are shown for some millisecond pulsars where no information is available about the viewing geometry. The triangles mark the expectation value, $\langle \rho \rangle$, with the upper bar indicating the 68 per cent statistical confidence level on the upper bound (see Kramer *et al.* 1998 for details). Assuming dipolar magnetic field lines, we have marked the region of opening angles that corresponds to the interior of a neutron star of 10 km radius. With the same assumption, one can translate an opening angle into an emission height in units of fraction of the light cylinder. This scale is indicated on the right border of the plot. The maximum possible value for ρ ($90°$) is marked by a dotted line.

(Equation (3.21)), we find

$$\rho = 1.5 \text{ radians} \left(\frac{r_{\mathrm{em}}}{R_{\mathrm{LC}}} \right)^{1/2} = 86° \left(\frac{r_{\mathrm{em}}}{R_{\mathrm{LC}}} \right)^{1/2} \qquad (3.30)$$

As for Equation (3.29) this expression is valid only for small angles, $\rho \lesssim 30°$ and, hence, $r_{\mathrm{em}} \lesssim 0.1 R_{\mathrm{LC}}$.

The observation that the widths of millisecond pulsar profiles are essentially independent of frequency (Section 1.1.2.4) suggests that the emission in these pulsars originates from the same height at multiple frequencies (Kramer *et al.* 1999b). This result is not surprising, since the radius of the light cylinder and, hence, the size of the magnetosphere is much smaller for millisecond pulsars (Equation (3.21)). This leaves very little space for the active emission region to spread out.

In addition to the frequency evolution of profile widths, RFM can also explain the appearance and disappearance of profile components (see, for example, Rankin (1983a)), the shift in component location at different frequencies (see, for example, Gangadhara and Gupta (2001)), and the observations that inner profile components appear to have steeper flux density spectra (Lyne & Manchester 1988; Kramer *et al.* 1994; Sieber 1997). As the emission cones become narrower at higher frequencies, the observer's line of sight cuts across a different part of the radio beam, modifying the observed pulse profile. Since aberration effects depend critically upon rotation speed – which increases with emission height – the beaming direction may differ slightly for profile components that are emitted at different altitudes. This effect may compete with a sweep-back of the field lines at high altitudes (Shitov 1983; see Section 3.3).

The frequency dependence of the beam opening angles is described well by a relationship of the form (Thorsett 1991; Xilouris *et al.* 1996)

$$\rho = A_1 f^{-q} + A_0, \tag{3.31}$$

which translates into a similar expression for the emission height

$$r_{\rm em} = B_1 \, f^{-p} + B_0 \tag{3.32}$$

with constants A_0, A_1, B_0, B_1 and q and RFM index p. The constant B_0 corresponds to the emission height at the highest radio frequencies. Recently, Mitra and Rankin (2002) showed that the opening angles of the outer edges of the profile follow the expected relationship while the angles corresponding to inner profile components appear to be frequency independent. Using the somewhat simpler form of Equation (3.32), without the offset B_0, Kijak and Gil (1998; 2003) combine data from several pulsars to derive

$$r_{\rm em}^{\rm KG} = 400 \pm 80 \, {\rm km} \left(\frac{f}{\rm GHz} \right)^{-0.26 \pm 0.09} \left(\frac{\dot{P}}{10^{-15}} \right)^{0.07 \pm 0.03} \left(\frac{P}{\rm s} \right)^{0.30 \pm 0.05}.$$
$$\tag{3.33}$$

In addition to this geometrical approach, one can use timing data obtained at several frequencies to look for delays in the expected arrival times of pulses caused by retardation, aberration effects or the impact of magnetic sweep-back near the light cylinder. So far, studies have only been able to place upper limits on the extent of the emission region

(Cordes & Stinebring 1984; Phillips 1992; Kramer *et al.* 1997). However, these limits are consistent with the geometrical results.

The emission heights determined from simple geometrical arguments also can be compared to those derived from polarisation measurements and fits to the position angle swing (see Section 3.4.4) or in very few cases from observations of scintillation effects (see, for example, Wolszczan and Cordes (1987)) (see Chapter 4). The general agreement between the geometrical method and the polarisation data (Blaskiewicz *et al.* 1991; von Hoensbroech & Xilouris 1997), combined with the success in explaining the frequency evolution of pulse profiles, gives strong support to the RFM model.

As a result of this success, RFM is embedded in a number of theoretical models. Ruderman and Sutherland (1975), for example, assumed that the radiation frequency is related to the local plasma frequency that decreases with distance from the surface (see Equation (3.19)). However, a radically different approach suggests that all emission is created at a single emission height and that the observations can be explained by the propagation of natural polarisation modes in the pulsar magnetosphere, (see, for example, Barnard and Arons (1986) and McKinnon (1997)). This model assumes that one polarisation mode can escape the magnetosphere directly, while the emission of the other mode is ducted along the field lines before it can also escape at a radius $r_{\rm esc}$ to reach the observer.

The existence of two simultaneously emitting polarisation modes is suggested by observed jumps in the position angle swing curve (see, for example, McKinnon and Stinebring (1988)) (see Section 1.1.5) and theory (see, for example, Barnard and Arons (1986) and Petrova (2001)). The radius $r_{\rm esc}$ may be related to the concept of a *polarisation limiting radius* which defines the region in the magnetosphere up to which the polarisation properties of the radio emission are affected (Barnard 1986).

While it is difficult to distinguish between $r_{\rm esc}$ and $r_{\rm em}$ from observations, differences that should be apparent for inner and outer profile components may have been observed by Mitra and Rankin (2002). For most practical purposes it may be sufficient to identify the term 'emission height' with the altitude at which the observed emission *leaves* the pulsar magnetosphere, regardless of whether it was created at that height or further below.

3.4.3 Beam structure

Komesaroff (1970) proposed that the radio emission is due to curvature radiation of particles flowing along the open field lines. As the emitted power increases with decreasing curvature radius, intensity would be largest for the last open field lines and should vanish along the magnetic axis, forming a centred hollow cone. When the line of sight of a distant observer cuts the cone, a two-component profile is observed, each component corresponding to one edge of the cone. A single component profile is observed if the emission cone is just grazed.

As seen in Chapter 1, some pulse profiles exhibit more than simply one or two components. In order to account for such cases, the hollow-cone model has been extended to include a pencil-like beam along the magnetic axis (Backer 1976), known as the *core* component (in contrast to *conal components*; Rankin 1983a). An additional modification to the model is to replace the simple hollow cone with a multiple cone structure nested around a central core component (Oster & Sieber 1976; Sieber & Oster 1977; Rankin 1993; Gil *et al.* 1993a). Different numbers of pulse components can be observed depending on the line of sight. (see Figure 3.6).

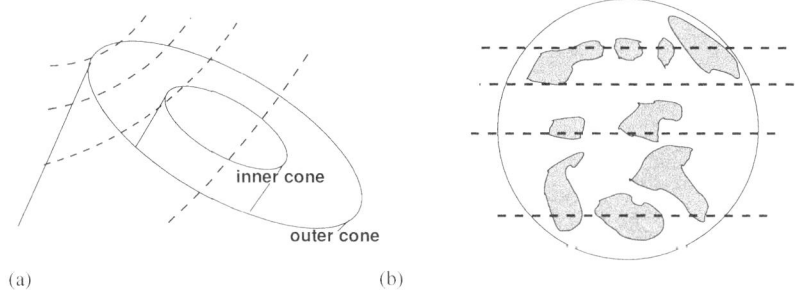

(a)   (b)

Fig. 3.6. Creation of multi-component pulse profiles in two competing beam models. (a) A nested cone structure (Rankin 1993; Gil *et al.* 1993). (b) A patchy beam structure (Lyne & Manchester 1988).

An alternative model suggests that the interior of the radio beam is filled randomly with discrete emitting regions and that only patches of the open field line region are active (Lyne & Manchester 1988). While a time-averaged nested cone structure may be justified physically by a model of plasma discharges occurring in a gap region above the polar cap (see Section 3.5.1), the model may have difficulties in explaining a number of very complex profiles observed for some pulsars (see Figure 1.2). In general, pulse profiles are not very complicated and those of normal and millisecond pulsars do not differ significantly in their complexity (Kramer *et al.* 1998); (see Section 1.1.2.1). Figure 3.7 shows a selection of normal and millisecond pulsar profiles. Can you guess which panel corresponds to which set of pulsars? See the footnote[2] at the bottom of this page for the answer.

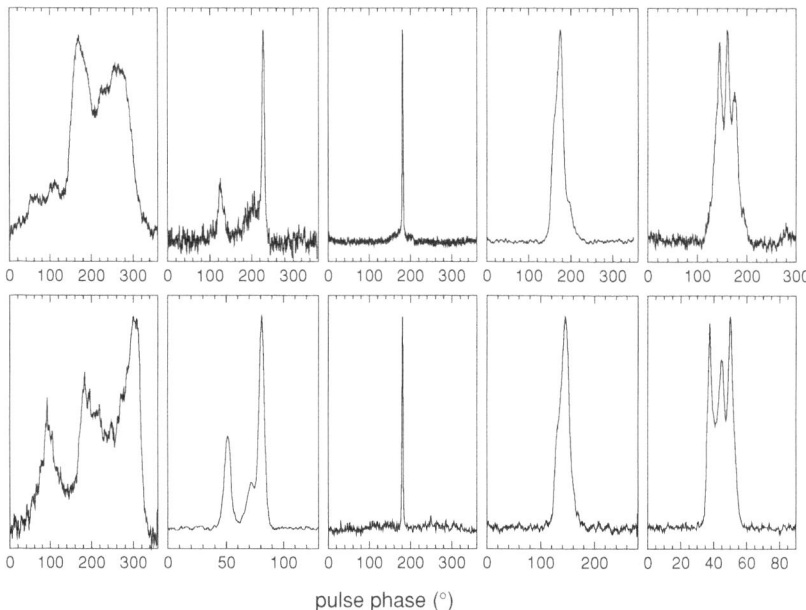

pulse phase (°)

Fig. 3.7. A comparison of selected profiles of normal and millisecond pulsars to demonstrate that both types of profiles generally exhibit similar complexity (Kramer & Xilouris 2000); (see text for details). All profiles are freely available as part of the EPN database (see Appendix 3).

If an only partly filled emission region is a good description for millisecond pulsar beams (Kramer *et al.* 1999a); (see Section 3.4.1), one would consider this as a patchy beam. However, unlike the nested cone/RFM approach, this model is usually unable to predict the profile evolution

2 The millisecond pulsar profiles are the ones shown in the upper panel! The pulsars are: (upper row) PSRs J0218+4218, J0621+1001, B1534+12, J1640+2224, J1730−2304, (lower row) PSRs B1831−04, B2045−16, B2110+27, B2016+28, B1826−17.

with frequency. The observations seem to provide some direct evidence for nested cones, as the opening angles for inner and outer profile components both follow $\rho \propto k P^{-1/2}$ but with different values for the scaling factor, $k_{\text{inner}} < k_{\text{outer}}$ (see, for example, Gil *et al.* (1993) and Rankin (1993)). However, the interpretation of the data remains controversial (see, for example, Han and Manchester (2001) and Kijak and Gil (2002)).

3.4.4 Rotating-vector model

Among the best arguments for the cone model is the behaviour of the linear polarisation observed in integrated profiles. Radhakrishnan and Cooke (1969) explained the S-shaped sweep of the polarisation position angle (PPA); (see Chapter 1) by simple geometrical arguments if the plane of the linearly polarised emission is determined by the direction of the magnetic field at the point of emission. When the beam sweeps across the observer, the projected direction rotates with the star and the measured PPA, Ψ, varies slowly at the outer wings of the profiles and changes rapidly at the profile centre (see Figure 3.8). The so-called *rotating vector model* (RVM) predicts the PPA swing as follows:

$$\tan(\Psi - \Psi_0) = \frac{\sin\alpha \ \sin(\phi - \phi_0)}{\sin(\alpha + \beta) \ \cos\alpha - \cos(\alpha + \beta) \sin\alpha \cos(\phi - \phi_0)}, \quad (3.34)$$

where, as before, ϕ is the rotational phase (pulse longitude), α is the magnetic inclination angle and β is the impact parameter. At the longitude of the fiducial plane, ϕ_0, the position angle $\Psi = \Psi_0$.

While fits of this model to the PPA swing allow in principle a determination of α and β, in practice the range of pulse longitudes ϕ with well-defined position angles usually is limited by the small pulse duty cycle. As a result, the available data do not always tightly constrain α and β. This limitation often prevents a distinction between the inner and outer lines of sight (see Section 3.4.1) which differ significantly in their $\Psi(\phi)$ behaviour for large longitudes $\phi - \phi_0$ (Narayan & Vivekanand 1982); (see Figure 3.8). In contrast, the steepest gradient of Equation (3.34) can be determined easily and is related to α and β as follows:

$$\left(\frac{d\Psi}{d\phi}\right)_{\text{max}} = \frac{\sin\alpha}{\sin\beta} \quad (3.35)$$

which is measured at the fiducial plane, $\phi = \phi_0$ (see Section 3.4.1). Apparently, the gradient is largest if the beam is cut centrally, i.e. $\beta = 0$.

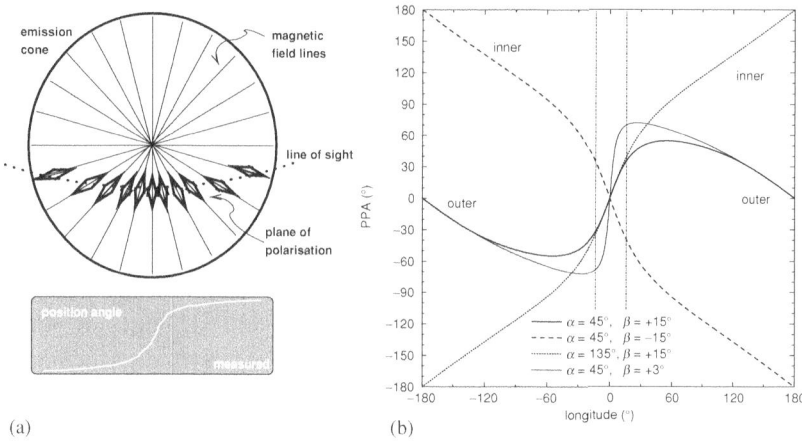

Fig. 3.8. Rotating-vector model of Radhakrishnan and Cooke (1969). (a) The top plot shows a pole-on view of the magnetic axis. The plane of linear polarisation is tied to the magnetic field lines resulting in the characteristic S-shaped curve shown on the lower left. The position angle of the linearly polarised emission, Ψ, is measured with respect to the projected direction of the magnetic axis and rotates throughout the pulse by at most $180°$. The steepest gradient is expected for the location of the fiducial plane. (b) The expected curves as a function of pulse longitude for a selection of inclination angles and impact parameters. Also, we have indicated the mean profile width of $30°$ for the pulsar population by the vertical lines. From this we note: (i) the steepest gradient increases as β decreases (see Equation (3.35)); (ii) that inner and outer lines of sight always have a characteristic form (i.e. a continual rise for the inner line and a roll-over for the outer line); (iii) when sampling only a small range of pulse longitudes, it is often not possible to discriminate between inner and outer lines of sight.

A shallow PPA swing curve, on the other hand, generally means that the beam is cut at its outer edge, i.e. the impact angle is large.

The simple RVM has been extended by including corrections due to aberration, retardation, magnetic sweep-back and the effects of plasma currents (Blaskiewicz *et al.* 1991; Hibschman & Arons 2001a). These effects can shift the PPA swing both in longitude ϕ and absolute value Ψ_0. Aberration, for example, advances the total intensity profile in phase, while the PPA curve is delayed. The resulting shift $\delta\phi$ between the profile and PPA swing $\delta\phi = \delta t/P = 4r_{em}/cP$ (Blaskiewicz *et al.* 1991). The dependence on the emission height provides a means to determine r_{em}. In almost all studied cases, the PPA is indeed delayed as expected and the resulting emission heights are consistent with (Blaskiewicz *et al.* 1991; von Hoensbroech & Xilouris 1997) or somewhat lower than the geometrically derived values (Mitra & Li 2004).

3.4.5 Energetics

A correct estimation of a pulsar's radio luminosity L is difficult for a number of reasons: (a) the radio emission is highly beamed; (b) we observe only a one-dimensional cut through an emission beam of unknown shape; (c) the flux density spectrum does not have the same shape for all pulsars (see Section 1.1.3); (d) reliable estimates of pulsar distances are difficult to obtain (see Section 1.2.1); (e) pulsar flux densities may be affected by scintillation (see Section 1.1.3). Assuming that the intensity distribution along the observer's cut through the emission cone is representative for the whole beam, for a pulsar at a distance d we may write

$$L = \zeta d^2 \int S(f)_{\text{peak}} \, df, \tag{3.36}$$

where the ζ is the solid angle illuminated by the pulsar beam. The fraction of the illuminated celestial sphere, $\zeta/4\pi$, is often referred to as the *beaming fraction*. The angle ζ is related to the radius of the emitting cone, ρ, as follows:

$$\zeta = \int_0^\rho 2\pi \sin \rho' \, d\rho' = 2\pi(1 - \cos \rho) = 4\pi \sin^2 \left(\frac{\rho}{2}\right). \tag{3.37}$$

If the beam geometry is known then ρ can be expressed in terms of the pulse width and period (see Section 3.4.1).

The peak flux density $S(f)_{\text{peak}}$ is simply the maximum intensity of the profile. To express this in terms of the more commonly quoted *mean flux density* S_{mean} (i.e. the integrated intensity under the pulse averaged over the full period), we define the equivalent width W_{eq} as the width of a top-hat pulse having the same area and peak flux as the true profile. It then follows that $S_{\text{mean}} = (W_{\text{eq}} S_{\text{peak}})/P$, where P is the pulse period.

The flux density spectrum is described usually (see Section 1.1.3) in terms of observing frequency f by a simple power law

$$S_{\text{mean}}(f) = S_{\text{mean}}(f_0) \left(\frac{f}{f_0}\right)^\xi, \tag{3.38}$$

where $S_{\text{mean}}(f_0)$ is the mean flux density measured at a frequency f_0.

Incorporating all these factors into Equation (3.36), we obtain

$$L = \frac{4\pi d^2}{\delta} \sin^2\left(\frac{\rho}{2}\right) \int_{f_1}^{f_2} S_{\mathrm{mean}}(f)\,\mathrm{d}f \tag{3.39}$$

$$= \frac{2\pi d^2}{\delta} (1 - \cos\rho)\, S_{\mathrm{mean}}(f_0)\, \frac{f_0^{-\xi}}{\xi + 1} \left(f_2^{\xi+1} - f_1^{\xi+1}\right), \tag{3.40}$$

where the pulse duty cycle $\delta = W_{\mathrm{eq}}/P$, and f_1 and f_2 describe the frequency range at which pulsars have been detected and studied, i.e. a range from a few tens of MHz (see, for example, Izvekova *et al.* (1981)) up to 87 GHz (Morris *et al.* 1997). Assuming typical values of $\delta \approx 0.04$, $\rho \approx 6°$ for $P \approx 1$ s (see, for example, Gould (1994)), $\xi \approx -1.8$ (Maron *et al.* 2000), $f_1 \approx 10^7$ Hz and $f_2 \approx 10^{11}$ Hz we find for a reference frequency $f_0 = 1400$ MHz:

$$L \simeq 7.4 \times 10^{27} \text{ erg s}^{-1} \left(\frac{d}{\mathrm{kpc}}\right)^2 \left(\frac{S_{1400}}{\mathrm{mJy}}\right) \tag{3.41}$$

where S_{1400} is the mean flux density measured at 1400 MHz. In a general case, the first factor needs to be computed from the observed pulse period, width and, ideally, the known geometry to derive ρ from α, β and W using Equation (3.27). In practice, this is not usually known reliably and we define the 'pseudoluminosities'

$$L_{400} \equiv S_{400} d^2 \quad \text{or} \quad L_{1400} \equiv S_{1400} d^2 \tag{3.42}$$

using the mean flux density measured at 400 or 1400 MHz, respectively, quoted in units of mJy kpc^2.

While pulsars are generally weak radio sources, the emission originates from very small regions, so that the estimated *brightness temperature*, T_{B}, is very large. The brightness temperature corresponds to the (thermodynamic) temperature of a black body radiating the same observed radio intensity in the classical Rayleigh–Jeans part of the Planck spectrum. The flux density S_{f} observed at a given frequency for a source of radius r at distance d is then given by

$$S_{\mathrm{f}} = \frac{2\pi r^2 f^2}{d^2} k_{\mathrm{B}} T_{\mathrm{B}}, \tag{3.43}$$

where k_{B} is Boltzmann's constant. If we receive a pulse of peak flux density S_{peak} with duration Δt from such a source, the maximum size of the source is given by the light travel time across it, i.e. $r < c\Delta t$. The corresponding brightness temperature

$$T_B = \frac{S_{\text{peak}}}{2\pi k_B} \left(\frac{f \Delta t}{d} \right)^2 \tag{3.44}$$

$$\simeq 10^{30} \text{ K} \left(\frac{S_{\text{peak}}}{\text{Jy}} \right) \left(\frac{f}{\text{GHz}} \right)^{-2} \left(\frac{\Delta t}{\mu s} \right)^{-2} \left(\frac{d}{\text{kpc}} \right)^2. \tag{3.45}$$

Inserting values observed for the giant pulses of the Crab pulsar (see, for example, Hankins *et al.* (2003)); (see Section 1.1.4.1), we obtain values well in excess of 10^{35} K. As such high brightness temperatures rule out all incoherent mechanisms, the pulsar emission must be a coherent process.

3.5 Acceleration gaps

Depending on the proposed radiation mechanisms (see Section 3.6), large magnetospheric plasma densities, exceeding the Goldreich–Julian density (see Equation (3.20)) by many orders of magnitude, are required. Many models predict such plasma multiplication due to the existence of magnetospheric *gap* regions. In these gaps the Goldreich–Julian condition of a co-rotating magnetosphere breaks down, the force-free state inside the gaps cannot be maintained and a residual electric field $E_\parallel$ must exist (see discussion of Equation (3.16)). Gaps therefore are expected in regions in which the magnetosphere is depleted of plasma. Two such regions have been identified in the literature: (a) in the open field line region above the magnetic polar cap ('polar gap'); (b) between the outer and inner field lines close to the light cylinder ('outer gap'). These regions are indicated in Figure 3.1.

3.5.1 Polar gap

Most radio emission models require a polar gap to provide a dense electron–positron pair plasma that is assumed to originate in a *pair cascade* in the gap, e.g. Sturrock (1971) and Ruderman and Sutherland (1975), and Figure 3.3. It is proposed that charged particles pulled from the surface are accelerated due to a large residual electric field (see Equation (3.18)), reaching relativistic energies ($\gamma \lesssim 10^7$). Moving along the curved magnetic field lines, these particles produce γ-ray photons either by curvature emission (see, for example, Ruderman and Sutherland (1975) and Arons (1983)) or by inverse Compton scattering on lower-energy photons, e.g. Daugherty and Harding (1986). Due to the presence of the strong magnetic field, the γ-ray photons can split, resulting in magnetic one-photon electron–positron pair creation if the photon energy exceeds twice the rest mass of an electron, i.e. $E_\gamma \geq 2m_ec^2$ (Erber 1966).

The new generation of particles may produce further photons and particles, leading to an avalanche of *secondary pair plasma* (Sturrock 1971). As a result, this pair cascade may serve to multiply the initial plasma density by a factor of 10 to 10^4, e.g. Hibschman and Arons (2001b) and Arendt and Eilek (2002). It is commonly believed that this secondary (or perhaps a created tertiary) plasma produces the observed radio emission at some distance from the pulsar surface. Detailed modifications to this basic picture include general relativistic effects (Muslimov & Tsygan 1992; Muslimov & Harding 1997); the effect of multipoles on the curvature radius of the magnetic field lines and their influence on the production of the γ-rays, e.g. Asseo and Khechinashvili (2002) and Zhang and Cheng (2003).

The pair cascade model implies that a lack of such processes will prevent pulsars from emitting radio emission. Assuming that the accelerating potential (see Equation (3.17)) needs to be larger than a critical value, its dependence on P and $\dot{P}$ provides a *death line* (see, for example, Chen and Ruderman (1993)) and an explanation as to why the lower-right part of the P–$\dot{P}$ diagram (see Figure 1.13) lacks the otherwise expected long-period pulsars. The discovery of the 8.5 s radio pulsar J2144–3933 by Young *et al.* (1999), however, provides an interesting challenge for most proposed death lines.

Pair cascade processes may also be prevented (Daugherty & Harding 1983) if the magnetic field strength exceeds the *critical magnetic field*

$$B_{\mathrm{crit}} = m_e^2 c^3/e\hbar = 4.4 \times 10^{13} \text{ Gauss}. \qquad (3.46)$$

For magnetic fields of such strengths, other processes may compete with magnetic photon splitting, changing the opacity of the gap region. A curvature photon may, for example, decay into a *positronium* – a bound electron–positron pair, cf. Daugherty and Harding (1983) and Usov and Melrose (1996). Alternatively, it might simply decay into two other gamma photons (Baring & Harding 1998). Such arguments have been used to explain the absence of radio emission from highly magnetised neutron stars called *magnetars* (Baring & Harding 1998; 2001) although these models have been challenged by the discovery of high magnetic field radio pulsars (Camilo *et al.* 2000a; McLaughlin *et al.* 2003); (see Chapter 9).

A pair cascade may not occur simultaneously across the whole po-lar cap but be localised in the form of discharges of small regions in the polar gap. Such *sparks* may produce columns of secondary or ter-tiary pair plasma that stream into the magnetosphere to produce the observed radio emission. As most parts of the polar gap region would be still depleted of plasma, the created charges would experience a drift that is different from the Goldreich–Julian co-rotation, resulting in a carousel-like rotation of the sparks about the magnetic axis. Ruder-man and Sutherland (1975) suggested that this spark motion and the progressively different positions of the associated plasma columns are responsible for the observed drift of sub-pulses (see Section 1.1.4.4). The spark model has been developed further also to explain a suggested nested cone structure of the pulsar beam (see Section 3.4.3) by a time-averaged circular motion of the sparks about the magnetic axis (Gil & Sendyk 2000; Gil & Mitra 2001). Alternatively, drifting sub-pulses may be explained by phenomena occurring outside the polar gap such as a 'diocotron instability' that may develop naturally in the plasma-filled magnetosphere (Spitkovsky & Arons 2002).

3.5.2 Outer gaps

The existence of gaps is also proposed for the outer magnetosphere near the location of the *null line* $\mathbf{\Omega} \cdot \mathbf{B} = 0$, which separates the space charges of different sign (see Equation (3.16) and Figure 3.3). Holloway (1973) pointed out that charges flowing out from this region would not be re-plenished from the surface, leaving an extended gap between oppositely charged surfaces. The modern form of the outer gap model was devel-oped by Cheng *et al.* (1986) and Romani (1996) and involves particle acceleration, high-energy emission and pair production. The relevant parameters for this region relatively close to the light cylinder differ quite significantly from those of the inner polar gaps. In particular, the local magnetic field is much weaker, so that the magnetic field strength at the light cylinder is a more important factor than the field strength at the surface. Generally, outer gaps are not expected to be associated with radio emission although exceptions may exist, e.g. in the Crab pul-sar (see Chapter 9). However, outer gaps are proposed to explain the high-energy emission of pulsars by curvature and synchrotron emission (see Section 3.7). In this case, the lower magnetic fields require particles with much higher γ-factors than for polar gap emission.

3.6 Radio emission

The nature of pulsar radio emission process is still unknown. Any proposed radiation process has not only to explain the coherency and high degree of polarisation of the emission, but also has to work efficiently over nearly 4 orders of magnitude in rotation period and 6 orders of magnitude in magnetic field, producing more or less similar emission characteristics. Similarly, pulsar emission is observed over about 10 octaves of radio frequencies (from ~ 100 MHz to 100 GHz), clearly requiring a very broad-band emission process.

We do not attempt to review the various proposed emission models, and will provide only a basic overview. In general, one can distinguish between antenna mechanisms (i.e. emission by bunches of particles), relativistic plasma emission and maser mechanisms, e.g. Melrose (2004). Most of them make use of a secondary pair plasma and place the origin of the pulsar radiation in the inner region of the magnetosphere.

In antenna mechanisms, N particles of charge q, confined in a volume of a size smaller than half of the emitted wavelength, will all radiate in phase, i.e. acting like a particle of charge Nq. The emitted power will then be N^2 times the power which is radiated by one single particle (rather than only N times the single particle power as in the incoherent case). Bunched curvature emission flowing along curved magnetic field lines was one of the first models proposed for pulsar emission (Komesaroff 1970; Sturrock 1971; Ruderman & Sutherland 1975), although curvature radiation is a rather inefficient process. Curvature emission in particular has been criticised (Lesch *et al.* 1998), while emission by bunches in general has been rejected on the ground that no mechanism exists that is fast enough to create the bunches and to maintain their shape for sufficiently long times, e.g. Melrose (1981; 1992).

The common feature of all models in the second class, involving relativistic plasma emission, is that they all require a certain kind of plasma instability. Apart from problems allowing the instability to grow sufficiently rapidly, the energy of the resulting plasma turbulence cannot, in many cases, escape directly. Instead, non-linear processes have to convert it first into a certain wave mode. The various models differ mainly in the proposed kind of plasma turbulence and the conversion processes used, e.g. Arons and Barnard (1986), Melrose (1992) and Asseo (1993).

Maser emission mechanisms proposed as a source for pulsar radiation can be interpreted as negative absorption. Again, various kinds of maser mechanisms have been suggested, e.g. the free electron maser (see, for example, Melrose (1989) and Rowe (1995)) or emission driven by the curvature-drift instability (see, for example, Kazbegi *et al.* (1991)). While the latter is again an indirect emission mechanism, maser mechanisms have generally the advantage over relativistic plasma emission in that the radiation can escape the magnetosphere directly.

3.7 High-energy emission

Likely emission processes for the observed non-thermal optical, X-ray and γ-ray emission are: (a) synchrotron emission; (b) curvature emission; (c) inverse Compton scattering. The location of this emission is thought to be either in the polar cap region (see, for example, Daugherty and Harding (1986) and Harding *et al.* (2002)) or in the outer gaps (see, for example, Cheng *et al.* (1986) and Romani (1996)). Both of these locations are shown schematically in Figure 3.1.

In the polar cap model, the observed emission is due to inverse Compton scattering of infrared photons by upward polar cap cascades. Resonant–inverse Compton scattering off thermal photons from the surface can be particularly effective (Zhang & Harding 2000). Polar cap models expect a cut-off of the high-energy spectrum due to γ-B-field absorption.

Outer gap models involve particle acceleration, pair production and high-energy emission near the light-cylinder. The observed emission is due to a combination of curvature radiation and synchrotron emission of downward propagating pairs. Inverse Compton scattering appears to be unimportant due to the low density of thermal photons from the stellar surface. The spectrum is curvature-radiation limited and should extend to higher energies than predicted by inner gap models (see, for example, Romani (1996)). By extending the observed γ-ray spectrum to a few tens of GeV with upcoming satellites like GLAST, it may be possible to distinguish between the outer gap and polar cap model.

In both families of models, the emission beam appears to be much wider than the radio beam. The most detailed calculations for the expected high-energy profiles and spectra have been made for the outer gap models (Romani & Yadigaroglu 1995; Romani 1996; Romani 2003; Cheng *et al.* 2000). Outer gap models are successful in predicting the prominent double-peaked profiles observed at high-energies (see Chapter 9) and in accounting for many details of the observed features by geometrical factors. On the other hand, the radio emission models suggest the existence of polar gap pair creation, which may then be naturally associated with the high energy emission. Both models are still discussed controversially.

3.8 Further reading

A detailed account of the properties of neutron stars is given in the books *Black Holes, White Dwarfs and Neutron Stars. The Physics of Compact Objects* by Shapiro and Teukolsky (1983) and *Compact Stars: Nuclear Physics, Particle Physics and General Relativity* by Glendenning (2000). Recent reviews about the characteristics of neutron star atmospheres, in particular in the presence of strong magnetic fields, were presented by Pavlov *et al.* (2002) and Zavlin and Pavlov (2002). Examples of detailed modern calculations can be found in Pons *et al.* (2002) and Braje and Romani (2002).

After the influential paper by Goldreich and Julian (1969), the concept of a pulsar magnetosphere has been developed by a large number of authors. The work is summarised and discussed in *Theory of Neutron Star Magnetospheres* by Michel (1991) and *Physics of the Pulsar Magnetosphere* by Beskin, Gurevich and Istomin (1993). Recent reviews updating these works are given by Michel and Li (1999) and Beskin (1999). The most recent progress is summarised by Spitkovsky (2004).

In a number of excellent reviews, Melrose (1992; 2000; 2004) discussed the problems related to the theoretical models proposed to explain the observed radio emission. These problems are not only concerned with the emission process, but also the aspects of wave propagation in the magnetosphere that also are discussed, for example, by Barnard and Arons (1986) and recently by Petrova (2001).

The two competing models put forward to account for high-energy emission of pulsars are developed in a number of publications, e.g. Daugherty and Harding (1986), Cheng *et al.* (1986), Romani and Yadigaroglu (1995) and Romani (1996).

4

Effects of the interstellar medium

In Chapters 1 and 2 we saw that the pulsar signal is affected in a number of ways (frequency dispersion, Faraday rotation, scintillation and scattering) as it propagates through the interstellar medium (ISM). This chapter presents the basic theoretical background in those areas necessary for planning and interpreting observations. We begin with a discussion of dispersion and Faraday rotation which can be understood by propagation through a homogeneous medium. Then we move on to describe scattering and scintillation effects caused by propagation through a more realistic model of the ISM which includes turbulence on a variety of length scales. As essentially perfect point sources, pulsars show scintillation effects that are absent in extended radio sources for the same reason that 'stars twinkle, and planets do not'. This has a number of important observational consequences that we discuss in detail.

4.1 Propagation through a homogeneous medium

The ISM is a cold, ionised plasma. Electromagnetic radiation from pulsars and other sources will experience a frequency-dependent index of refraction as they propagate through the ISM. Neglecting, for now, small corrections due to the Galactic magnetic field, the refractive index

$$\mu = \sqrt{1 - \left(\frac{f_{\rm p}}{f}\right)^2},\qquad(4.1)$$

where f is the wave (observing) frequency and the plasma frequency

$$f_{\rm p} = \sqrt{\frac{e^2 n_{\rm e}}{\pi m_{\rm e}}} \simeq 8.5\,{\rm kHz}\left(\frac{n_{\rm e}}{{\rm cm}^{-3}}\right)^{1/2}.\qquad(4.2)$$

Here n_e is the electron number density, while e and m_e are the charge and mass of an electron respectively. For the ISM, we find that typically $n_e \sim 0.03$ cm^{-3} (see, for example, Ables and Manchester (1976)), so that $f_p \simeq 1.5$ kHz. We note from Equation (4.1) that a wave will not propagate if $f < f_p$. This fact is sometimes used to constrain the plasma density in a pulsar magnetosphere or the location where the radio emission leaves the magnetosphere (Ruderman & Sutherland 1975).

4.1.1 Dispersion

From Equation (4.1) we note that $\mu < 1$. It follows therefore that the group velocity of a propagating wave $v_g = c\mu$ is less than the speed of light c. Consequently, the propagation of a radio signal along a path of length d from the pulsar to Earth will be delayed in time with respect to a signal of infinite frequency by an amount

$$t = \left(\int_0^d \frac{dl}{v_g} \right) - \frac{d}{c}. \tag{4.3}$$

Substituting $v_g = c\mu$, and noting that $f_p \ll f$ to approximate μ, we find

$$t = \frac{1}{c} \int_0^d \left[1 + \frac{f_p^2}{2f^2} \right] dl - \frac{d}{c} = \frac{e^2}{2\pi m_e c} \frac{\int_0^d n_e \, dl}{f^2} \equiv \mathcal{D} \times \frac{\text{DM}}{f^2}, \tag{4.4}$$

where the *dispersion measure*

$$\text{DM} = \int_0^d n_e \, dl \tag{4.5}$$

is usually expressed in cm^{-3} pc and the *dispersion constant*

$$\mathcal{D} \equiv \frac{e^2}{2\pi m_e c} = (4.148808 \pm 0.000003) \times 10^3 \text{ MHz}^2 \text{ pc}^{-1} \text{ cm}^3 \text{ s}. \tag{4.6}$$

Note that the uncertainty in $\mathcal{D}$ is determined by the uncertainties in e and m_e. Throughout the rest of this book, we will often write the approximate numerical factor of 4.15 for convenience, but recommend using the full precision in $\mathcal{D}$ in practice[1]. With this in mind, the delay between two frequencies, f_1 and f_2 both in MHz, is

$$\Delta t \simeq 4.15 \times 10^6 \text{ ms } \times (f_1^{-2} - f_2^{-2}) \times \text{DM}. \tag{4.7}$$

---

1 Note that our definition is the reciprocal of that introduced by Manchester and Taylor (1972), i.e. $1/\mathcal{D} \equiv 2.41 \times 10^{-4}$ MHz^{-2} pc cm^{-3} s^{-1}.

From a measurement of the pulse arrival time at two or more different frequencies, we can infer the DM along the line of sight to the pulsar. The DM then can be used to estimate the distance to a pulsar by numerically integrating Equation (4.5) assuming a model for the Galactic electron density distribution, n_e.

Following earlier relatively simple symmetric distributions for n_e (see, for example, Lyne *et al.* (1985)), Taylor and Cordes (1993) included the effects of spiral arm structure in their model. The most recent 'NE2001' model (Cordes & Lazio 2002a,b) was shown in Figure 1.9 and incorporates the largest number of pulsars with independent distance measurements used for calibration. It models the electron density distribution as the sum of several components, i.e. a thin and thick disk component, features of the local ISM, the spiral arms and contributions from the Galactic centre region. Additionally, it includes 'voids' and 'clumps' in n_e in order to account for specially identified features, e.g. the Gum nebula.

4.1.2 Faraday rotation

The frequency-dependent delay of the signal also can be expressed in terms of phase rotations. Compared to a pulse observed at infinite frequency (i.e. no dispersion), the pulse phase at a frequency f is observed to lag in phase by $\Delta\Psi = -kd$, where $k = 2\pi/\lambda$ is the wavenumber and λ is the wavelength. For a cold, now magnetised plasma:

$$k(f) = \frac{2\pi}{c}\mu f = \frac{2\pi}{c}f\sqrt{1 - \frac{f_p^2}{f^2} \mp \frac{f_p^2 f_B}{f^3}}, \qquad (4.8)$$

and the Galactic magnetic field along the line of sight, $B_{\parallel}$ enters the computations via the cyclotron frequency

$$f_B = \frac{eB_{\parallel}}{2\pi m_e c} \simeq 3\,\mathrm{MHz}\left(\frac{B_{\parallel}}{\mathrm{G}}\right). \qquad (4.9)$$

For the Galaxy, $B_{\parallel} \sim 1\mu\mathrm{G}$, so that we expect $f_B \sim 3\,\mathrm{Hz}$ for the ISM.

The last term in Equation 4.8 reflects the different propagation speeds for the left (upper '−' sign) and right-hand (lower '+' sign) circularly polarised waves[2] in a magnetised medium. The time corrections to the group velocity for an unmagnetised medium are $\lesssim 1$ ns and, hence, negligible. However, this minute difference in propagation speeds is the

---

2 The polarisation state of an electromagnetic wave is defined in Appendix 1.

cause of Faraday rotation. The differential phase rotation between left and right circular polarisations

$$\Delta\Psi_{\text{Faraday}} \simeq \int_0^d (k_{\text{R}} - k_{\text{L}})\, dl, \tag{4.10}$$

where k_{L} and k_{R} are the wavenumbers of the left and right circularly polarised wave, respectively. Since $f \gg f_{\text{p}}$ and $f \gg f_{\text{B}}$, we find:

$$\Delta\Psi_{\text{Faraday}} = \frac{e^3}{\pi m_e^2 c^2 f^2} \int_0^d n_e B_{||}\, dl. \tag{4.11}$$

In addition, as the polarisation position angle (PPA), (see Chapters 3 and 7) is periodic on π rather than 2π for phase, the change in PPA

$$\Delta\Psi_{\text{PPA}} = \Delta\Psi_{\text{Faraday}}/2 \equiv \lambda^2 \times \text{RM}, \tag{4.12}$$

where λ is the wavelength, and we define the *rotation measure*

$$\text{RM} = \frac{e^3}{2\pi m_e^2 c^4} \int_0^d n_e B_{||}\, dl. \tag{4.13}$$

The measurement of both RM and DM gives the average magnetic field strength along the line of sight weighted by electron density:

$$\langle B_{||} \rangle \equiv \frac{\int_0^d n_e B_{||}\, dl}{\int_0^d n_e\, dl} = 1.23\,\mu\text{G} \left(\frac{\text{RM}}{\text{rad m}^{-2}} \right) \left(\frac{\text{DM}}{\text{cm}^{-3}\,\text{pc}} \right)^{-1}. \tag{4.14}$$

As we discuss below, the electron density n_e is not homogeneous throughout the Galaxy. Regions of enhanced electron density (e.g. HII regions along the line of sight) produce an unequal weighting and can contribute significantly to the measurement of $\langle B_{||} \rangle$. Also, if the direction of the magnetic field changes, represented by a sign change in the contributing $B_{||}$, the average value, $\langle B_{||} \rangle$, is a poor reflection of the true situation. An interpretation of derived magnetic field strengths therefore should be made with care (see, for example, Mitra *et al.* (2003) and Chapter 7).

4.2 Propagation through a turbulent medium

In the discussion so far, we have described the propagation of pulsar signals assuming a homogeneous ISM. In reality, the electron density is not homogeneous but shows variations in concentration on a wide range of length scales. In addition to temporal variations in dispersion measures (see Figure 4.1), these changes in concentration also distort and scatter the pulse shape. A relative motion between the pulsar, the scattering medium and the observer leads to the phenomenon of interstellar scintillation (ISS) which manifests itself as intensity variations on various timescales. The theoretical description of these phenomena depends to some degree on the assumed form of the turbulence spectrum of the ISM. We keep intentionally the discussion of these effects simple, describing only the case for the so-called *Kolmogorov spectrum* and assuming that the effects of the scattering medium can be described by the *thin screen* model (Scheuer 1968) introduced in Chapter 1.

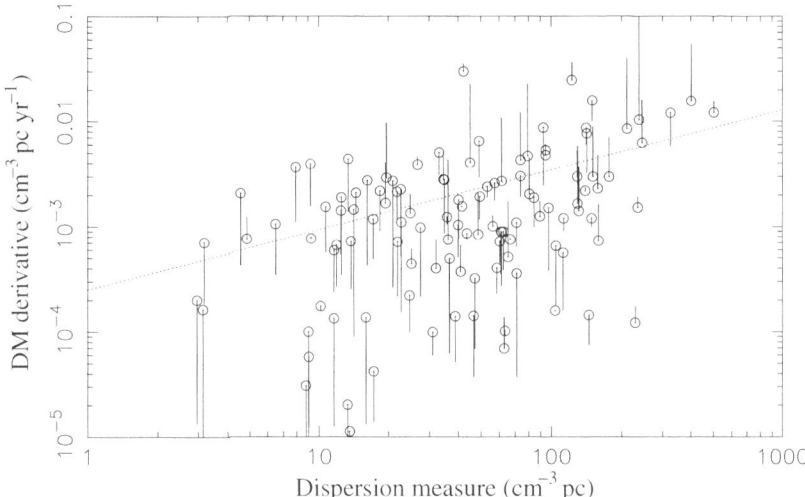

Fig. 4.1. The magnitude of temporal dispersion measure variations plotted against dispersion measure as presented by Hobbs *et al.* (2004). The dotted line suggests a relationship $|d(\mathrm{DM})/dt| \simeq 0.0002\sqrt{\mathrm{DM}}$ cm^{-3} pc yr^{-1} although the scatter is large (~ 1 order of magnitude). The vertical lines at the end of each point give an indication of the variation of $d(\mathrm{DM})/dt$ expected due to each pulsar's transverse velocity. Figure provided by George Hobbs. See also Phillips and Wolszcan (1991) and Backer *et al.* (1993b).

Although the turbulent medium extends through the whole space between the pulsar and the observer, the thin screen model concentrates these irregularities midway along the propagation path. In spite of its simplicity, this model is capable of explaining the basic effects of scattering and scintillation described in Chapter 1. Other assumptions about the thickness and location of the screen modify the scaling laws of the observed parameters with frequency and distance (see Appendix 2) to some extent. However, for most practical purposes, the basic theory presented below is sufficient to plan and understand observations of pulsars.

4.2.1 Interstellar scattering basics

For simplicity, let us assume first that the inhomogeneities distorting the wave consist of variations in the electron density n_e with a single typical size, a, as shown in Figure 4.2. These distortions perturb the phases

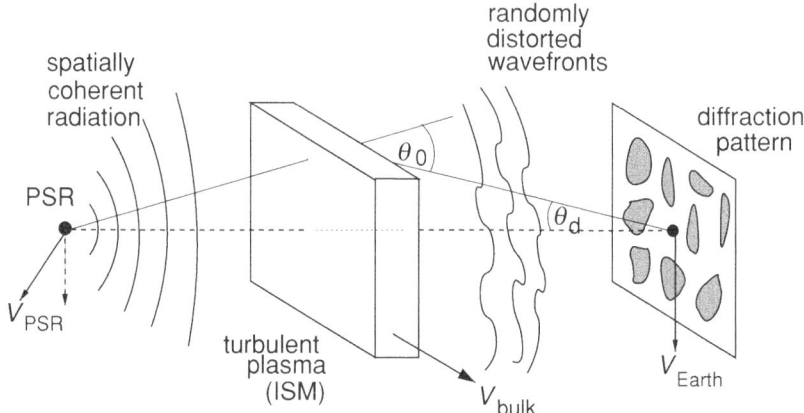

Fig. 4.2. Sketch showing inhomogeneities in the ISM that result in the observed scattering and scintillation effects discussed in this chapter. The initially spatially coherent electromagnetic radiation from the pulsar is distorted by a thin screen of irregularities of various scales. The resulting randomly distorted waves are bent by an angle θ_0 forming a scatter-broadened image of radius θ_d (see text). Scintillation is produced as the randomly distorted wavefronts form an interference pattern at the location of a distant observer. Figure adapted from an original version (Cordes 2002) provided by Jim Cordes.

of a propagating wave due to changes in the refractive index, $\Delta\mu$. After propagation through an inhomogeneity of length a, a wave of frequency f has experienced a phase shift $\delta\Phi = \Delta k\, a$. With $k = (2\pi/c)\,\mu\, f$, using Equations (4.1) and (4.2) we obtain

$$\Delta k = \frac{2e^2}{c\, m_e\, f}\, \Delta n_e. \tag{4.15}$$

The corresponding phase shift is then

$$\delta\Phi = \Delta k\, a \approx \frac{2e^2}{m_e c}\, \frac{a\Delta n_e}{f}. \tag{4.16}$$

After traversing a distance d between the pulsar and the Earth, a ray has encountered a total of about d/a such irregularities in n_e, resulting in a root mean square phase variation

$$\Delta\Phi \approx \sqrt{\frac{d}{a}}\,\delta\phi = \frac{2e^2}{m_e c}\frac{\sqrt{ad}\Delta n_e}{f}. \tag{4.17}$$

We can view the effect of this phase difference $\Delta\Phi$ as a bending of the wavefront by an angle θ_0 at a screen of scale a located midway between the Earth and the pulsar, where

$$\theta_0 \approx \frac{\Delta\Phi/k}{a} \approx \frac{\Delta\Phi c}{2\pi a f} = \frac{e^2}{\pi m_e}\frac{\Delta n_e}{\sqrt{a}}\frac{\sqrt{d}}{f^2}. \tag{4.18}$$

As a result of this bending (Figure 4.2) the observer sees a scatter-broadened image as a diffuse disk centred around the pulsar with an angular radius

$$\theta_d = \theta_0/2 \approx \frac{e^2}{2\pi m_e}\frac{\Delta n_e}{\sqrt{a}}\frac{\sqrt{d}}{f^2}, \tag{4.19}$$

and an angular intensity distribution

$$I(\theta)d\theta \propto \exp(-\theta^2/\theta_d^2)\,2\pi\theta d\theta. \tag{4.20}$$

Rays received by the observer at an angle θ arrive slightly later than those that travelled undeflected. The geometric time delay

$$\Delta t(\theta) = \frac{\theta^2 d}{c} \tag{4.21}$$

can be used to derive the observed intensity as a function of time:

$$I(t) \propto \exp(-c\Delta t/(\theta_d^2 d)) \equiv e^{-\Delta t/\tau_s}, \tag{4.22}$$

where

$$\tau_s = \frac{\theta_d^2 d}{c} = \frac{e^4}{4\pi^2 m_e^2}\frac{\Delta n_e^2}{a}\,d^2\,f^{-4}. \tag{4.23}$$

As a result, a sharp single pulse emitted by the pulsar is measured by the observer as a one-sided exponential function with a *scattering time-scale* τ_s. Any more complicated pulse shape will appear as a convolution of the pulse with the exponential. The larger the scattering disk, the longer the time delay and tail of the exponential.

This simple picture of multi-path propagation reproduces the basic features of the observed pulse scattering and predicts a strong dependence with frequency and distance to the pulsar ($\tau_s \propto f^{-4}d^2$); (see

Chapter 1). A more detailed comparison of this model with the data requires some modifications that we discuss in Section 4.2.3.

4.2.2 Interstellar scintillation basics

The short-term intensity variations seen for many pulsars and interpreted as interstellar scintillation (see Figure 1.10) can be understood in terms of the simple model we have applied so far to scattering. The signals received over the time τ_s show a variety of phases given by $\delta\Phi \sim 2\pi f \tau_s$. Interference of the signals produces an interference pattern at the observer's plane. This pattern of patches with enhanced and reduced intensity moves over the observer due to relative motions between the pulsar, the scattering medium and the observer (see Figure 4.2). As a result, we observe changes in intensity with a timescale Δt that depend upon the relative velocity.

Interference can occur only if the phases of the waves do not differ by more than about 1 radian. As the phases depend on frequency, there is a limitation in bandwidth of the interfering waves, i.e. waves of frequencies outside a *decorrelation bandwidth*, also referred to as the *scintillation bandwidth* Δf, will not contribute. The condition for interference is therefore

$$2\pi\Delta f \tau_s \sim 1, \tag{4.24}$$

which implies the scaling $\Delta f \propto 1/\tau_s \propto f^4$.

Scintillation therefore produces a pattern of intensity variation in both frequency and time. This pattern can be measured in form of a *dynamic spectrum*, a two-dimensional image of pulse intensity as a function of observation time and frequency (see Section 7.4.4 and Figure 7.9). A *scintle* is an enhanced region of flux density in this frequency time plane. Its size in frequency, scintillation bandwidth Δf, is usually defined as the half-width at half-maximum of the autocorrelation function of the spectrum. Similarly, the scintle size in time, the scintillation timescale Δt, can be measured as the half-width at $1/e$ along the time axis (Cordes 1986); (see Section 7.4.4 for more details).

4.2.3 Modelling turbulence in the interstellar medium

While the simple picture presented above describes the basic phenomena of interstellar scattering and scintillation, in reality the variations in electron density in the ISM show a distribution of scales rather than a single size. In order to interpret the observations in more detail, we now replace the single linear size a by a distribution of length scales that can be characterised by a spatial wavenumber spectrum. A commonly used description (see, for example, Rickett (1990)) is the extended power law model:

$$P_{n_e}(q) = \frac{C_{n_e}^2(z)}{(q^2 + \kappa_0^2)^{\beta/2}} \exp\left[-\frac{q^2}{4\kappa_i^2}\right],\qquad(4.25)$$

where $q = 1/a$ is the magnitude of the three-dimensional wavenumber, the 'inner' and 'outer' scales of the turbulence are κ_i and κ_o, respectively, and $C_{n_e}^2(z)$ denotes the strength of the fluctuations along a given line of sight. The integral

$$\mathrm{SM} = \int_0^d C_{n_e}^2(z)\,\mathrm{d}z \qquad(4.26)$$

provides a measure of the electron density fluctuations along the line of sight and is known as the *scattering measure* (see, for example, Cordes *et al.* (1991)). Although SM can be determined from pulse broadening and scintillation measurements and from angular diameter measurements of Galactic and extra-galactic sources, different observables correspond to different lines of sight weighting for $C_{n_e}^2$, resulting in different values of SM for the same line of sight (see, for example, Cordes & Lazio (2002a)). We concentrate our discussion on scintillation parameters and pulse broadening measurements.

The inner and outer scales κ_i and κ_o correspond to cut-offs in scale sizes. For wavenumbers q in the range $\kappa_o \ll q \ll \kappa_i$, Equation (4.25) simplifies to a power law model with a spectral index β, i.e.

$$P_{n_e}(q) = C_{n_e}^2 q^{-\beta}. \qquad(4.27)$$

The study of other turbulent media suggest that energy cascades down from larger to smaller scales, resulting in a *Kolmogorov spectrum* with $\beta = 11/3$. Most observations show that this is also the case for the ISM (even over 10 orders of magnitude in wavenumber, i.e. 10^8 m $< 1/q < 10^{18}$ m); (Armstrong *et al.* 1995), although deviations are seen along some lines of sight. A recent review by Gupta (2000) summarises the arguments for and against a pure Kolmogorov spectrum. For the remainder of this chapter, we assume for simplicity that the interstellar turbulence can indeed be described by Equation (4.27) and $\beta = 11/3$.

For the wavenumber spectrum of Equation (4.27), scattering time and decorrelation bandwidth depend upon observing frequency as follows:

$$\tau_s \propto f^{-\alpha}, \quad \Delta f_{\mathrm{DISS}} \propto f^{\alpha}, \tag{4.28}$$

where $\alpha = 2\beta/(\beta - 2)$ (see, for example, Lee and Jokipii (1975)). For a Kolmogorov spectrum, where $\beta = 11/3$, we have $\alpha = 4.4$ rather than 4 in the thin screen model presented above. Also note that we have added the subscript 'DISS' (diffractive interstellar scintillation) to the scintillation bandwidth as we will find below (Section 4.2.5.2) that different branches of scintillation exist, of which 'diffractive' is relevant here.

In our simple picture we had highlighted also the relationship between τ_s and Δf_{DISS} which we now write in a more general form:

$$2\pi \tau_s \Delta f_{\mathrm{DISS}} = C_1. \tag{4.29}$$

Although the 'constant' C_1 is, indeed, close to unity as was the case for Equation (4.24) (Rickett 1977), its value changes slightly for different geometries and models of the turbulence wavenumber spectrum. For a Kolmogorov spectrum, $C_1 = 1.16$ (see, for example, Lambert and Rickett (1999)). In principle, a simultaneous determination of Δf_{DISS} by scintillation measurements and τ_s by pulse broadening observations can be used to test this relationship. However, it was emphasised by Cordes and Rickett (1998) that the $1/e$ decay time determined from scattered pulse profiles (as we introduced this quantity in Equation (4.22)) can deviate somewhat from the mean delay time of the scattered wave signal (as given in Equation (4.21)).

4.2.4 Interstellar scintillation revisited

The different sizes of the electron density inhomogeneities in the ISM contribute differently to the observed phenomena. Depending also on the distance to the pulsar, three different kinds of scintillation can be observed. While we can distinguish between *weak* and *strong* scintillation, strong scintillation has two different branches: (a) *diffractive*; (b) *refractive* scintillation. In order to discuss the differences between these regimes, it is useful first to introduce a number of related quantities.

As before, we consider the waves as passing through a thin scattering screen mid-way between us and the pulsar. The screen disturbs the phases of the waves so that they are scattered into an angular spectrum of $(e^{-1/2})$ width θ_d (cf. Equation (4.19)). At some distance d the phase modulation caused by the screen produces an amplitude modulation and hence a scintle pattern in the plane of the observer. Whether weak or strong scintillation is observed depends upon the size of the total phase perturbations at this distance. We can determine the regime by considering the size of the circular region on the scattering screen centred around the source at which the phase differences are less than or equal to 1 radian. The size of this region is known as the *field coherence scale* $s_0 = 1/(k\,\theta_d)$. In the simple model with a single scale size, we find $s_0 \propto fd^{-0.5}$. For the Kolmogorov spectrum of turbulence, the dependencies are modified slightly so that $s_0 \propto f^{1.2}d^{-0.6}$.

We can compare the coherence scale to the *Fresnel scale*, l_F known from diffractive optics when we observe an interference pattern in the near field of, for example, an illuminated circular aperture (see, for example, Smith and Thompson (1988)). In this case, the radius of the first *Fresnel zone*

$$l_F = \sqrt{\frac{d}{k}} \simeq 1.2 \times 10^9 \text{ m} \left(\frac{d}{\text{kpc}}\right)^{1/2} \left(\frac{f}{\text{GHz}}\right)^{-1/2}. \qquad (4.30)$$

For the case $s_0 \gg l_F$, the phase fluctuations within l_F are very small. At an observer's distance d for which this conditions holds, intensity variations are caused only by the small phase perturbations within the Fresnel zone that lead to weak scintillation. The Fresnel scale therefore is the dominating length scale for weak scintillation.

At larger distances, we leave eventually the near field and enter the regime of Fraunhofer diffraction in which larger phase perturbations of many radians across the screen are effective. This occurs when $s_0 \ll l_F$. On this scale the effective screen also contains many points with stationary phase. This leads to multi-path propagation, observed as pulse scatter broadening. In terms of geometric optics, we reach a distance at which rays of different parts of the screen can cross, leading to large intensity variations and, hence, strong scintillation.

As indicated, strong scintillation consists of two branches: (a) diffractive; (b) refractive scintillation. While diffractive scintillation is that described in our simple picture above – caused by interference between components of the angular spectrum – the refractive branch corresponds to the larger sizes of the scattering disk. This defines the *refractive scale*

$$l_R = d\theta_d = \frac{d}{ks_0} = \frac{l_F^2}{s_0}. \qquad (4.31)$$

Refractive scintillation causes small-amplitude intensity variations from large-scale focusing and defocusing of the radiation.

4.2.5 Observable manifestations of the scintillation regimes

Now we compare the observational manifestations of the three different types of scintillation. In addition to scintillation bandwidth and timescale, two useful quantities are: (a) modulation index; (b) scintillation strength. The *modulation index*

$$m = \sigma_S / \langle S \rangle , \tag{4.32}$$

where σ_S is the standard deviation of the observed flux densities and $\langle S \rangle$ is their mean. The *scintillation strength*, u, is defined (Rickett 1990) and related to observing frequency f and the observable scintillation bandwidth of strong (diffractive) scintillation, Δf_{DISS} as follows:

$$u \equiv \frac{l_{\mathrm{F}}}{s_0} = \sqrt{\frac{f}{\Delta f_{\mathrm{DISS}}}} \propto f^{-1.7} \, d^{1.1}. \tag{4.33}$$

This expression can be used to define the transition between weak scintillation ($u < 1$) and strong scintillation ($u > 1$).

4.2.5.1 Weak scintillation

As discussed above, weak scintillation occurs when $s_0 \gg l_{\mathrm{F}}$ or when $u < 1$ and phase perturbations are small at the distance of the observer. A relative motion of the pulsar with a speed V_{ISS} moves the scintle pattern produced in the plane of the observer. For a fixed observer, this results in an intensity modulation as the bright and weak intensity patches pass by. The timescale for this intensity variation is determined by the speed of the scintle pattern and, hence, that of the pulsar (see, for example, Rickett (1990)),

$$\Delta t_{\mathrm{weak}} = \frac{l_{\mathrm{F}}}{V_{\mathrm{ISS}}} \simeq 3.35 \, \mathrm{h} \left(\frac{d}{\mathrm{kpc}} \right)^{1/2} \left(\frac{f}{\mathrm{GHz}} \right)^{-1/2} \left(\frac{V_{\mathrm{ISS}}}{100 \, \mathrm{km \, s^{-1}}} \right)^{-1}. \tag{4.34}$$

The expected modulation index $m_{\mathrm{weak}} \sim 0.1 - 0.3$ can be related to the scintillation strength as follows:

$$m_{\mathrm{weak}} = \sqrt{u^{5/3}} = \left(\frac{f}{\Delta f_{\mathrm{DISS}}} \right)^{5/12} \propto f^{-1.4} \, d^{0.9}, \tag{4.35}$$

which is in good agreement with the observations (see, for example, Malofeev *et al.* (1996) and Kramer *et al.* (2003b)). The scintillation bandwidth

$$\Delta f_{\mathrm{weak}} \sim f \tag{4.36}$$

and therefore is large. We can use the condition $u = 1$ to estimate the transition frequency, f_c, from weak to strong scintillation. Given the frequency and distance dependencies from Equation (4.33) we find that

$$f_c \equiv f(u = 1) \propto d^{0.65}, \qquad (4.37)$$

which is again consistent with the observations (Malofeev *et al.* 1996). We will derive a numerical estimate in the next section discussing strong, diffractive scintillation (see Figure 4.3), but in general we expect a transition from strong to weak scintillation at a frequency of a few GHz for a pulsar at a distance of about 1 kpc. As an example, for PSR B1133+16 shown in Figure 1.10 the transition frequency is estimated to be at 2 GHz (Kramer *et al.* 2003b). Hence, most pulsar observations are carried out in the strong scintillation regime. As we will see later, at the transition frequency, f_c, deep modulations can occur.

4.2.5.2 Diffractive scintillation

Diffractive interstellar scintillation (DISS) is in the strong scintillation regime and occurs when $s_0 \ll l_F$ or when phase perturbations are large at the distance of the observer and $u > 1$, and is recognised as strong-intensity variations in both time and frequency. These become most obvious in dynamic spectra, as introduced in Section 4.2.2, where the timescale, Δt_{DISS}, and frequency scale, Δf_{DISS}, can be measured readily. As for weak scintillation, we expect also the timescale to be related to the pulsar speed. For DISS, the dominating length scale is the field coherence scale, s_0, so that

$$\Delta t_{\mathrm{DISS}} = \frac{s_0}{V_{\mathrm{ISS}}} \propto f^{1.2}\, d^{-0.6}. \qquad (4.38)$$

The measurement of the scintillation timescale can therefore be used to determine the transverse speed of a pulsar. We will discuss this application in more detail in Sections 4.2.6 and 7.4.4.

The decorrelation bandwidth, Δf_{DISS}, is related to the scattering timescale via Equation (4.29), which for a Kolmogorov spectrum becomes

$$\Delta f_{\mathrm{DISS}} = \frac{1.16}{2\pi \tau_s} \simeq 185\,\mathrm{Hz}\left(\frac{\tau_s}{\mathrm{ms}}\right)^{-1}. \qquad (4.39)$$

As τ_s increases with distance, we expect Δf_{DISS} to decrease accordingly. A rough estimate (Cordes *et al.* 1985)

$$\Delta f_{\mathrm{DISS}} \sim 11\,\mathrm{MHz}\left(\frac{f}{\mathrm{GHz}}\right)^{4.4}\left(\frac{d}{\mathrm{kpc}}\right)^{-2.2} \qquad (4.40)$$

was derived from scintillation measurements and a normalisation of the turbulence spectrum (see Equation (4.27)) for a given line of sight. This is achieved by determining $C_{n_e}^2$ from observables. For a Kolmogorov spectrum, Cordes *et al.* (1990) found

$$C_{n_e}^2 = 0.002 \text{ m}^{-20/3} \left(\frac{f}{\text{GHz}} \right)^{3.67} \left(\frac{d}{\text{kpc}} \right)^{-1.83} \left(\frac{\Delta f_{\text{DISS}}}{\text{MHz}} \right)^{-0.83} . \quad (4.41)$$

Rather than using the estimate from Equation (4.40), it is preferable wherever possible to obtain a measurement of either Δf_{DISS} or τ_s. The other quantity can be computed from Equation (4.39). In fact, this result implies also that very often only one of these two quantities is measurable, depending on observing frequency and available spectral resolution. Only at intermediate frequencies may both quantities be determined.

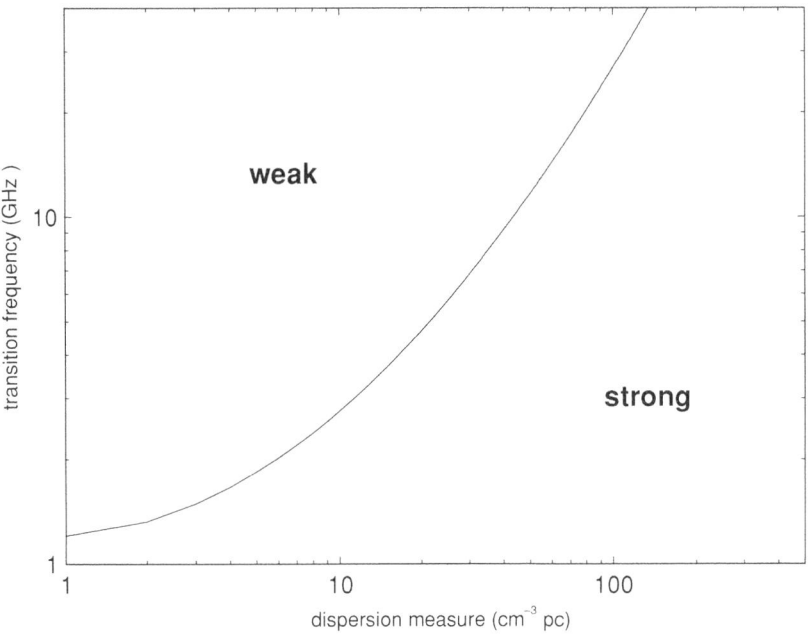

Fig. 4.3. Estimated frequency for the transition from strong to weak scintillation as a function of the observable dispersion measures (see text).

If a measurement of neither Δf_{DISS} nor τ_s is available, we can use also an empirical relationship between τ_s and the dispersion measure. A number of authors have investigated this relationship (see, for example. Ramachandran *et al.* (1997) and Cordes and Lazio (2002b)). We show the most recent compilation of measurements by Bhat *et al.* (2004) in Figure 4.4. Fitting these data to a parabola in log(DM), Bhat *et al.* (2004) find

$$\log \tau_s = -6.46 + 0.154 \log(\text{DM}) + 1.07(\log \text{DM})^2 - 3.86 \log f, \quad (4.42)$$

where the units are ms for τ_s, cm^{-3} pc for DM and GHz for f. Note that the fitted frequency exponent is 3.86 rather than the expected value of 4.4 for Kolmogorov turbulence. This was noticed by Löhmer *et al.* (2001) and may arise from large-scale refraction or from the truncation of scattering screens transverse to the Earth-pulsar line of sight.

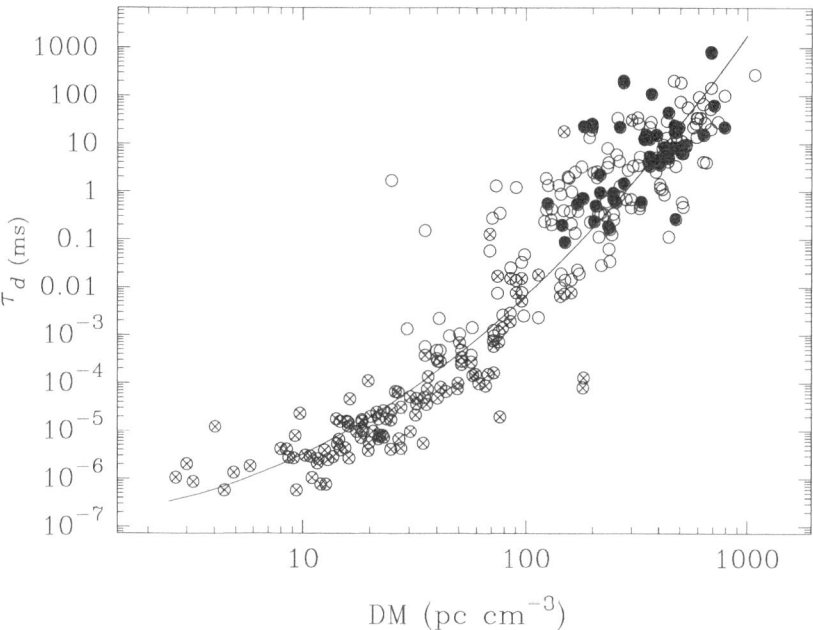

Fig. 4.4. Pulsar scatter-broadening times as a function of dispersion measure (Bhat *et al.* 2004). The solid curve represents the best fit (see Equation (4.42)). A large scatter about this fit is noticeable. Figure provided by Ramesh Bhat.

Using Equations (4.33), (4.39) and (4.42), and setting $u = 1$, we can make an estimate of the transition frequency from strong to weak scintillation, f_c. The result is shown in Figure 4.3. Given the large scatter around the fitted curve in Equation (4.42), a more reliable estimate is obtained using actual measurements of τ_s or Δf_{DISS} for a given pulsar.

For diffractive scintillation, the intensity is strongly modulated and the observed flux density of a pulsar can differ significantly from its intrinsic flux density. The number of scintillation maxima that are averaged together in time and frequency determines the strength of the modulation. If the observing bandwidth $\Delta f_{\mathrm{obs}} \lesssim \Delta f_{\mathrm{DISS}}$, only one scintle is sampled at any instant and the modulation is up to 100 per cent between successive observations. Averaging in either time, so that $\Delta t_{\mathrm{obs}} \gg \Delta t_{\mathrm{DISS}}$, or frequency, so that $\Delta f_{\mathrm{obs}} \gg \Delta f_{\mathrm{DISS}}$ (see, for example, Figure 1.10), is required in order to obtain a reliable flux density measurement of the pulsar.

We can estimate the modulation index from the number of scintles sampled in a $\Delta f_{\mathrm{obs}} \times \Delta t_{\mathrm{obs}}$ plane. Following Cordes and Lazio (1991), the number of scintles sampled in time and frequency are, respectively:

$$N_t \approx 1 + \kappa \frac{\Delta t_{\mathrm{obs}}}{\Delta t_{\mathrm{DISS}}} \quad \text{and} \quad N_f \approx 1 + \kappa \frac{\Delta f_{\mathrm{obs}}}{\Delta f_{\mathrm{DISS}}}, \qquad (4.43)$$

where κ is empirically determined to lie in the range 0.1–0.2. The observed modulation index is then

$$m_{\mathrm{DISS}}^{\mathrm{obs}} = \frac{m_{\mathrm{DISS}}}{\sqrt{N_t N_f}} = \frac{1}{\sqrt{N_t N_f}}. \qquad (4.44)$$

If the scintles are large compared to Δt_{obs} and Δf_{obs}, large flux modulation is observed, i.e. $m_{\mathrm{DISS}}^{\mathrm{obs}} \sim 1$. If the scintles are small, as is the case for highly dispersed distant pulsars, as $\Delta f_{\mathrm{DISS}} \propto d^{-2.2}$, the observed modulation index becomes small and we obtain stable flux density measurements. In this case, only a weak and slow modulation is observed due to refractive scintillation.

4.2.5.3 Refractive scintillation

Although diffractive scintillation was recognised soon after the discovery of pulsars (Lyne & Rickett 1968), it was not until 1982 that an additional refractive branch of strong scintillation was noticed. Sieber (1982) realised that the timescales of long and weak intensity modulations, which had been considered to be intrinsic to the pulsar, increased with source distance. Rickett *et al.* (1984) proposed that these were due to refractive interstellar scintillation (RISS) and a number of theoretical studies followed (see, for example, Romani *et al.* (1986)). Today, RISS is also accepted as an explanation for intensity variations observed in active galactic nuclei (see, for example, Rickett (1990), Rickett *et al.* (1995), Jauncey *et al.* (2001), Dennett-Thorpe and de Bruyn (2002)). These sources are too extended to show DISS, but are small enough to exhibit RISS, which takes place when $l_R \gg l_F$. Given Equation (4.31), this implies $l_F \gg s_0$, indeed confirming that RISS occurs in the strong scintillation regime.

The basic principles of RISS can be understood in terms of geometric optics and the effect of the focusing and defocusing of rays from the scattering screen (see, for example, Romani *et al.* (1986)), leading to clear predictions for the modulation index, m_{RISS} and scintillation timescale Δt_{RISS}. Observational results, however, indicate that there are significant deviations from theoretical predictions (see, for example, Gupta (2000)). We will nevertheless list the expected basic relationships.

As RISS involves the much larger refractive scale, its timescale

$$\Delta t_{RISS} = \frac{l_R}{V_{ISS}} = \frac{l_F^2}{s_0 \, V_{ISS}} = \frac{l_F^2}{s_0^2} \frac{s_0}{V_{ISS}} \tag{4.45}$$

$$= u^2 \, \Delta t_{DISS} = \frac{f}{\Delta f_{DISS}} \, \Delta t_{DISS} \propto f^{-2.2} \, d^{1.6} \tag{4.46}$$

is much longer than the other regime. This expression conveniently relates the DISS and RISS quantities, enabling us to estimate the expected timescales. As expected from Sieber's (1982) observations, Δt_{RISS} *increases* with distance, d, in contrast to Δt_{DISS}.

We can relate the modulation index to the scintillation strength as (see, for example, Romani *et al.* (1986) and Rickett (1996))

$$m_{RISS} = u^{-1/3} = \left(\frac{l_F}{s_0}\right)^{-1/3} = \left(\frac{\Delta f_{DISS}}{f}\right)^{1/6} \propto f^{0.57} \, d^{-0.37}. \tag{4.47}$$

The modulation index increases towards the transition frequency, f_c, contributing to the strong intensity variation due to DISS. As a result, near $u \approx 1$, the total variance in intensity

$$m_{total}^2 \approx m_{DISS}^2 + m_{RISS}^2 + m_{DISS} m_{RISS}. \tag{4.48}$$

can be greater than unity (Rickett 1990).

Refractive scintillation also is observed in dynamic spectra with a variety of phenomena such as bands of features drifting in frequency and time, long-term changes in the decorrelation bandwidth Δf_{DISS}, crisscross patterns with overlapping slopes of opposite signs and also periodic patters referred to as 'fringes'. For a summary of these phenomena see, for example, Rickett (1990) or Gupta *et al.* (1994). Their study can be used not only to constrain the turbulence wavenumber spectrum on large scales, but also to resolve a pulsar magnetosphere when systematic shifts in the periodic spectra are interpreted as multiple imaging events from different parts of the pulse profile (Wolszczan & Cordes 1987; Smirnova & Shishov 1989). The derived emission heights are typically larger than those estimated from other methods (see Chapter 3), but recent results produce more consistent numbers (Gupta *et al.* 1999).

Recently, the loosely organised criss-cross patterns in the dynamic spectra have been shown to produce faint but distinct features in the 'secondary spectrum', representing a Fourier transform of the dynamic spectrum (Stinebring *et al.* 2001; Hill *et al.* 2003). These features appear in the form of 'arcs' that can be used to study deviations from the simple thin-screen model and to probe the structure of the ISM on ~ 1 AU size scales. The curvature of the arc is related to the location of the screen along the line of sight. Multiple arcs indicate the presence of multiple thin screens. We will discuss secondary spectra and their applications further in Chapter 7.

4.2.5.4 Summary

We summarise the behaviour of the modulation in all three scintillation regimes in Figure 4.5 as a function of frequency and dispersion measure. The various scaling laws for the observable parameters are summarised in Appendix 2.

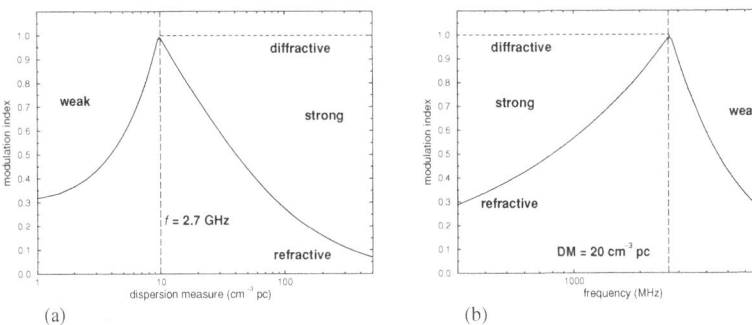

Fig. 4.5. The modulation indices due to weak scintillation, strong diffractive and strong refractive scintillation as a function of dispersion measure (a) and observing frequency (b). These are estimates using the quoted relationships for an observing frequency of 2.7 GHz (a) and a DM of 20 cm^{-3} pc (b). This frequency often is considered to be ideal for high-precision pulsar timing as it strikes a compromise between the effects of 'interstellar weather' and the steep pulsar spectra.

4.2.6 Scintillation speeds

The relationship between the scintillation timescales and the relative motion of the pulsar allows the determination of the pulsar speed transverse to the line of sight, V_{ISS} (see Figure 4.2). Since diffractive parameters usually are easier to measure, this method uses observations of DISS to infer V_{ISS}. From Equation (4.38) we see that $V_{\text{ISS}} = s_0/\Delta t_{\text{DISS}}$. Since $s_0 = 1/k\theta_{\text{d}}$, using our definition of τ_{s} (Equation (4.23)) and Equation (4.29), we find $s_0 \propto (d\Delta f_{\text{DISS}})^{1/2}/f$ (see also, for example, Gupta *et al.* (1994)). Inserting this into Equation (4.38), we obtain

$$V_{\text{ISS}} = A \left(\frac{d}{\text{kpc}} \right)^{1/2} \left(\frac{\Delta f_{\text{DISS}}}{\text{MHz}} \right)^{1/2} \left(\frac{f}{\text{GHz}} \right)^{-1} \left(\frac{\Delta t_{\text{DISS}}}{\text{s}} \right)^{-1} , \quad (4.49)$$

where the constant A depends again on the geometry, the location of the scattering screen and the form of the turbulence spectrum. Its numerical value varies slightly among different authors. Gupta (1995) quotes $A = 3.85 \times 10^4$ km s^{-1}, while Cordes and Rickett (1998) derive $A = 2.53 \times 10^4$ km s^{-1}. In spite of these uncertainties, the numbers are in good agreement (Nicastro *et al.* 2001) with proper motion measurements using pulsar timing (Chapter 8) or interferometric measurements (Chapter 9). In Chapter 7 we show how to measure the scintillation timescale and bandwidth and apply this method to observations of binary pulsars.

4.3 Further reading

There are a number of useful reviews on the propagation effects caused by the ISM. Cordes (2002) describes their impact on pulsar observations. Han (2004) gives a summary of RM measurements and their implications for the Galactic magnetic field. See also the recent work by Mitra *et al.* (2003) demonstrating possible caveats in such analyses. It is useful to put these Galactic measurements in the context of observations of magnetic fields in other Galaxies (see, for example, Beck (2002)).

Specialised reviews on propagation issues in the ISM can be found in Rickett (1977; 1990; 1996), Narayan (1992) and Gupta (2000). Further details can be found in Cordes *et al.* (1985; 1986), Armstrong *et al.* (1995) and Cordes and Lazio (1991). While we have assumed a Kolmogorov spectrum for the interstellar turbulence, those wishing to go further should consult Lambert and Rickett (1999), Gupta (2000) and Cordes and Lazio (2001) for different geometries and turbulence spectra. Observational evidence for non-Kolmogorov turbulence was presented by Löhmer *et al.* (2001) and Bhat *et al.* (2004). For observable deviations from the thin-screen model, e.g. by extended or multiple screens, see Lambert and Rickett (1999; 2000) or Cordes and Lazio (2001).

Interstellar scintillation and scattering also can be used to study special regions in the Galaxy. Recent work on the local ISM and the so-called 'Local Bubble' can be found in Bhat *et al.* (1998). Lazio and Cordes (1998a,b) studied the scattering effects in the very interesting Galactic Centre region (see also Section 6.5.6). Mitra and Ramachandran (2001) studied the effect of scattering in the direction of the Gum nebula.

Work on scintillation speeds has been presented by numerous authors (see, for example, Gothoskar and Gupta (2000), Johnston *et al.* (1998), Nicastro and Johnston (1995) and Gupta (1995)). Details of the comparison between scintillation speeds and pulsar proper motions can be found in Lyne and Smith (1982), Cordes (1986), Gupta (1995), Cordes and Rickett (1998), Deshpande and Ramachandran (1998) and most recently by Nicastro *et al.* (2001). If all model parameters are known from a combination of proper motion, scintillation and scattering measurements, the distance of a pulsar can be determined (see, for example, Deshpande and Ramachandran (1998)).

The important aspects of 'interstellar weather' on pulsar timing observations are addressed by Foster and Cordes (1990), Backer and Wong (1999) and recently by Backer and Ramachandran (2004). As pointed out by Backer and Wong (1999), the frequency dependence of the scattering angle and path geometry of the propagating signals means that each observing frequency is probing a different volume of the ISM. As a consequence, the column density of electrons depends to some extent on frequency, so that the DM is not an accurately definable quantity at any instant, contributing to the observed DM variations (see Figure 4.1).

We have not discussed the rarely observed *extreme scattering events*. These were first seen in observations of extra-galactic sources (Fiedler *et al.* 1987) and later seen in timing observations of PSR B1937+21 (Cognard *et al.* 1993). Romani *et al.* (1987), Fiedler *et al.* (1994) and Walker (2001) give a theoretical description of these events.

4.4 Available resources

An extensive source on modelling the Galactic electron density distribution is provided by the authors of the latest model: 'NE2001' (Cordes & Lazio 2002a,b). Also, they cover some aspects of propagation effects in the inhomogeneous ISM and provide software that implements their electron density model to convert between dispersion measure and distance. Additional parameters related to their model predictions for the fluctuations in the electron density and scintillation bandwidths and timescales are also computed. Software to estimate pulsar distances using alternative electron density models (see, for example, Gomez, Benjamin and Cox (2001)) are also available. On the book web site (see Appendix 3) we provide links to the various software, as well as a single program to compare distance estimates for a variety of models.

5

Instrumentation for pulsar observations

Pulsar observations place higher demands on data acquisition systems than most other astronomical observations. High time resolution is necessary to search for millisecond and rapidly rotating pulsars (Chapter 6), to resolve microsecond and nanosecond scale structure in individual pulses (Chapter 7) and to obtain high-precision arrival times (Chapter 8). Since pulsars are generally weak sources, observations covering a wide bandwidth are desirable to maximise sensitivity. As discussed in Chapter 4, the frequency-dependent refractive index of the interstellar medium means that pulses emitted at higher radio frequencies travel faster and arrive earlier than those emitted at lower frequencies. For a finite receiver bandwidth, this dispersion process broadens the pulse so that its signal to noise ratio (S/N) is reduced. Frequency resolution therefore is vital to combat the effects of dispersion. We distinguish between two main approaches to de-dispersion: (a) incoherent devices (filterbanks and correlators) where the phase content of the signal is not recorded; (b) coherent devices (baseband recorders and coherent de-dispersers) which make full use of the incoming signal to properly recover the true pulse shape.

5.1 Example observational set-up

Before discussing the various devices for dispersion removal, we begin with an overview of the signal path from the reception of raw voltages by the telescope to the input of a signal into a data acquisition device. The example shown in Figure 5.1 contains all the essential elements used in single dish pulsar observations discussed here.

After being brought to a focus by the antenna, the radiation is sampled

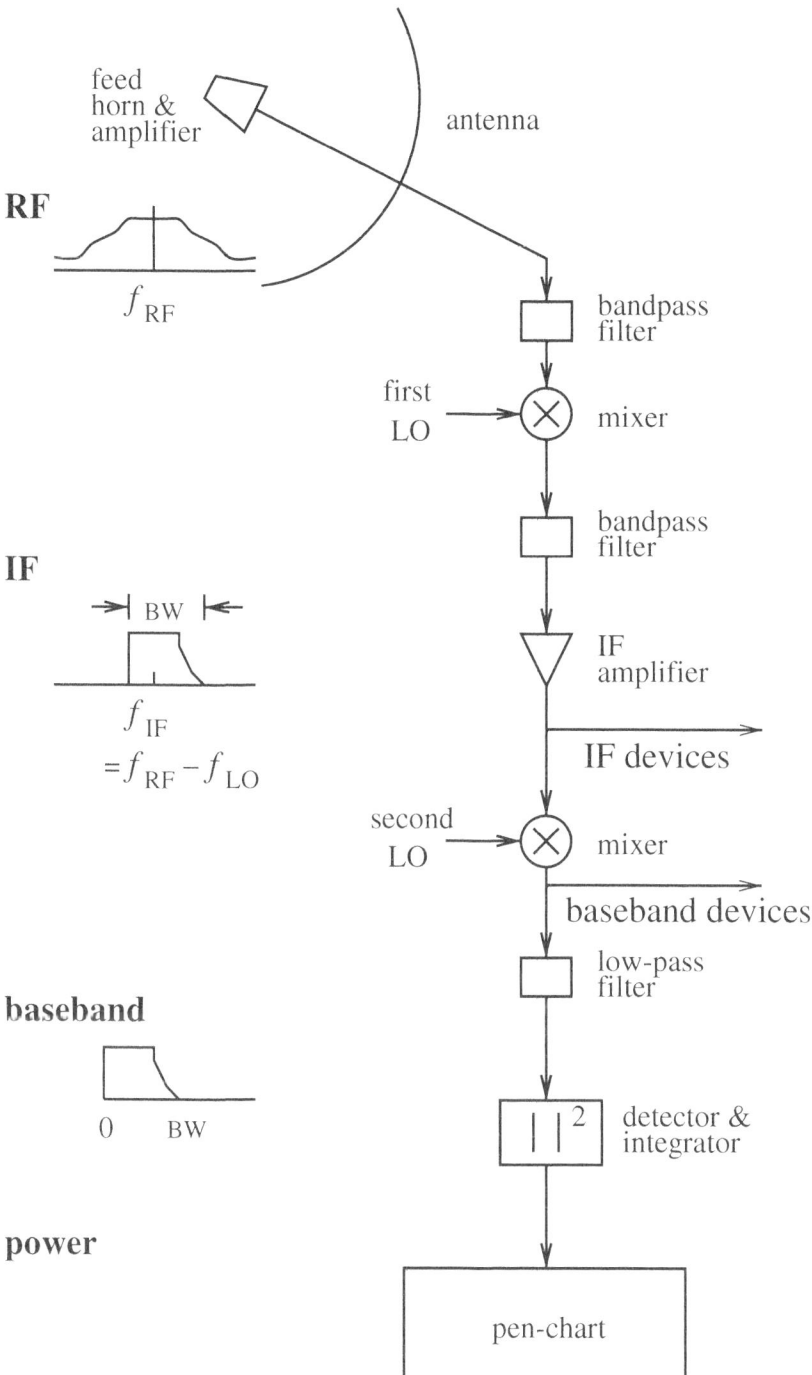

Fig. 5.1. Schematic showing the signal path in a single-dish radio telescope. See text for a discussion of the various stages. Figure adapted from an original version (Bhattacharya 1998) provided by Dipankar Bhattacharya.

by a wave-guide *feed*, usually having two receptors that sample orthogonal senses of polarisation, usually dual linear or dual circular (see Appendix 1). We refer, throughout this chapter, to these orthogonal inputs as *polarisation channels*, and in illustrations show the signal path from just a single channel. The weak radio signal then is amplified by a low-noise amplifier that has a specific frequency response within a band centred at a radio frequency (RF) which we denote by f_{RF}. The amplified signal then is passed through a bandpass filter so that any harmonics from out-of-band interference signals (e.g. from communication devices, TV stations etc.) are rejected. At this point, the RF signal usually is converted to a lower frequency for two reasons: (a) it is more efficient to transmit a low-frequency signal (i.e. cable losses are lower); (b) much of the data-acquisition hardware which we will discuss in later sections accept signals in a certain frequency range.

Frequency conversion is achieved by a device known as a *mixer* that beats the RF signal with a monochromatic signal of frequency f_{LO} provided by a *local oscillator* (LO). The result is an *intermediate frequency* (IF) signal at the sum and difference of f_{RF} and f_{LO}. Although the sum of the frequencies can be used, most of the time we require the difference frequency for *down conversion*, so that $f_{IF} = f_{RF} - f_{LO}$. Usually, $f_{LO} < f_{RF}$, so that the resulting IF (known as the *upper sideband*) has the same frequency sense as the RF, i.e. an increase in IF corresponds to an increase in RF. The *lower sideband* case, where $f_{LO} > f_{RF}$, produces a frequency inversion, i.e. an increase in IF corresponds to a decrease in RF) and is used less frequently.

The IF signals either may be sent directly to data acquisition devices such as filterbanks or correlators (see Sections 5.2.1 and 5.2.2) or converted to *baseband*, with a further mixing process as shown before being sent to baseband sampling devices (see Section 5.3). As a real-time signal and interference monitor, the signal is traditionally low-pass-filtered and detected (squared) so that the resulting power is plotted on a pen chart recorder. The pen chart recorder was, inadvertently, the first 'pulsar backend' used serendipitously by Jocelyn Bell-Burnell to discover pulsars back in 1967. Sadly, these wonderful devices are becoming less common in observatories nowadays as modern data acquisition systems discussed below provide enhanced functionality.

5.2 Incoherent de-dispersion

The simplest way to compensate for the effects of pulse dispersion is to split the incoming frequency band into a large number of independent *frequency channels* using a spectrometer, and apply appropriate time delays to each channel, so that the received pulses arrive at the output of each channel at the same time. This is carried out using either dedicated hardware, or nowadays more commonly, software. In either case, the appropriate time delays should be calculated by recalling the dispersion relationship (see Equation (4.7)) in which the difference in arrival times Δt between a pulse received in a channel with frequency f_{chan} relative to some reference frequency f_{ref} (often chosen to be the centre frequency of the observed band) can be written as simply

$$\Delta t \simeq 4.15 \times 10^6 \text{ ms } \times (f_{\text{ref}}^{-2} - f_{\text{chan}}^{-2}) \times \text{DM}. \tag{5.1}$$

In this equation, as before, the frequencies are measured in MHz and the dispersion measure DM (cm^{-3} pc) is the integrated column density of free electrons along the line of sight. The exact value of the constant in this expression is given in Chapter 4 (see also Appendix 2). As shown in Figure 5.2, the appropriately delayed frequency channels then can be added together to produce a *de-dispersed time series*.

It is apparent from Figure 5.2 that this *incoherent de-dispersion* process is limited by the width of the individual frequency channels that inherently retain a small dispersion delay. Rearranging Equation (5.1) for a finite bandwidth Δf about some centre frequency f, we find, for the practical case in which $f \gg \Delta f$, that the dispersive delay across a frequency channel of width Δf, for the same units as above, is given by

$$t_{\text{DM}} \simeq 8.3 \times 10^6 \text{ ms } \times \text{DM} \times \Delta f \times f^{-3}. \tag{5.2}$$

Careful choice of the channel bandwidths therefore is required to ensure that t_{DM} does not become a significant fraction of the pulse period. At 430 MHz, most current spectrometers have $\Delta f \sim 0.1$ MHz. This translates to $t_{\text{DM}} \simeq 10\,\mu s$ DM^{-1} and, in the absence of additional interstellar scattering, is adequate to detect millisecond pulsars with DMs of up to 100 cm^{-3} pc. The strong dependence on observing frequency means that broader channel bandwidths may be used at higher frequencies. For example, a 512×0.5 MHz channel spectrometer currently in use at Parkes has $t_{\text{DM}} = 1.5\,\mu s$ DM^{-1} at 1400 MHz. Although the dispersion limit usually dominates time resolution considerations, we note that the fundamental resolution for a filter channel of width $\Delta \nu$ is $1/\Delta \nu$. For $\Delta \nu = 0.5$ MHz, this corresponds to a resolution of $2\mu s$.

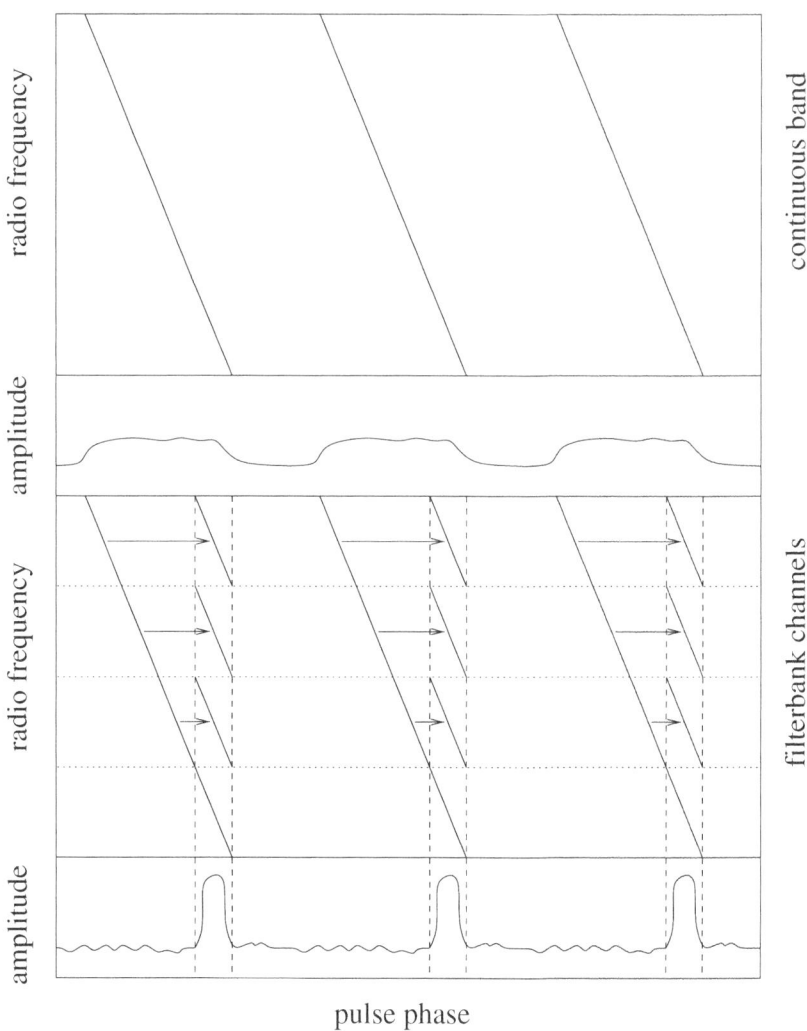

pulse phase

Fig. 5.2. Pulse dispersion and the process of de-dispersion. Simply detecting the pulse over a finite bandwidth results in a significantly broadened profile (top panel). Dividing the passband into smaller bandwidth channels and applying the appropriate delay to each channel considerably reduces the broadening and increases the pulse S/N ratio (lower panel).

5.2.1 Analogue filterbank spectrometers

The simplest and most widely used data acquisition device for incoherent de-dispersion is the analogue filterbank spectrometer. This device per-

forms the task shown in Figure 5.2, in which a broad-band signal is split
up into adjacent frequency channels that are passed through narrow-
band filters tuned (e.g. using a simple LC circuit) to accept frequencies
only within a specific range. Each frequency channel is then digitised
separately.

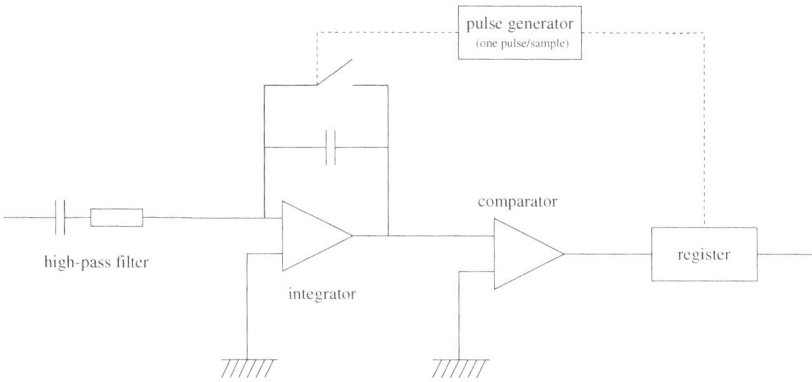

Fig. 5.3. Schematic showing a one-bit digitisation scheme to sample filterbank
data. Figure design provided by Andrew Lyne.

The simplest digitisation scheme is one-bit sampling. Shown schematically in Figure 5.3, the incoming narrow-band signal passes through the
RC circuit with a time constant of typically ~ 1 s. This effectively filters
out low-frequency receiver variations. The integrator outputs the running mean of this signal, which is compared with 0 V by the comparator.
Sampling is achieved by clocking the output of the comparator with the
register once per sampling interval, t_{samp}, and synchronously closing the
switch on the integrator as shown. The resulting output is then either 1
or 0 depending on whether the running mean is positive or negative. It
is worth noting that since the integrator effectively convolves the signal
with a top-hat function of width t_{samp}, it acts as a low-pass filter rejecting most frequencies greater than $1/t_{\mathrm{samp}}$. This is particularly useful,
since it negates the need for specialised anti-aliasing filters developed
in earlier implementations of these systems (see, for example, Brinklow
(1989)).

At first sight, single-bit digitisation appears to be an overly crude approach to sampling. However, particularly for search projects in which
the sampled data from narrow channel bandwidths often are dominated
by noise with only a small contribution from the signal, it can be shown

(see, for example, Brinklow (1989)) that the loss of sensitivity relative to perfect sampling is only $(1 - \sqrt{2/\pi}) \simeq 20\%$[1] For large search projects, this relatively minor reduction in sensitivity is often well worth the saving in data storage and reduction. In addition, unlike multi-bit schemes, one-bit digitisation is extremely robust against strong impulsive bursts of interference. Indeed pulsar surveys made using single-bit digitisers sampling the filterbanks at the Parkes and, to a lesser extent, Jodrell Bank radio telescopes have been responsible for the discovery of about two-thirds of all known pulsars (Lyne 2003).

For polarisation studies, the four Stokes parameters from filterbanks can be obtained using an adding polarimeter that we do not discuss here as such systems are generally complicated to calibrate (for details see, for example, von Hoensbroech and Xilouris (1997)).

5.2.2 Autocorrelation spectrometers

An alternative means of obtaining a filterbank-style output is the auto-correlation spectrometer. Shown schematically in Figure 5.4, this device multiplies the signal with a delayed version of itself. The resulting set of *lagged products* in the time domain then can be used to calculate the power in the radio frequency domain. Consider a voltage, $v(t)$, and its complex conjugate, $v^*(t)$, from a radio telescope as a function of time t. The Wiener–Khinchin theorem (see, for example, Rohlfs and Wilson (2000)) states that the autocorrelation function

$$R(\tau) = \lim_{T \to \infty} \frac{1}{T} \int_0^T v(t) v^*(t + \tau) \, \mathrm{d}t \qquad (5.3)$$

and the power spectrum are a Fourier transform pair, i.e.

$$P(f) = \frac{1}{2\pi} \int_{-\infty}^{+\infty} R(\tau) e^{-2\pi i f \tau} \, \mathrm{d}\tau. \qquad (5.4)$$

Normally, the lagged products are stored on disk for subsequent off-line processing to produce $P(f)$. Note that the number of synthesised frequency channels is equal to the number of lags recorded. Since this number can be varied easily, autocorrelation spectrometry provides variable channel bandwidths and is therefore much more flexible than an analogue filterbank where the width and number of filters are fixed.

---

1 Extending Brinklow's (1989) analysis to multi-bit sampling of weak signals, we find that the degradation in S/N to be about 14 and 3 per cent, respectively, for three- and nine-level sampling.

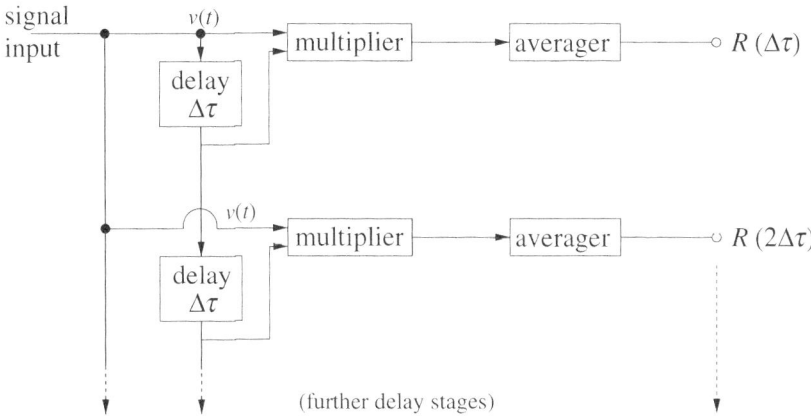

Fig. 5.4. Block diagram of an autocorrelation spectrometer. The input signal here from a single polarisation channel is split and multiplied by delayed versions of itself for a number of different delay times $\Delta\tau$. The resulting signals are passed to an accumulator that outputs the averaged lagged products once per data sampling interval. As mentioned in the text, this approach can be extended easily to form the cross-correlation products for two orthogonal polarisation channels. Figure adapted from an original design by Jon Hagen.

The above discussion assumes autocorrelation of a single polarisation channel. In practice, we sample two orthogonal channels obtaining a spectrum $P(f)$ for each. Denoting these two channels by A and B and the resulting power spectra of the autocorrelation functions by P_{AA} and P_{BB}, the total intensity $I = P_{\mathrm{AA}} + P_{\mathrm{BB}}$. In addition to autocorrelating each channel, the cross-correlation products also can be formed easily, i.e. P_{AB} and P_{BA}. This provides a very straightforward way to obtain all four Stokes parameters I, Q, U and V (see Appendix 1) for polarimetric observations. For a dual-linear feed:

$$\begin{bmatrix} I \\ Q \\ U \\ V \end{bmatrix} = \begin{bmatrix} P_{\mathrm{AA}} + P_{\mathrm{BB}} \\ P_{\mathrm{AA}} - P_{\mathrm{BB}} \\ 2\,\mathrm{Re}[P_{\mathrm{AB}}] \\ 2\,\mathrm{Im}[P_{\mathrm{BA}}] \end{bmatrix}. \tag{5.5}$$

while for observations with a dual-circular feed:

$$\begin{bmatrix} I \\ Q \\ U \\ V \end{bmatrix} = \begin{bmatrix} P_{\mathrm{AA}} + P_{\mathrm{BB}} \\ 2\,\mathrm{Im}[P_{\mathrm{BA}}] \\ 2\,\mathrm{Re}[P_{\mathrm{AB}}] \\ P_{\mathrm{BB}} - P_{\mathrm{AA}} \end{bmatrix}. \tag{5.6}$$

This is far simpler than the corresponding formation of the Stokes parameters from the analogue filterbanks. The Stokes parameters formed from the auto- and cross-correlation products need to be properly calibrated prior to analysis and interpretation. Details of these calibration procedures are discussed in Chapter 7.

As was also necessary for the filterbank data, the auto- and cross-correlation products need to be digitised. Most early correlators used a one-bit (two-level) scheme. Most devices currently in use apply three-level, or, for high dynamic range observations, nine-level sampling. A secondary effect of finite sampling is to bias the measured correlation functions. While S/N losses are unavoidable, the bias can be corrected for. This correction was first developed for one-bit sampling by van Vleck and Middleton (1966) who showed that, for a measured correlation coefficient r, the unbiased correlation coefficient $\hat{\rho} = \sin(\pi r / 2)$. Unfortunately, the generalisation of this so-called *van Vleck correction* to multi-bit sampling becomes rapidly non-trivial. The three-level sampling case (see, for example, Hagen and Farley (1973)) requires a non-analytic solution to the integral

$$r = \frac{1}{\pi} \int_0^{\hat{\rho}} \left[\exp\left(\frac{-(\varepsilon/\sigma)^2}{1+x} \right) + \exp\left(\frac{-(\varepsilon/\sigma)^2}{1-x} \right) \right] \frac{\mathrm{d}x}{\sqrt{(1-x^2)}}, \quad (5.7)$$

where ε is the digitiser threshold and σ is the root mean square voltage. Kulkarni and Heiles (1980) have considered the three-level case in detail and derived a number of useful approximations to the van Vleck formula for the auto- and cross-correlation products.

5.3 Coherent de-dispersion systems

The deleterious effects of interstellar dispersion can be completely removed by making use of the phase of the incoming voltage from the telescope. This technique – known as *coherent de-dispersion* – was pioneered by Hankins and Rickett (1975). After measuring the complex voltage induced in the telescope feed by the incoming electromagnetic radiation, $v(t)$, this method recovers the intrinsic complex voltage as it originated from the pulsar, $v_{\mathrm{int}}(t)$. This then is transformed into a real signal that retains the instrumental (Nyquist) time resolution without being affected by the dispersive effects of the interstellar medium[2].

---

2 Multi-path scattering, discussed in Chapters 1 and 4, is a different effect that can not be removed in a direct manner. Scattering therefore affects both incoherently and coherently dispersed pulse profiles. Methods to restore the unscattered pulse

The superiority of the coherent de-dispersion scheme versus incoherent de-dispersion is demonstrated clearly for PSR B1937+21 in Figure 1.2.

5.3.1 Principles of coherent de-dispersion

Coherent de-dispersion exploits the fact that the modification of the propagating signal by the interstellar medium can be described as the work of a 'phase-only' filter, or *transfer function H*. This relationship becomes particularly simple in the frequency domain. For a signal centred on a frequency f_0 and with a limited bandwidth, Δf, we can write

$$V(f_0 + f) = V_{\text{int}}(f_0 + f)H(f_0 + f), \tag{5.8}$$

where $V(f)$ and $V_{\text{int}}(f)$ are the corresponding Fourier transforms of the raw voltages $v(t)$ and $v_{\text{int}}(t)$. These are non-zero only for $|f| < \Delta f/2$, i.e.

$$v(t) = \int_{f_0-\Delta f/2}^{f_0+\Delta f/2} V(f)e^{i2\pi ft}\mathrm{d}f \tag{5.9}$$

$$v_{\text{int}}(t) = \int_{f_0-\Delta f/2}^{f_0+\Delta f/2} V_{\text{int}}(f)e^{i2\pi ft}\mathrm{d}f. \tag{5.10}$$

From our discussion of the propagation of radio waves in the interstellar medium in Chapter 4, we recall that the delay in the interstellar medium can be represented as phase rotations that depend on frequency and path length travelled, $\Delta\Phi = -k(f_0 + f)d$, where $k(f)$ is the wavenumber and d is the distance to the pulsar (see Chapter 4). Therefore, the transfer function becomes simply

$$H(f_0 + f) = e^{-ik(f_0+f)d}. \tag{5.11}$$

The idea is now to determine H for a given pulsar and to apply its inverse to the measured, Fourier-transformed voltage. Once this is done, the result is transformed back into the time domain to obtain the desired de-dispersed signal.

In order to obtain a practical expression for the transfer function, we recall the relationship from Chapter 4 between the wavenumber and

shape by deconvolution are discussed by Kuzmin and Izvekova (1993), Löhmer *et al.* (2001) and Bhat, Cordes and Chatterjee (2003).

frequency (see Equation (4.8)). This becomes

$$k(f_0 + f) = \frac{2\pi}{c}(f_0 + f)\sqrt{1 - \frac{f_p^2}{(f_0 + f)^2} \mp \frac{f_p^2 f_B}{(f_0 + f)^3}}, \tag{5.12}$$

where, as before, f_p and f_B are respectively the plasma and cyclotron frequencies. In Chapter 4, we saw that for the interstellar medium $f_p \sim$ 2 kHz and $f_B \sim 3$ Hz. Therefore, when the observing frequency f is above a few hundred MHz, we find the sizes of the last two terms in the above expression to be of the order of 10^{-10} and 10^{-18}, respectively. We can approximate, therefore, the above expression by keeping only the first terms in a Taylor expansion, resulting in

$$k(f_0 + f) \approx \frac{2\pi}{c}(f_0 + f)\left[1 - \frac{f_p^2}{2(f_0 + f)^2}\right]. \tag{5.13}$$

While we ignored here the usually minute difference in propagation speed for left- and right-hand circularly polarised signals, it may be necessary in special cases to use different transfer functions for different polarisation channels, which may also mean that we should keep higher-order terms of the Taylor expansion.

Inserting the above expression for the wavenumber into the transfer function, we find

$$H(f_0 + f) = e^{-i\left\{\frac{2\pi}{c}(f_0 + f)\left[1 - \frac{f_p^2}{2(f_0 + f)^2}\right]\times d\right\}} \tag{5.14}$$

$$= e^{i\frac{2\pi}{c}d\left(f_0 + f - \frac{f_p^2}{2(f_0 + f)}\right)}. \tag{5.15}$$

Finally, using the identity

$$\frac{1}{f_0 + f} = \frac{1}{f_0} - \frac{f}{f_0^2} + \frac{f^2}{(f_0 + f)f_0^2}, \tag{5.16}$$

we arrive at the desired result:

$$H(f_0 + f) = e^{-i\frac{2\pi}{c}d\left\{\left(f_0 - \frac{f_p^2}{2f_0}\right) + \left(1 + \frac{f_p^2}{f_0^2}\right)f - \frac{f_p^2}{2(f + f_0)f_0^2}f^2\right\}}. \tag{5.17}$$

Here, we have sorted the terms that are independent of, linear and quadratic in f, respectively. The first term represents an arbitrary, constant phase offset that cannot be determined as it is lost during the later (square-law) signal detection. The second term depends on f and, according to the shift-theorem (Bracewell 1998), corresponds to a delay in the time domain. We can ignore this term simply by shifting the arrival time by the appropriate amount. The last term causes phase rotations within the band that are quadratic in f. These need to be unwound in order recover the original pulsar signal. Taking only this term into account, we obtain the transfer function relevant for coherent de-dispersion

$$H(f_0 + f) = e^{+i\frac{2\pi}{c}d\frac{f_p^2}{2(f+f_0)f_0^2}f^2}. \tag{5.18}$$

This expression still includes the distance, d, and the (average) plasma frequency, f_p. We can replace these quantities by the directly observable dispersion measure, DM. Since the dispersion constant, $\mathcal{D}$, defined in Equation (4.6) is related to the plasma frequency by

$$\mathcal{D} = \frac{f_p^2}{2cn_e}, \tag{5.19}$$

where n_e is the average electron density along the line of sight, we can identify the expression

$$\frac{f_p^2 d}{2c} = \text{DM} \times \mathcal{D} \tag{5.20}$$

which we use to write the transfer function as

$$H(f_0 + f) = e^{+i\frac{2\pi\mathcal{D}}{(f+f_0)f_0^2}\text{DM}f^2}. \tag{5.21}$$

5.3.2 Baseband sampling

By applying to the inverse of the transfer function, H^{-1}, to the sampled voltage data, the originally emitted voltage is recovered. In order to do this, we have to modify the phases of the complex Fourier components of $V(f_0 + f)$ in the way described by H^{-1}. This requires sampling and digitisation of $v(t)$, such that both amplitude and phases are measured using a method called *baseband sampling*.

The bandwidth-limited complex voltage, $v(t)$, can be written as a combination of its (real) amplitude, $a(t)$, a time-varying phase term, $\phi(t)$, and the carrier wave centred at f_0,

$$v(t) = a(t)e^{i\phi(t)}e^{i2\pi f_0 t}. \tag{5.22}$$

Now we mix the signal with that of a local oscillator (LO) of frequency f_{LO}, to produce a signal $\mathcal{I}$. A second signal $\mathcal{Q}$ is produced that uses the same LO but with a phase shift of 90° (or $\pi/2$). We can represent the derived two signals as being the result of the multiplication with the cosine and sine branch of the mixing carrier wave,

$$\mathcal{I}(t) \quad = \quad a(t)e^{i\phi(t)}e^{i2\pi f_0 t}\cos(2\pi f_{\text{LO}}t) \tag{5.23}$$

$$\mathcal{Q}(t) \quad = \quad a(t)e^{i\phi(t)}e^{i2\pi f_0 t}\sin(2\pi f_{\text{LO}}t). \tag{5.24}$$

Using the Euler identities, we obtain

$$\mathcal{I}(t) \quad = \quad \frac{1}{2}a(t)e^{i\phi(t)}\left\{ e^{i2\pi(f_0+f_{\text{LO}})t} + e^{i2\pi(f_0-f_{\text{LO}})t} \right\} \tag{5.25}$$

$$\mathcal{Q}(t) \quad = \quad \frac{1}{2i}a(t)e^{i\phi(t)}\left\{ e^{i2\pi(f_0+f_{\text{LO}})t} - e^{i2\pi(f_0-f_{\text{LO}})t} \right\}. \tag{5.26}$$

The next step is to low-pass-filter the signal, i.e. to allow only frequencies with $f \leq \Delta f/2$ to propagate further. This removes the frequency parts corresponding to $f_0 + f_{\text{LO}}$, hence,

$$\mathcal{I}(t) = \frac{1}{2}a(t)e^{i\phi(t)}e^{i2\pi(f_0-f_{\text{LO}})t} = \frac{1}{2}a(t)e^{i(2\pi(f_0-f_{\text{LO}})t+\phi(t))} \tag{5.27}$$

$$\mathcal{Q}(t) = \frac{i}{2}a(t)e^{i\phi(t)}e^{i2\pi(f_0-f_{\text{LO}})t} = \frac{i}{2}a(t)e^{i(2\pi(f_0-f_{\text{LO}})t+\phi(t))} \tag{5.28}$$

If we choose the LO frequency such that it is in the centre of our band (i.e. $f_{\text{LO}} = f_0$), the bandpass $[f_0 - \Delta f/2; f_0 + \Delta f/2]$ is shifted to $[-\Delta f/2; +\Delta f/2]$, which is called *baseband*. For the digitised real parts of the signals, we obtain

$$I(t) \quad \equiv \quad \text{Re}(\mathcal{I}(t)) = \frac{1}{2}a(t)\cos(\phi(t)) \tag{5.29}$$

$$Q(t) \quad \equiv \quad \text{Re}(\mathcal{Q}(t)) = -\frac{1}{2}a(t)\sin(\phi(t)), \tag{5.30}$$

giving us access to both the amplitude and phase of the complex voltage $v(t)$. The signals $I(t)$ and $Q(t)$ can be viewed as the real and imaginary part of $v(t)$. This form of baseband mixing with $f_{\text{LO}} = f_0$ is shown in Figure 5.5 and provides *complex sampled* data. Another way of viewing it is to consider $I(f)$ and $Q(f)$ as providing both positive and negative frequencies of the bandpass relative to f_0 in the Fourier domain, $[-\Delta f/2; +\Delta f/2]$.

The data rate required for Nyquist sampling I or Q is $2 \times \Delta f/2 = \Delta f$. As we have to sample both signals I and Q, the total data rate is $2\Delta f$. Choosing f_{LO} to coincide with the upper or lower edge of our observing frequency range, as shown in Figure 5.5, produces a passband at $[-\Delta f, 0]$ when $f_{\text{LO}} = f_0 + \Delta f/2$ (lower-sideband down conversion) or one at $[0, \Delta f]$ when $f_{\text{LO}} = f_0 - \Delta f/2$ (upper-sideband down conversion). For these cases, we have to sample only one signal, but the data rate must

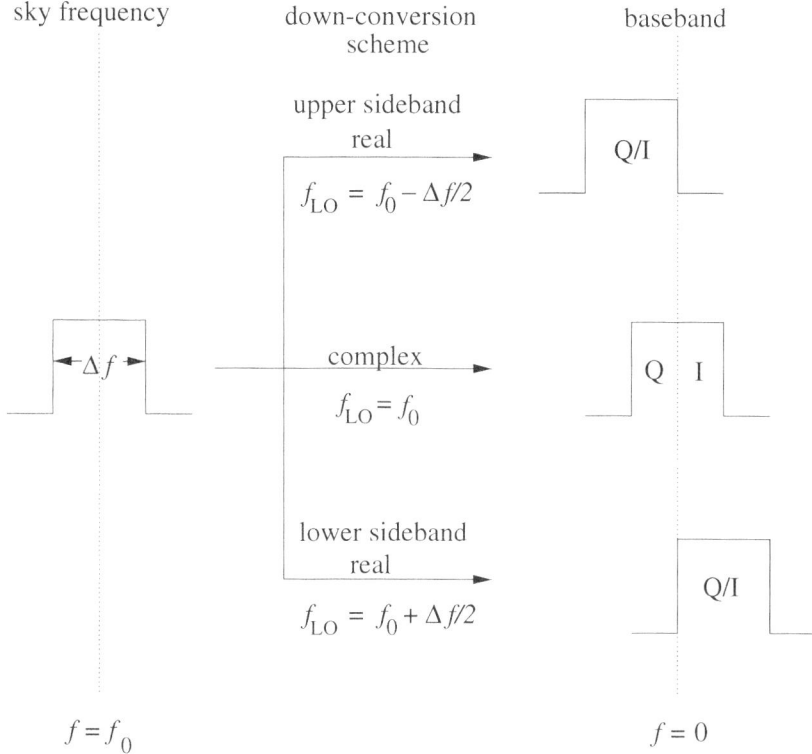

Fig. 5.5. Schematic showing various possible baseband mixing schemes depending on the placement of the local oscillator (see text).

be higher, $2\Delta f$. While this is the same as the total rate as for complex sampled data, the demand on the data-acquisition hardware in terms of speed is higher. Moreover, since we do not sample orthogonal signals, an additional Fourier transform is needed to recover the relative phase information that is needed for polarimetry (Shrauner 1997). Both factors usually lead to the implementation of complex sampling rather than real sampling.

This shift of our passband to baseband does not affect the Fourier transforms of our voltages, so that we can finally apply

$$V_{\text{int}}(f) = V(f)H^{-1}(f_0 + f). \tag{5.31}$$

In practical implementations, the inverse transfer function is combined with a taper, T, that is chosen such that it emphasises anti-aliasing in the low-pass filtering. The combination of both:

$$C(f_0, f) \equiv T(f)H^{-1}(f_0 + f) \qquad (5.32)$$

is known as the *chirp function*.

5.3.3 Implementation

The multiplication of the chirp function with the measured voltage in the Fourier domain corresponds to a convolution in the time domain (see, for example, Bracewell (1998)). This is important for considering the length of the voltage data required for de-dispersion. We certainly need a length of at least the dispersion delay between the upper and lower edge of our bandpass, t_{DM} (see Equation (5.2)). For a bandwidth Δf that is complex sampled at the Nyquist rate, the number of samples needed is $n_{DM} = t_{DM}\Delta f$ for each of I and Q. However, since a discrete convolution of each point of a time series of length n depends on $n/2$ points both before and after it, we need to pad our voltage series with another $n_{DM}/2$ samples at the beginning and at the end. Therefore, the shortest data set we can coherently de-disperse must be at least $2n_{DM}$ samples long. For efficiency, longer data sets are usually chosen, in particular since past constraints such as memory size become increasingly less important today. In any case, the 'wings' of length $n_{DM}/2$ at the beginning and end of each data set need to be ignored after the convolution.

A typical implementation of coherent de-dispersion is given by the following recipe:

(i) Mix signals of the observing frequency to baseband, digitise both I and Q by sampling each at a rate Δf.

(ii) Take a set of data of length n samples.

(iii) Compute a discretely sampled chirp function for n samples.

(iv) Fourier transform the data set of n points and multiply the resulting Fourier components with the elements of the chirp function.

(v) Inverse Fourier transform the result back into the time domain.

(vi) Ignore the wings of $n_{DM}/2$ samples at the beginning and end of the time series and save the remaining $n - n_{DM}$ points as output.

(vii) Take the next $n - n_{DM}$ points from the input data and add them to last n_{DM} points of the current set. Proceed with step (iii) on the new set of n samples.

By this cyclic procedure shown also in Figure 5.6, a continuous coherently de-dispersed time-series is formed.

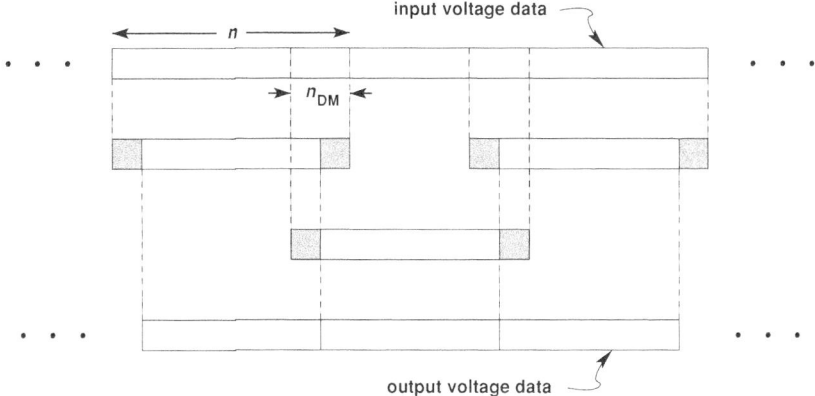

Fig. 5.6. Schematic showing the overlap-save procedure in coherent de-dispersion. The input data are split into overlapping sections of n points. Each section is Fourier-transformed separately and the chirp function is applied before Fourier transforming each segment back into the time domain. Points from the overlapping regions shown in grey are discarded before re-combining the resulting independent segments into a single time series. Figure adapted from an original version (van Straten 2003) provided by Willem van Straten.

Two practicalities to this general scheme are worth noting for completeness: (a) in order to avoid discarding the first $n_{\mathrm{DM}}/2$ samples in the output data, the initial data block is zero-padded with a dummy wing of $n_{\mathrm{DM}}/2$ samples; (b) additional steps to mitigate radio-frequency interference in both the frequency and time domains (see Section 7.5 for details) can be implemented before and after steps (iv) and (v).

5.3.4 Polarimetry

The application of coherent de-dispersion based on complex-sampled data has the additional important benefit that all four Stokes parameters are readily accessible if two orthogonal polarisation channels are sampled by the feed. In a similar way to the correlators described earlier, we can obtain the four Stokes parameters I, Q, U and V from the signals from two orthogonal linear polarisation channels X and Y as follows:

$$\begin{bmatrix} I \\ Q \\ U \\ V \end{bmatrix} = \begin{bmatrix} |X|^2 + |Y|^2 \\ |X|^2 - |Y|^2 \\ 2\,\mathrm{Re}(X^*\,Y) \\ 2\,\mathrm{Im}(X^*\,Y) \end{bmatrix}, \tag{5.33}$$

where $*$ indicates a complex conjugate. All signals can be easily formed by simple computations in memory during step (vi) of the recipe above. For the case of left and right-handed signals from circular feeds L and R, we have:

$$
\begin{bmatrix} I \\ Q \\ U \\ V \end{bmatrix} = \begin{bmatrix} |L|^2 + |R|^2 \\ 2\,\mathrm{Re}(L^*\,R) \\ 2\,\mathrm{Im}(L^*\,R) \\ |R|^2 - |L|^2 \end{bmatrix}.
\tag{5.34}
$$

Calibration of the Stokes parameters will be discussed in Chapter 7.

5.3.5 Baseband recorders

It was not possible, until recently, to perform the required number of floating-point operations necessary for coherent de-dispersion on-line in software. Unless dedicated hardware were built to perform the calculations (see below), a common solution has been to write the sampled data to a fast-recording tape medium. The data then can be 'played back' subsequently and analysed by a computer off-line. The advantage of this technique is that the software can be improved constantly and data re-analysed. A disadvantage, however, is that one is observing 'blindly' as no immediate feed-back about the quality of the observations is available. Moreover, very often the observing bandwidth had been limited by speed (or price!) of the available tape media. For this reason, some systems are now equipped with a large number of hard disks in which the data are saved for subsequent processing.

5.3.6 On-line coherent de-dispersers

On-line systems provide immediate access to the data and therefore are the most flexible and efficient in terms of observing flexibility. Three main types have been developed over the past years.

In a hardware solution, dedicated processor chips perform the computations in real time. This can be achieved by employing a digital filterbank, so that only small pieces of baseband need to be processed by each chip. In some implementations, the de-dispersion is not achieved by applying the chirp in the frequency domain, but by performing a convolution in the time domain. A disadvantage of some of these systems has been that the de-dispersed data cannot be dumped fast enough, so that a number of pulses have to be integrated to keep the data rate low. Single-pulse observations therefore are not possible. As for all on-line systems, the baseband data cannot be played back for a second re-processing.

The recent advance in processor speed and the development of easy-to-build *Beowulf clusters* has triggered the birth of a new branch of on-line de-disperser. In one implementation, the data are stored on disks before they are picked up and analysed by a large number of PC processors. For small dispersion measures and bandwidths, this process is sustained in quasi real time. If even further processing power is available, the data can be sent directly to processors without being written to disk in an immediate stage. This can be facilitated by splitting the band into smaller chunks that are easier to process.

In order to get an idea of how the computational requirements scale, consider processing a segment of data of length comparable to the dispersion delay across an observing band Δf. From the dispersion relation, the observation time for this segment $t_{\rm obs} \propto {\rm DM} \times \Delta f$. Complex sampling of the data requires a sampling interval $t_{\rm samp} = 1/\Delta f$. The resulting number of samples per fast Fourier transform (FFT) $n_{\rm FFT} = t_{\rm obs}/t_{\rm samp} \propto {\rm DM} \times \Delta f^2$. Since the time to compute each FFT $t_{\rm FFT} \propto n_{\rm FFT} \log_2(n_{\rm FFT})$, (see, for example, Press *et al.* (1992)) then the ratio of computational time to observing time $t_{\rm FFT}/t_{\rm obs} \propto \Delta f \log_2({\rm DM}\,\Delta f^2)$.

5.3.7 Software filterbanks

The flexibility of modern coherent de-dispersion machines becomes apparent for observing modes other than coherent de-dispersion. For example, by taking short Fourier transforms of the complex-sampled data followed by the computation of the power spectrum, it is relatively straightforward to modify the software in order to simulate an incoherent filterbank. Sorting the samples accordingly, a simple filterbank output is achieved. Performing all the operations in software gives complete flexibility of the channelisation of the data, unlike the simple analogue filterbanks discussed in Section 5.2.1.

In a variation of this approach, one takes the Fourier components of the coarse filterbank channels and performs coherent de-dispersion on them rather than computing simply the power spectrum.

In both cases, care should be taken to avoid contamination of the data of one coarse frequency channel by signals from the others; this is known as *spectral leakage* and occurs because a short Fourier transform is not very efficient in separating the frequency components. Further details on software filterbanks can be found in Jenet *et al.* (1997).

5.4 Further reading

Most of the techniques to sample and process the pulsar signal presented in this chapter were originally described in the landmark paper by Hankins and Rickett (1975). Many of these ideas are covered in a slightly different style in the very useful review by Bhattacharya (1998).

Further discussion of filterbanks can be found in Jacoby and Anderson (2001) and Lyne (2003). Although remarkably successful over the years, and still in use at many observatories, analogue filterbanks are becoming somewhat dated. One aspect of the rapid development of digital signal processing technology that is starting to be applied to pulsar research is the so-called *digital filterbank* (Foster *et al.* 1996; Backer *et al.* 1997). Based on finite impulse response filters and high-speed Fourier transform chips, much like the correlators, digital filterbanks offer a much more flexible output than their analogue counterparts.

Although often hardware-specific, a number of useful references describing pulsar data acquisition systems often provide a good background in some of the techniques and technology implemented. The Princeton University pulsar group has been active in this area for many years with a number of generations of incoherent devices (see, for example, the filterbank-based 'Mk III' system described by Stinebring *et al.* (1992)), or more recently the 'Mk IV' baseband recording system (Shrauner 1997; Stairs *et al.* 2000a; Ord 2002). Other recent developments include the Caltech Baseband Recorder at Arecibo (Jenet et al. 1997), PuMa, a baseband sampler in use at the Westerbork telescope (Voûte *et al.* 2002) and the CPSR/CPSR2 recording systems at Parkes (van Straten 2003). Beowulf cluster approaches to data reduction either at or close to real-time observing are now becoming more common. At the time of writing (February 2004), a number of systems are in development at Arecibo (ASP; Arecibo Signal Processor), Green Bank (GASP; Green Bank Astronomical Signal Processor) and Jodrell Bank (COBRA; Coherent Online Baseband Receiver for Astronomy), (Joshi *et al.* 2003).

5.5 Available resources

On the book web site (Appendix 3), we maintain an up-to-date table of existing pulsar data acquisition systems that are available for use (either freely or by arrangement with the owners) for pulsar observations at most of the large radio telescopes around the world. Where available, links to contact people, relevant documentation, data formats and software are provided. For various reasons, each data acquisition system writes data in its own local format. While attempts to standardise the format of the processed data from these machines are being made, e.g. the EPN format (Lorimer *et al.* 1998) and PSRFITS (Hotan *et al.* 2004), raw data are invariably written in non-standard formats that are unique to each system. Some attempts at generic data recognition are being made: the SIGPROC, PRESTO and PSRCHIVE packages (Lorimer (2001), Ransom (2001) and Hotan *et al.* (2004), respectively) are capable of reading raw data in a variety of formats. These packages are now freely available and can be obtained via the links on the web site.

6

Finding new pulsars

Pulsar searching is conceptually a simple process – the detection of dispersed pulses in noisy data. The first pulsars were discovered serendipitously by visual inspection of the total power output from a radio telescope (Hewish *et al.* 1968). However, only a small fraction of the 1700 pulsars currently known are strong enough to be discovered via their individual pulses. The vast majority of known pulsars, and most that still await discovery, are faint objects which require sensitive telescopes and innovative techniques to reveal their periodic nature. From the discussion in Chapters 1 and 2, the motivation for probing deeper into this population is to discover exotic pulsars (e.g. those in binary systems) and to better characterise the Galactic distribution and evolution of neutron stars.

Since the early days of pulsar astronomy, a lot of effort has gone into developing sophisticated algorithms to maximise the sensitivity and efficiency of the pulsar search process. A summary of most of the resulting techniques we explore in this Chapter is presented in Figure 6.1. We begin by describing the main components of the 'standard' frequency domain radio pulsar search procedure which involves de-dispersion, Fourier transformation and candidate selection. We then move on to extensions of this approach to searches for short-period binary pulsars. Searches in the time domain are becoming increasingly popular; we discuss the fast-folding algorithm and single-pulse searches in this context. Virtually all searches of radio data now need to combat the ever-increasing levels of interference present; we discuss briefly time- and frequency-domain mitigation techniques. We conclude with an overview of tried and tested strategies required to optimise a search for the various types of pulsars.

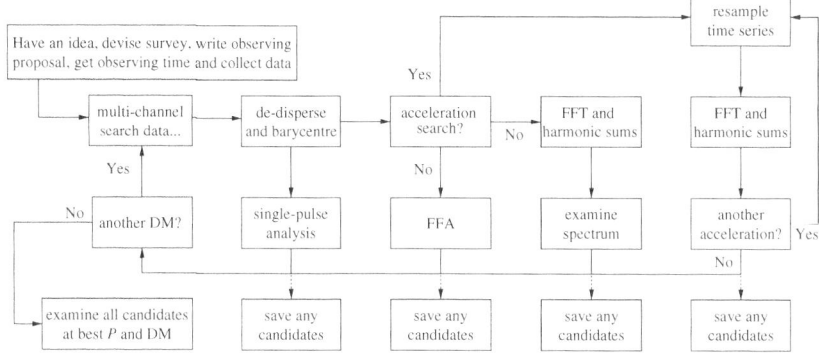

Fig. 6.1. Flow diagram summarising the main steps in a pulsar search and the most commonly used algorithms. Frequency domain acceleration searches and radio frequency excision techniques are not shown here (for clarity) but are described in detail in Sections 6.2.2 and 6.4, respectively.

6.1 Standard search procedure

We begin by describing the most commonly used procedure to find a periodic signal of unknown pulse period and dispersion measure (DM). The data are first de-dispersed to form a number of time series spanning a wide range of trial DM values. Each time series then can be independently searched for the presence of periodic signals. The standard procedure is to Fourier transform the time series and search the resulting amplitude or power spectra for significant features. The best candidates from the analysis are saved and the whole process is repeated for another trial DM. After processing all of the time series in this way, a list of pulsar candidates is compiled and the de-dispersed data are folded modulo each candidate period for further inspection.

6.1.1 The de-dispersion stage

Now we expand on these steps in detail, beginning with a description of the basic algorithms for dispersion removal, optimal choice of trial DM step and an efficient de-dispersion scheme.

6.1.1.1 Simple de-dispersion

Considering the raw data as a two-dimensional array of time samples and frequency channels, we write the jth time sample of the lth frequency channel as $\mathcal{R}_{jl}$. For n_{chans} frequency channels, the jth sample of the de-dispersed time series, $\mathcal{T}_j$, is then

$$\mathcal{T}_j = \sum_{l=1}^{n_{\text{chans}}} \mathcal{R}_{j+k(l),l}, \tag{6.1}$$

where $k(l)$ is the nearest integer number of time samples corresponding to the dispersion delay of the lth frequency channel relative to some reference frequency. Labelling the lth frequency channel of the data by f_l, and recalling the dispersion relation (see Equation (4.7)), we can write $k(l)$ in terms of the trial DM, the data sampling interval t_{samp} and the channel frequencies as follows:

$$k(l) = \left(\frac{t_{\text{samp}}}{4.15 \times 10^6 \, \text{ms}}\right)^{-1} \left(\frac{\text{DM}}{\text{cm}^{-3} \, \text{pc}}\right) \left[\left(\frac{f_l}{\text{MHz}}\right)^{-2} - \left(\frac{f_1}{\text{MHz}}\right)^{-2}\right]. \tag{6.2}$$

Here we have assumed that the channel ordering in $\mathcal{R}$ starts at the highest frequency ($l = 1$) and proceeds in descending frequency order. The channel frequencies are therefore given by

$$f_l = f_1 - (l-1)\Delta f_{\text{chans}}, \tag{6.3}$$

where Δf_{chans} is the channel bandwidth.

6.1.1.2 Choice of dispersion step size

Some consideration needs to be made in order to choose the most appropriate interval between trial DM values. This should not be so large that a real pulsar with a true DM lying between two trial values is significantly broadened and sensitivity is lost. Conversely, the interval should not be too small that computing power is wasted on producing and searching de-dispersed time series that are virtually identical for neighbouring trial DMs. In order to quantify this, consider a simple top-hat pulse of intrinsic width W_{int}. De-dispersion at an incorrect trial DM value that differs by ΔDM from the true value broadens the pulse across the entire bandwidth. When $\Delta f \ll f$, the resulting effective pulse width

$$W_{\text{eff}} = \sqrt{W_{\text{int}}^2 + (k_{\text{DM}} \times |\Delta\text{DM}| \times \Delta f/f^3)^2}, \tag{6.4}$$

where $k_{\text{DM}} = 8.3 \times 10^6$ ms if pulse widths are measured in ms and, as usual, DM is in units of cm^{-3} pc and the bandwidth Δf and centre frequency f are in MHz. In Appendix 1, we derive the observing sensitivity to pulse signals. Ignoring constant system-dependent factors in

Equation (A1.22) of Appendix 1, we find a simple relationship between signal to noise (S/N) ratio, effective pulse width W_{eff} and period P:

$$\text{S/N} \propto \sqrt{\frac{P - W_{\text{eff}}}{W_{\text{eff}}}}. \tag{6.5}$$

Combining the above two equations, we can calculate the response of a data acquisition system to incorrect de-dispersion. This is shown in Figure 6.2, in which we plot S/N versus trial DM for a variety of different pulse periods relative to the S/N value for a true DM of 50 cm^{-3} pc. As expected, the choice of DM step becomes critical when attempting to detect dispersed pulsars with periods below a few hundred ms.

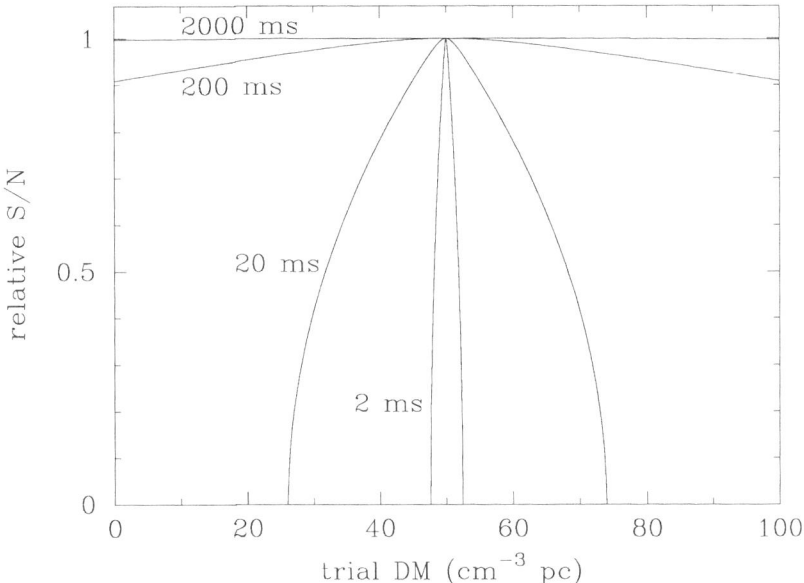

Fig. 6.2. Relative S/N as a function of trial DM for a hypothetical data acquisition system spanning an 8 MHz band centred at 430 MHz. Each curve corresponds to a different pulse period as indicated. In each case the true DM value is assumed to be 50 cm^{-3} pc and the intrinsic pulse duty cycle (W_{int}/P) is assumed to be 5 per cent. Effects of scattering and dispersion across filterbank channels are assumed to be negligible.

A sensible choice of DM step is to set the delay between the highest and lowest frequency channels equal to the data sampling interval. Again, starting with the dispersion relation, the ith DM value can be written in terms of the total bandwidth Δf (MHz), centre frequency f

(MHz) and sampling time t_{samp} (ms) as

$$DM_i = 1.205 \times 10^{-7} \, cm^{-3} \, pc \, (i-1) t_{samp} (f^3/\Delta f). \qquad (6.6)$$

The case $i = 1$ corresponds to the 'zero DM' time series – simply combining all the frequency channels without any time delays. This time series is used primarily to identify sources of interference (see Section 6.4). When $i = n_{chans} + 1$, the so-called 'diagonal DM' value is reached. At this DM, the total delay across the band is equal to $n_{chans} \times t_{samp}$ and the broadening across an *individual frequency channel* is equal to t_{samp}.

Above the diagonal DM the effective time resolution starts to become dominated by the broadening in the individual channels. Usually when $i = 2n_{chans}$ or $i = 3n_{chans}$, adjacent time samples are added together so that the rest of the processing requires fewer computations. Since dispersion broadening is now the dominant effect, halving the effective time resolution does not impact on the sensitivity to short-period pulsars. The entire process now can be repeated on the new coarser data to produce time series with higher dispersion measures out beyond the now higher value of the diagonal DM until the desired DM limit is reached. Typically, we expect DMs in excess of 1000 cm^{-3} pc for surveys along the Galactic plane and $\lesssim 50$ cm^{-3} pc for high Galactic latitudes. As a general rule, we recommend using a reasonable model of the electron density distribution (see, for example, Section 4.4) to estimate the maximum DM for the lines of sight sampled by the survey and multiply this by a factor of two to account for uncertainties in the electron density model.

6.1.1.3 Tree de-dispersion

The simple de-dispersion process described above is rather computationally expensive, since it requires n_{chans}^2 floating-point operations to de-disperse every n_{chans} time samples. Inspired by the fast-folding algorithm developed by Staelin (1969); (see Section 6.3.1), Taylor (1974) proposed a more efficient de-dispersion scheme known as the 'tree algorithm'. As shown in Figure 6.3, the tree algorithm derives its name from the fact that it can be built from successively smaller components which all start from simple two-channel 'branches'. For example, a four-channel tree can be built from a pair of two-channel branches; an eight-channel tree derives from two four-channel branches etc. As a result, the tree algorithm requires a base-two number of frequency channels.

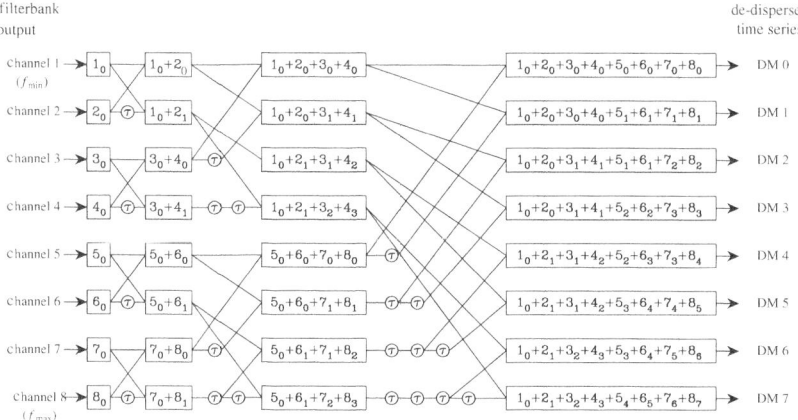

filterbank output

de-dispersed time series

Fig. 6.3. Taylor's tree algorithm for de-dispersion showing the eight-channel case. Figure provided by Bernd Klein.

Rather than requiring n_{chans}^2 operations, the identification of redundant and repeated operations reduces the overall computational requirement to $n_{\text{chans}} \log_2 n_{\text{chans}}$ operations. In general, the resulting number of time series produced by the tree algorithm is equal to the input number of frequency channels. When the time delay is set to be equal to t_{samp}, the DM steps are the same as in Equation (6.6).

An implicit assumption in the tree de-dispersion process is that the dispersion delay across the frequency band is linear. While the exact form of the dispersion law (see Equation (4.7)) is a quadratic function of frequency, the relatively narrow bandwidths employed mean that this assumption is often adequate. For some searches, however, the data need to be corrected before the tree algorithm can be used. For example, in the Parkes multibeam survey (Manchester *et al.* 2001) where $f = 1374$ MHz and $\Delta f = 288$ MHz, the ninety-six channel filterbank data were padded by an additional thirty two channels across the band. This 'linearisation' process changes the effective channel frequency slightly so that the dispersion delay more closely approximates a straight line. In this case, the total number of channels becomes a base-two number as required.

In practice, the additional computations required to linearise the data before the tree stage, as well as the computational input/output requirements demanded by the tree algorithm in producing multiple de-dispersed time series, mean that it may not always be optimal to implement a tree algorithm. Benchmarking brute-force de-dispersion schemes against tree de-dispersion is often advisable during the planning phases of a pulsar survey.

6.1.2 Barycentric correction for long time series

Most radio pulsar search data are of relatively short (< 30 min) duration. Over this time, the effects of the rotation of the Earth and its motion around the Sun can be safely ignored. For deep radio searches (see Sections 6.5.4 and 6.5.5), and searches at X-ray and γ-ray wavelengths carried out with orbiting satellites, the relative motion between the observatory and the target becomes important and should be corrected prior to attempting a periodicity search. As for pulsar timing applications (see Chapter 8) the standard approach is to refer the observed (topocentric) data collected at the telescope to the solar system barycentre (SSB) which, to a very good approximation, is an inertial reference frame.

To transform a topocentric time series to the SSB, first we must delay or advance the start time of the observation, t_{start}, appropriately to match the arrival time of the first sample at the SSB, $t_{\mathrm{start,SSB}}$. This is identical to the correction applied to pulse arrival times described in detail in Chapter 8. The arrival time of subsequent samples then needs to be monitored so that it does not differ significantly from the expected arrival time at the SSB. Specifically, we compare the arrival time of the ith sample $t_i = t_{\mathrm{start,SSB}} + (i-1)\,t_{\mathrm{samp}}$ with the arrival time corrected for the relative motion between the observatory and the SSB: $\tau_i = t_{\mathrm{start,SSB}} + (i-1)\,t_{\mathrm{samp,SSB}}$. The corrected sampling time $t_{\mathrm{samp,SSB}}$ is a variable quantity due to the continually changing relative motion and can be calculated using a pulsar timing program such as TEMPO (see Chapter 8). The correction then proceeds by adding or subtracting whole samples from the time series so that $|\tau_i - t_i| < t_{\mathrm{samp}}$. Added samples are usually chosen to be the mean value of the time series.

6.1.3 Periodicity searches using the Fourier transform

Given a de-dispersed time series $\mathcal{T}_j$ that, if necessary, has been appropriately barycentred, we need an algorithm to search it for the presence of periodic signals. One of the most efficient and widely used techniques is to take the Fourier transform of the time series and examine the Fourier (frequency) domain. In the following subsections, we review the salient properties of Fourier transform for pulsar searching.

6.1.3.1 The discrete Fourier transform

Since the de-dispersed time series T_j is a set of N independently sampled data points, rather than the continuous form of the Fourier transform, we compute the discrete Fourier transform (DFT). By definition, the kth Fourier component of the DFT

$$\mathcal{F}_k = \sum_{j=0}^{N-1} T_j \exp(-2\pi i jk/N), \tag{6.7}$$

where $i = \sqrt{-1}$ and N is the number of elements in the time series. Throughout this chapter, we will assume that all Fourier components $\mathcal{F}_k$ have been normalised by the factor $(N\overline{T_j^2})^{1/2}$ (see, for example, Ransom *et al.* (2002)). As we shall see later, this turns out to be convenient when estimating significance levels of pulsar candidates. For equally-spaced data in the time domain with a sampling interval t_{samp}, the frequency of the kth Fourier component $\nu_k = k/(Nt_{\text{samp}}) = k/T$, where T is the length of the observation and $1 < k < N/2$. The width of each frequency 'bin' therefore is simply $1/T$ and the highest frequency component is the Nyquist frequency $\nu_{\text{Nyq}} = 1/(2t_{\text{samp}})$.

It is apparent, from Equation (6.7), that the computation of an N-point DFT requires N^2 floating-point operations. For modern applications, in which we deal often with time series for which $N \geq 2^{25}$, this can become very time consuming, even for current-day computers. Fortunately, there are two ways to reduce the computation time. The first of these is the fast Fourier transform (FFT), which requires only $N \log_2 N$ operations to transform an N-sample time series. A further improvement comes from the fact that the time series is a set of purely real numbers. As a result, while the highest frequency in the Fourier domain is nominally $1/t_{\text{samp}}$, the DFT is symmetric about the Nyquist frequency, ν_{Nyq}. Fourier components above ν_{Nyq} are just the complex conjugates of their low-frequency counterparts:

$$\mathcal{F}_{N-k} = (\mathcal{F}_k)^*, \tag{6.8}$$

where the asterisk denotes the complex conjugate. This redundancy can be exploited (see, for example, Press *et al.* (1992)) to calculate the DFT of two N-point real data sets simultaneously, or a single data set of length $N/2$.

6.1.3.2 Searching for periodic signals in the Fourier domain

Displaying either the amplitudes ($\mathcal{A}_k = |\mathcal{F}_k|$) or powers ($\mathcal{P}_k = |\mathcal{F}_k|^2$) of the Fourier components as a function of frequency is an extremely sensitive means of revealing a periodic signal. This is illustrated in Figure 6.4, in which we have sampled a 25 Hz sine wave in the presence of purely Gaussian noise with a standard deviation that is 3 times the amplitude of the signal. As a result, the sinusoidal signal is essentially undetectable by examining the time domain. The 'power spectrum' in the Fourier domain, however, shows a clearly visible line at 25 Hz.

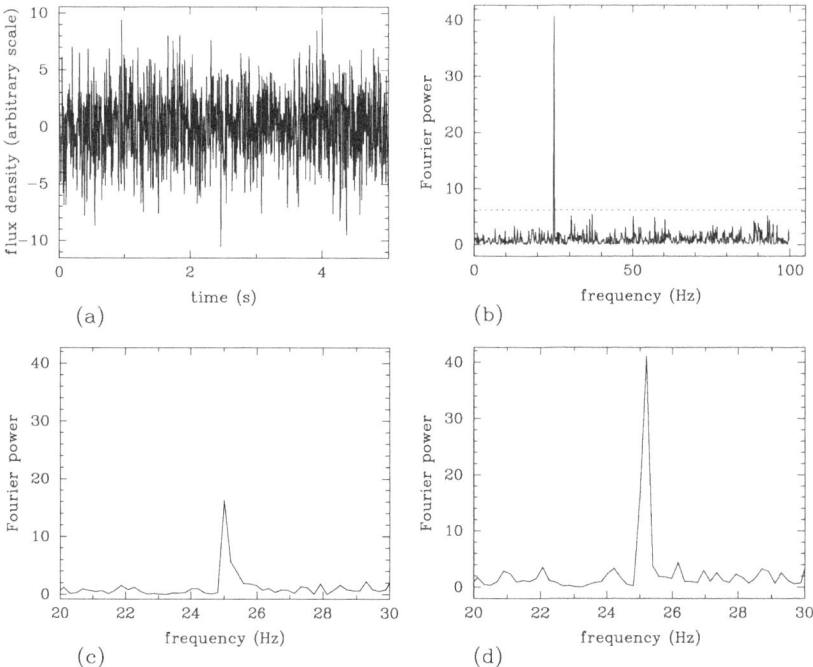

Fig. 6.4. (a) A noisy time series containing a 25 Hz signal; (b) the power spectrum of the DFT of this time series (i.e. $|\mathcal{F}_k|^2$ as a function of k). The dashed line shows the detection threshold based on the number of independent estimates of spectral power; (c) an expanded view of the spectrum showing the DFT response to a 25.125 Hz signal in which the effects of scalloping reduce the power by about 60 per cent; (d) recovery of the power by Fourier interpolation (see text).

One limitation of the DFT is that its frequency response is not uniform and is in fact only ideal for signals that match exactly the centre frequencies of the Fourier bins. As mentioned above, each bin in the frequency domain is characterised by its central frequency $\nu_k = k/T$

and width $\Delta\nu = 1/T$. The result of this 'scalloping' effect is a reduction in sensitivity to signals away from ν_k. Consider a signal of frequency $\nu_r = r/T$, where r is now a real number such that ν_r lies within the frequency range of bin k, i.e. $|\nu_k - \nu_r| < \Delta\nu/2$. Starting from the definition of the DFT, it can be shown (see, for example, Ransom *et al.* (2002)) that

$$\mathcal{F}_r = \mathcal{F}_k \operatorname{sinc}[\pi(k-r)]. \tag{6.9}$$

In the extreme case, when the signal frequency falls midway between two bins ($|k - r| = 0.5$), the Fourier amplitude is reduced by 36 per cent and the corresponding power by 60 per cent. This reduction can be seen in Figure 6.4(d) where the frequency of the original 25 Hz sinusoid has been changed to 25.125 Hz.

The main methods to recover this loss in sensitivity are zero padding, Fourier domain interpolation and interbinning. Zero padding – simply the addition of zero-valued elements to the time series – is the most straightforward procedure. Since this process adds no noise or signal into the time domain data (i.e. Parseval's theorem; (see Bracewell (1998)) the Fourier components are unaffected by this operation. This increases N and reduces $\Delta\nu$ accordingly. While this reduces the effects of scalloping somewhat, it does, of course, mean that the required FFTs are larger. For long time series, this can become computationally too costly.

When the expected frequency of the signal is known, or has been found from a preliminary search, the signal can essentially be recovered completely by correlating the nearest integer bin to r, which we denote by $[r]$, with the inverse of the sinc response, i.e.

$$\mathcal{F}_r \simeq \sum_{k=[r]-m/2}^{[r]+m/2} \mathcal{F}_k \exp(-i\pi(r-k)) \operatorname{sinc}[\pi(r-k)]. \tag{6.10}$$

Typical values for the range of bins around r might be $m = 32$ (see Ransom *et al.* (2002) for further details).

A less computationally expensive version of *Fourier interpolation* is known as 'interbinning', in which the power at half-integer frequencies can be estimated from just the two neighbouring bins (i.e. $m = 2$ in the above expression). In this case, the half-integer power $\mathcal{P}_{k+\frac{1}{2}} = |\mathcal{F}_{k+\frac{1}{2}}|^2$, where

$$\mathcal{F}_{k+\frac{1}{2}} \simeq \frac{\pi}{4}(\mathcal{F}_k - \mathcal{F}_{k+1}). \tag{6.11}$$

Using this simple technique, the Fourier amplitude of a signal lying midway between two independent bins is reduced by only 7 per cent.

A related approach also in use is to compare the power in each Fourier bin with those of its two nearest neighbours and replace it according to

$$\mathcal{P}_k = \max \left\{ \frac{|\mathcal{F}_{k-1} + \mathcal{F}_k|^2}{2}, |\mathcal{F}_k|^2, \frac{|\mathcal{F}_k + \mathcal{F}_{k+1}|^2}{2} \right\}. \qquad (6.12)$$

As a result, a signal split between two bins is combined into a single bin. This method is used to great effect in Figure 6.4(d). One result of all these schemes is that the resultant Fourier components are not independent quantities. This turns out to have a relatively small effect on the noise statistics discussed later.

6.1.3.3 Removing low-frequency noise

Our example time series and its Fourier transform in Figure 6.4 assumed purely Gaussian noise. The Fourier spectrum of Gaussian noise is 'white', i.e. the Fourier power is distributed uniformly over the entire frequency range. Well-behaved white noise is ideal, because the estimation of the significance level of any signal is relatively simple (see below). Although time series obtained from real pulsar search data closely resemble Gaussian noise, fluctuations in the receiver and/or data acquisition systems often manifest themselves via a significant low-frequency or 'red noise' component when viewed in the Fourier domain. An example of this is shown in Figure 6.5.

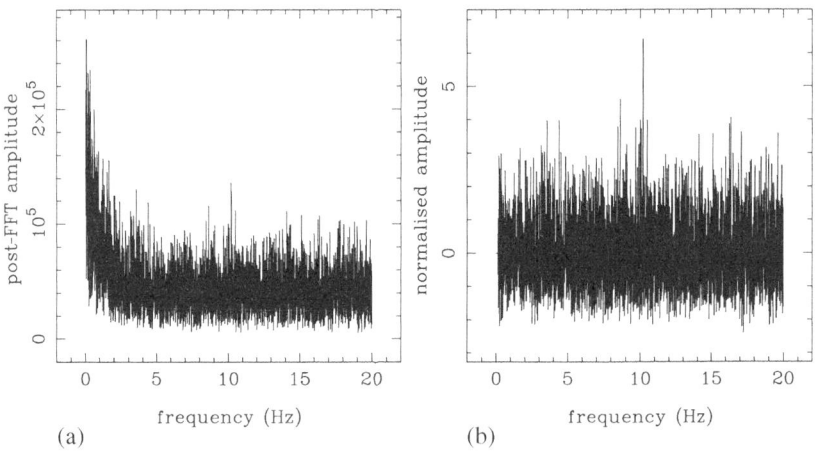

(a)   (b)

Fig. 6.5. (a) Amplitude spectrum from data collected using the Parkes telescope. (b) Spectrum after a whitening procedure has been applied to remove the red noise component. The whitened spectrum then has been normalised so that it has a zero mean and unit root mean square (see text).

Before attempting to estimate significance levels of any signal present in these data, it is standard practice to whiten the spectrum so that the response to noise is as uniform as possible. The most common technique in use is to break the spectrum up into a number of contiguous pieces, calculating the mean and root mean square value for each one. Care must be taken at this stage to avoid biasing the mean and root mean square values from outlying points, and often the median is used rather than the mean. Subtracting a running median and normalising the local root mean square will result in the whitened spectrum having a zero mean and unit root mean square. With this normalisation scheme, the S/N ratio of any spectral feature is simply its amplitude.

6.1.3.4 Increasing sensitivity to narrow pulses

Our discussion so far has considered purely sinusoidal signals that appear in the Fourier domain as a single line at the fundamental frequency of the sinusoid. In reality, however, the pulsed signals we are trying to detect have a duty cycle (i.e. the pulse width divided by the period) that is typically only a few per cent. In the Fourier domain, the power from such a narrow pulse is distributed between the fundamental frequency and a significant number of harmonics.

In order to estimate the number of harmonics present, consider the time series as a train of top hat pulses of width W spaced by the pulse period P. In the time domain, this can be viewed as a single top hat function convolved with a train of delta functions separated by P (i.e. the Shah or comb function). In the Fourier domain (see, for example, Bracewell (1998)) this convolution is just the product of the Fourier transforms of the two functions. The Fourier transform of the top hat is proportional to $\mathrm{sinc}(\pi f W)$, that has a first null when the Fourier frequency $f = 1/W$. The Shah function Fourier transforms to another Shah function with spacing of $1/P$. The resulting Fourier transform of our pulse train is simply a series of delta functions harmonically spaced by $1/P$ with amplitudes that are bounded by the envelope of the sinc function. Taking the extent of the harmonics as being roughly the width of the sinc envelope to its first null, the number of harmonics is roughly P/W, i.e. the reciprocal of the pulse duty cycle.

For typical pulse duty cycles of order 5 per cent, the expected number of harmonics is of order 20, as shown in the example spectrum for PSR B2303+30 (Figure 6.6(a)). In order to take full advantage of the power contained in these harmonics, a technique known as 'incoherent harmonic summing', first devised by Taylor and Huguenin (1969), is

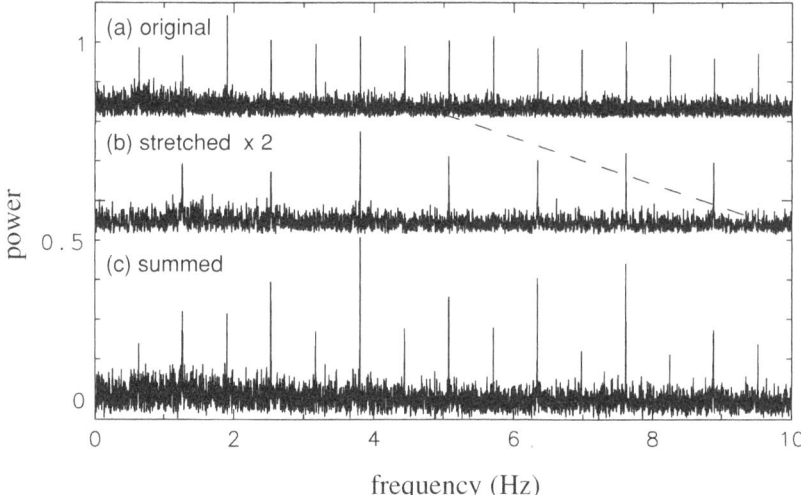

Fig. 6.6. The process of harmonic summing illustrated in the power spectrum of an observation of PSR B2303+30 collected with the Ooty radio telescope at 327 MHz (see text for further details). Figure provided by Dipankar Bhattacharya.

used. In the example shown in Figure 6.6, the lower half of the original spectrum is stretched by a factor of 2 (Figure 6.6(b)) and then added to the original unstretched spectrum. As a result, all second harmonics are added to their corresponding fundamentals. Although this summation process increases the noise in the folded spectrum by a factor of $\sqrt{2}$, the amplitudes of two harmonics add directly. For two harmonics of roughly equal power, the net gain in S/N is of order $\sqrt{2}$. By repeating this process several times, and taking care to add in odd-numbered harmonics, the S/N to a narrow duty cycle pulse increases significantly.

A good illustration of the improvement gained by harmonic summing is shown in Figure 6.7 which is the result of an analysis assuming idealised pulses described by Ransom *et al.* (2002). This shows the various duty cycle regimes in which harmonic summing is effective. While the single harmonic sensitivity is adequate for duty cycles wider than 30 per cent, it is clear for narrower pulses that the harmonic summing schemes are essential in order to retain full sensitivity. Most pulsar search codes produce a total of five spectra that are searched independently: the spectrum of the DFT itself, and four subsequent harmonically folded versions which contain the sum of the first two, four, eight, and sixteen

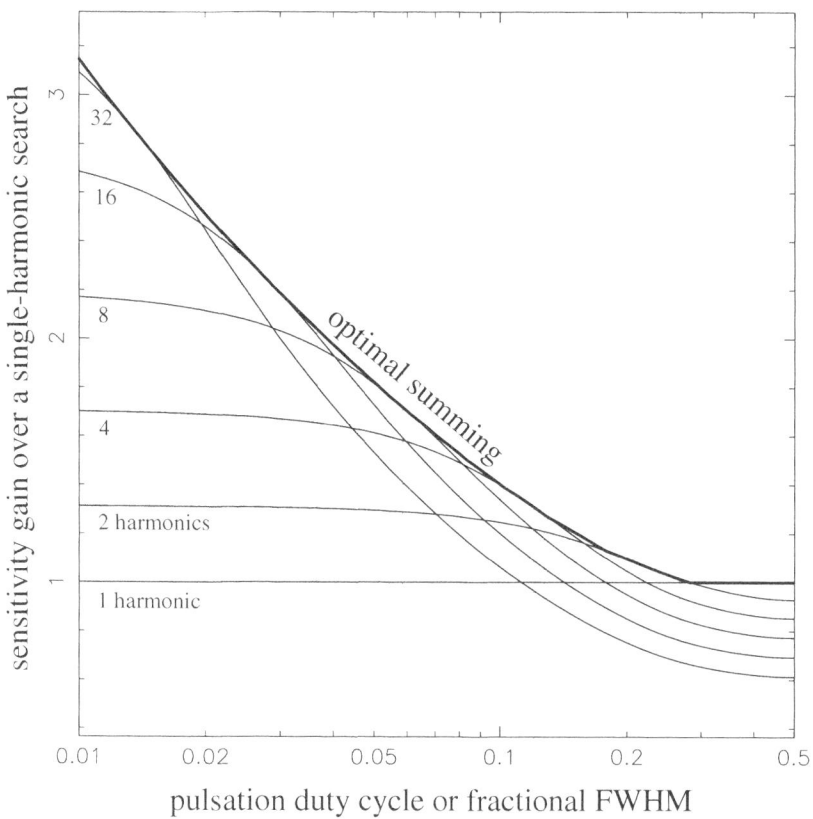

sensitivity gain over a single-harmonic search

32

16

8

4

2 harmonics

1 harmonic

optimal summing

pulsation duty cycle or fractional FWHM

Fig. 6.7. An illustration of the gain in sensitivity due to harmonic summing (thin curves) over a single-harmonic DFT analysis (horizontal line) as a function of pulse duty cycle. The thick solid line shows the optimal harmonic sum. Figure and analysis provided by Scott Ransom.

harmonics, respectively. As Figure 6.7 shows, this choice guarantees that most of the range of pulse duty cycles are searched with optimal sensitivity. However, for duty cycles less than 2 per cent, some improvement in sensitivity can be made by summing up to thirty-two harmonics.

6.1.3.5 False-alarm probabilities and S/N ratios

In order to calculate the significance of signals in the Fourier domain, we need to know the response of the DFT to random noise. In the ideal case[1] for a time series containing pure Gaussian noise, the probability

---

1 As discussed earlier, in reality the noise does not follow strictly a Gaussian distribution. Also, the Fourier interpolation discussed above means that the amplitude

density function (PDF) of the real and imaginary parts of the Fourier components also follow a Gaussian PDF. In the spectral analysis, we deal either with Fourier power (the sum of the squares of the real and imaginary components) or the amplitude (square root of the power). In general, the sum of the squares of n independent variables with Gaussian PDFs is the χ^2 distribution with n degrees of freedom. In our case, where $n = 2$, the powers follow an exponential PDF that can be integrated easily to show that the probability that the power in an individual bin $\mathcal{P}_k$ exceeds some threshold $\mathcal{P}_{\min}$ is simply proportional to $\exp(-\mathcal{P}_{\min})$. This is sometimes called the *false-alarm probability*, i.e. the chance of a candidate being the result of noise fluctuations rather than a real event. For the single harmonic shown in Figure 6.4(b), the normalised power level is ~ 40. The corresponding false-alarm probability $p_{\text{false}} = \exp(-40) \simeq 4 \times 10^{-18}$!

For the more general case, in which harmonic folding has been performed m times, the PDF follows a χ^2 distribution with $2m$ degrees of freedom. The corresponding false-alarm probability can be written as

$$p_{\text{false}}(\mathcal{P} > \mathcal{P}_{\min}) = \sum_{j=0}^{m-1} \frac{(\mathcal{P})^j}{j!} \exp(-\mathcal{P}_{\min}). \tag{6.13}$$

Here, it is assumed that the powers have been normalised by the factor $N\overline{\mathcal{T}_j^2}$ where, as before, N is the number of samples in the time series $\mathcal{T}_j$ (see, for example, Ransom *et al.* (2002)).

We can estimate a reasonable detection threshold for any given search based on the false-alarm probability of a single event and the number of trials. For the sinusoidal signal in Figure 6.4, the number of points in the time series is 1024, resulting in 512 separate samples of power in the Fourier domain. The expected number of false alarms above a given power threshold $\mathcal{P}_{\min}$ is simply $512 \exp(-\mathcal{P}_{\min})$. Setting this number to be less than one event, it follows that $\mathcal{P}_{\min} = -\ln(1/512) \simeq 6.2$. This threshold power level is shown by the dashed line in Figure 6.4(b).

For analyses that work with Fourier amplitudes ($\mathcal{A}$) rather than powers, it is more common to quote S/N thresholds. Since S/N values are calculated from the Fourier amplitudes by subtracting a mean value ($\overline{\mathcal{A}}$) and dividing by the local root mean square ($\sigma_{\mathcal{A}}$) the false-alarm proba-

and power values are not strictly independent quantities. Fortunately, the departure from a χ^2 distribution due to these effects is small and can be ignored safely.

bility that the S/N will exceed a threshold S/N$_{\mathrm{min}}$ is given by

$$p_{\mathrm{false}}(\mathrm{S/N} > \mathrm{S/N}_{\mathrm{min}}) = \exp(-[\sigma_{\mathcal{A}}\,\mathrm{S/N}_{\mathrm{min}} + \overline{\mathcal{A}}]^2). \qquad (6.14)$$

Integrating the exponential PDF in terms of amplitude rather than power, we find $\overline{\mathcal{A}} = \sqrt{\pi/4}$ and $\sigma_{\mathcal{A}} = 1 - \pi/4$. Using these values, we can repeat the above threshold calculation in terms of S/N. In general, for a search with n_{trials} independent S/N estimates, a suitable threshold is:

$$\mathrm{S/N}_{\mathrm{min}} = \frac{\sqrt{\ln[n_{\mathrm{trials}}]} - \sqrt{\pi/4}}{1 - \pi/4} \simeq \frac{\sqrt{\ln[n_{\mathrm{trials}}]} - 0.88}{0.47}. \qquad (6.15)$$

For the 512-sample DFT in Figure 6.4, the corresponding threshold S/N would be 3.4. In a real pulsar search, the number of trials is much greater than 512, and S/N$_{\mathrm{min}}$ is appropriately larger. For example, in a recent Arecibo drift-scan survey (Lorimer *et al.* 2004a) an analysis used 492 trial DM values, each of which produces 2^{18} Fourier components which, for the harmonic summing analysis, were analysed independently five times. As a result, $n_{\mathrm{trials}} \sim 6.4 \times 10^8$ and S/N$_{\mathrm{min}} \simeq 7.7$. In practice, because of radio-frequency interference (see Section 6.4), a slightly higher threshold is required (for example, in their search, Lorimer *et al.* (2004a) set S/N$_{\mathrm{min}} = 8$).

6.1.3.6 Reconstructed profiles

No use of the phase information in the Fourier components is made when forming amplitude or power spectra. This information can be used to eliminate 'signals' from the harmonic summing analysis that result from random superposition of noise features. Unlike the harmonics produced by a train of pulses, random superposition will result in a random phase relationship between the 'harmonics'. This can be tested by taking the inverse DFT of the harmonics to form a pulse profile. To ensure that this 'reconstructed profile' is entirely real, the DFT is doubled in size to include the complex conjugates of each harmonic used. If the Fourier components are truly phase related, this 'reconstructed' profile should have a S/N in the time domain that is comparable to the spectral S/N. A spurious candidate often will show up at this stage with a profile that has a much reduced S/N.

6.1.3.7 Two-dimensional Fourier analysis

The procedure discussed so far is essentially a one-dimensional Fourier analysis that repeats on different de-dispersed time series. An alternative approach is to consider the raw data as a two-dimensional array of time samples and frequency channels. The two-dimensional Fourier transform of such an array is the space of fluctuation frequency and dispersion delay. A dispersed periodic signal plotted in this phase space appears as a set of harmonically spaced dots in a graph of delay versus frequency. The slope of the dots is proportional to the dispersion measure. Interpolating along lines of constant dispersion measure then would produce amplitude or power spectra that can be analysed with the harmonic summing techniques described above. Although this method has been implemented (see, for example, Camilo *et al.* (1996)), and is in principle more computationally efficient than the one-dimensional approach, the simplicity of brute-force one-dimensional de-dispersion and Fourier transforms usually is preferred.

6.1.3.8 Discontinuities in the time series

The techniques outlined so far assume the time series to be continuous. Unforeseen interruptions in the observation (e.g. due to parking of the telescope in high winds or power outages) can result in a number of discontinuous time series. While these data can be analysed separately, the ideal solution is to perform a coherent Fourier transform over the entire observation. This is achieved by appropriately zero padding the missing time samples.

6.1.4 Candidate selection

The result of all the de-dispersion and Fourier transform stages described above is a list of candidate periods and S/N ratios for all harmonic folds and DMs. A real pulsar usually will appear many times in this list at a variety of S/N values, with the maximum ideally being at the DM that is closest to the true DM value. At this stage, the standard practice is to de-disperse and fold the raw data at the candidate period and DM, and produce diagnostic plots for visual inspection. The details of the folding process are discussed in Chapter 7.

A typical pulsar candidate is shown in Figure 6.8. This example is the original discovery observation of PSR J1842–0415, one of four pulsars discovered in a search using the 100 m Effelsberg radio telescope (Lorimer *et al.* 2000). The integrated profile (top left) has a well defined narrow pulse that appears consistently throughout the integration (bottom left) and in all frequency channels (bottom right). The S/N of the integrated profile should compare well with the S/N in the amplitude or

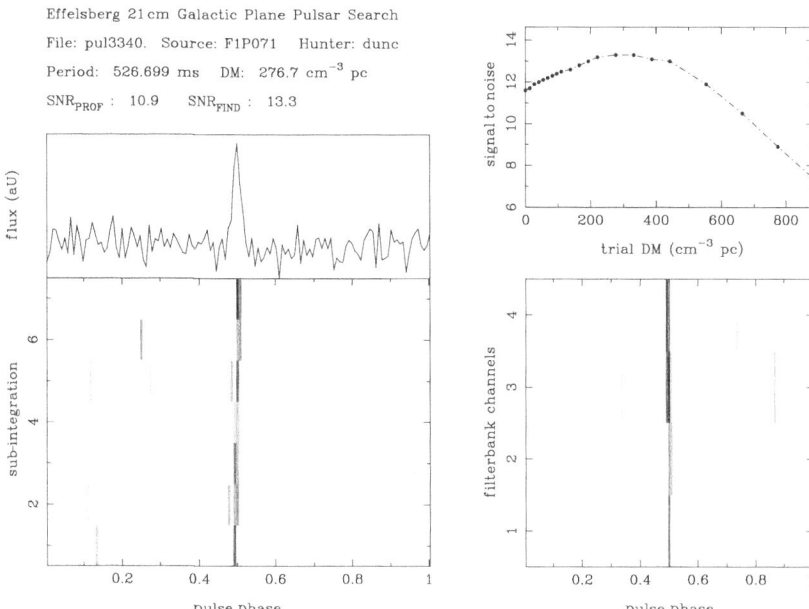

Fig. 6.8. Sample search code output for PSR J1842–0415, the first pulsar discovered using the Effelsberg telescope. See text for further details.

power spectrum and (if calculated) the reconstructed S/N. In addition to the folded profiles, perhaps the most useful diagnostic is S/N versus trial DM shown on the top right as a clear peak at a non-zero DM. As we showed in Section 6.1.1 and Figure 6.2, every pulsar produces a characteristic shape in the S/N–DM plane that depends primarily on the pulse width, period, observing frequency and bandwidth. Good agreement between the observed and theoretical S/N–DM responses is an essential test that any viable pulsar candidate should pass.

6.2 Searches for pulsars in binary systems

Although the Fourier transform is extremely good at finding periodic signals, as noted by a number of authors (see, for example, Middleditch and Priedhorsky (1986) and Johnston and Kulkarni (1991)), the frequency domain analyses discussed above have reduced sensitivity to pulsars in short-period binary systems. The effect of binary motion is to cause a change in the apparent pulse frequency during the integration, spreading the emitted signal power over a number of neighbouring Fourier bins. As a result, the sharpness of the spectral features and, hence, the S/N ratio and sensitivity of the search are reduced significantly. All is not lost, however, since there are now a number of techniques in use to compensate for, and in some cases fully recover, the loss of sensitivity due to binary motion. These techniques described below generally work optimally under certain conditions and often demand significant computational resources.

6.2.1 Time domain resampling

One approach to signal recovery is to resample the time series to the rest frame of an inertial observer with respect to the pulsar. A straightforward periodicity search of the corrected time series then would detect the pulsar as if it were a solitary object without loss of sensitivity. The correction to the time series is a simple application of the Doppler formula to relate a time interval in the pulsar frame, τ, to the corresponding interval in the observed frame, t:

$$\tau(t) = \tau_0(1 + V_l(t)/c), \qquad (6.16)$$

where $V_l(t)$ is the radial velocity of the pulsar along the line of sight, c is the speed of light and we have neglected terms in (v/c) higher than first order. The constant τ_0 is used for normalisation purposes (see, for example, Camilo *et al.* (2000b)). Given a functional form for $V_l(t)$, the re-sampling process proceeds by calculating the time intervals in the new frame from Equation (6.16). New sample values then are created based on a linear interpolation running over the original time series (see also Middleditch and Kristian 1984). Alternatively, the correction can be carried out by adding or removing samples to compensate for the relative phase drift between τ and t as described in Section 6.1.2 for the barycentre correction.

If the orbital parameters of the binary system are known (e.g. if a search is being made for a pulsar in a known binary system) $V_l(t)$ can be calculated from Kepler's laws (see Section 8.3.1.2) and the effects of orbital motion can be removed entirely from the time series. Following unsuccessful attempts on other systems (see, for example, Prince *et al.* (1991)), this technique may well prove fruitful following the discovery of the 2.7 s pulsar companion in the J0737–3039 system (Lyne *et al.* 2004). While this pulsar did not require an exhaustive search, given the stunning applications of J0737–3039 (see Chapter 2), deeper searches for companions in other systems should be carried out.

For the purposes of a blind search, in which the orbital parameters are a priori unknown, assuming a Keplerian model for $V_l(t)$ would require a five-dimensional search of all the parameter space. In practice, computing requirements demand a simpler solution. Although dropping orbital eccentricity and longitude of periastron would allow a three-dimensional search for circular orbit binaries, the simplest model is to assume a constant orbital acceleration a_l during the integration, i.e. $V_l(t) = a_l t$. The so-called 'acceleration search' then can be carried out on time series corrected assuming different trial values of a_l in order to cover a region of acceleration space.

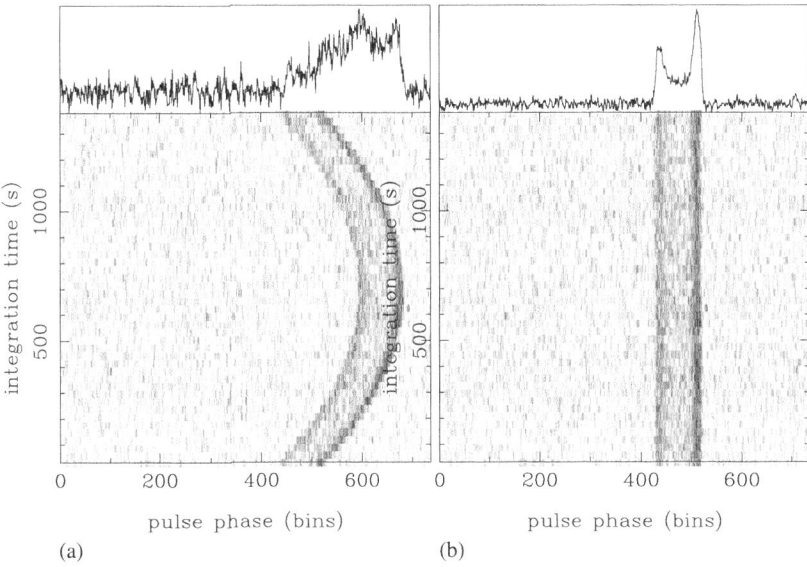

(a)   (b)

Fig. 6.9. (a) Folded pulse profiles as a function of time for a 22 min Arecibo observation of PSR B1913+16 showing the effects of a changing apparent pulse period. (b) The same time series now folded assuming $a_l = -16$ m s^{-2}.

An example of the improvement from the use of an acceleration search is shown in Figure 6.9 for a 22 min observation of the original binary pulsar B1913+16. Although the pulsar is strong enough to be detectable in the observation without any acceleration searching, folding the data at the nominal period from the search results in the heavily smeared profile shown on the left. A search in acceleration space shows a much stronger detection at $a_l \sim -16$ ms^{-2}. Folding the acceleration-corrected time series as shown on the right effectively has removed the deleterious effects of Doppler smearing and the true pulse shape is seen with greater significance.

Some care is required in choosing an appropriate acceleration step size in these searches to compromise between unnecessary processing incurred by over-sampling the parameter space and the loss of sensitivity caused by under-sampling. To quantify this, we need to calculate the number of Fourier bins an accelerated signal will occupy if no correction is applied. Applying the Doppler equation again, the observed pulse frequency $\nu(t) = \nu_0(1 - V_l(t)/c)$, where ν_0 is the true spin frequency of the pulsar in its rest frame and $V_l(t)$ is the line of sight velocity as before. Under the simple assumption that the acceleration approximation is valid (i.e. $V_l(t) = a_l t$), we find that the corresponding frequency drift $|\dot{\nu}| = a_l \nu_0/c$. For an observation of length T where the width of a Fourier bin $\Delta\nu = 1/T$, the number of frequency bins drifted by the signal

$$N_{\mathrm{drift}} = \dot{\nu}T/\Delta\nu = a_l \nu_0 T^2/c. \qquad (6.17)$$

For an acceleration search, the step size Δa_l should be such that the corresponding number of frequency bins drifted $\Delta N_{\mathrm{drift}} < 1$, i.e. $\Delta a_l < cP/T^2$, where P is the spin period. As an example, in 17.5 min analyses of data from the globular cluster 47 Tucanae, Camilo *et al.* (2000b) chose $\Delta a_l = 0.3$ m s^{-2} which guarantees that any pulsars with $P > 2$ ms do not drift by more than one spectral bin.

6.2.2 Frequency-domain techniques

For relatively small numbers of samples ($< 2^{23}$), time domain acceleration searches have been used to good effect to search for pulsars in globular clusters (see, for example, Anderson (1992) and Camilo *et al.* (2000b)). For longer data sets, however, the need to FFT repeatedly after each correction to the time series means that the computational time becomes dominated by FFTs that are very similar to one another. A more efficient approach developed by Ransom *et al.* (2002) is to work entirely in the frequency domain so that only one FFT need be carried out per DM trial.

6.2.2.1 Correlation method

The response of the DFT to a signal of varying frequency can be thought of as the idealised response to a stationary signal convolved with a finite impulse response (FIR) filter which spreads the power over a number of spectral bins. In a similar manner to coherent de-dispersion (Chapter 5) where an inverse chirp function is applied to the Fourier coefficients, the correlation technique applies an inverse FIR filter (the complex conjugate of the Fourier response) which effectively 'sweeps up' the power into a single Fourier bin. If the Fourier response can be written as $\mathcal{F}_{k-r_0}$, where $|k - r_0|$ is the frequency offset of the kth bin with respect to a reference frequency bin r_0, the corrected Fourier component

$$\mathcal{F}_{r_0} = \sum_{k=r_0-m/2}^{k=r_0+m/2} \mathcal{F}_k \mathcal{F}_{r_0-k}^*. \tag{6.18}$$

The exact form of the template $\mathcal{F}_{r_0-k}^*$ is a phase rotation term and a set of Fresnel integrals which are functions of the centre frequency bin r_0 and its derivative $\dot{r}$. Calculating this template for a range of $\dot{r}$ values is equivalent to the time domain acceleration correction, but is computationally much cheaper. This technique was first used to discover PSR J1807–2459, a 1.7 h binary in NGC 6544 (Ransom *et al.* 2001).

6.2.2.2 Stack/slide searches

Both the time-domain resampling and correlation techniques allow a fully coherent search in acceleration space. For large-scale searches (e.g. the Parkes multibeam survey (Manchester *et al.* 2001)), even the correlation approach becomes computationally expensive. A faster acceleration search used for this purpose that works in the frequency domain is the 'stack/slide' technique, which follows the orbital motion of a binary pulsar by breaking the integration up into a number of contiguous subsets, each of which is Fourier-transformed separately. This approach has two benefits: (a) the width of a Fourier bin in each segment is larger than for the entire observation; (b) an accelerated signal has less time to drift in frequency during each segment. As a result, the number of Fourier bins through which an accelerated signal will drift is reduced by the square of the number of segments used (see also Equation (6.17)).

Analogous to the time delays applied to compensate for dispersion (see Chapter 5), the effects of a changing frequency can be compensated for by applying frequency shifts (slides) to each spectra before adding (stacking) them together. The resulting stacked spectrum then can be searched for periodicities in the usual way. For an observation of length T split into n_{seg} segments, the change in frequency between two segments due a constant acceleration is simply $\nu_0 a_1 T/(n_{\text{seg}} c)$ Hz. While this simple incoherent stacking is less sensitive than a fully coherent acceleration search (typically 20 per cent according to Faulkner

(2004)) it offers a considerable improvement over the standard search. An excellent example is the recent discovery of the 7.7 hr binary pulsar J1756–2251 (Faulkner *et al.* 2004). This highly relativistic system was completely missed in the standard search analyses and only detectable as a result of the stack/slide technique.

6.2.2.3 *Phase-modulation searches*

For orbital periods comparable to, or much less than, the integration time, the use of one-dimensional acceleration searches is clearly far from optimal, since the orbital motion can no longer be approximated as $V_1(t) = a_1 t$. Although acceleration derivatives can be added to the search, this soon becomes computationally prohibitive. As an alternative approach, Jouteux *et al.* (2002) and Ransom *et al.* (2003a) developed a 'phase-modulation search' which becomes optimal when the observation encompasses several complete orbits. This is often the case for globular cluster searches with typical integration times of 2–10 h.

The phase-modulation search utilises the fact that the amplitude or power spectrum of an observation covering several orbits of a binary system has a characteristic shape imprinted by the constantly changing signal frequency. A high S/N example of this is shown in Figure 6.10(a). Ransom *et al.* (2003a) demonstrated that the imprint can be described by a family of regularly spaced Bessel functions forming a set of sidebands about the spin frequency of the pulsar. The beauty of this result is that the spacing of the sidebands is simply the orbital period. A straightforward DFT of the region of interest will detect the orbital period (Figure 6.10(c)). Just like the search for signals in the time domain, the DFT is an extremely sensitive means of detecting the periodicity in the sidebands, even when they are not readily apparent (as shown in Figure 6.10(b) and (d)).

Once the orbital period is known, the orbital semi-major axis and epoch of ascending node can be determined from the width and phases of the sidebands. For weak pulsars, this often requires specialised techniques to make full use of the phase information of the Fourier components. If the orbit is circular, these three parameters can be used by the correlation technique to determine the appropriate template to recover the power as before. Full details of this procedure can be found in Ransom *et al.* (2003a). A number of current searches of globular clusters are now routinely utilising the phase-modulation technique and may perhaps soon break the 96-min record held for the shortest radio pulsar binary (Camilo *et al.* 2000b).

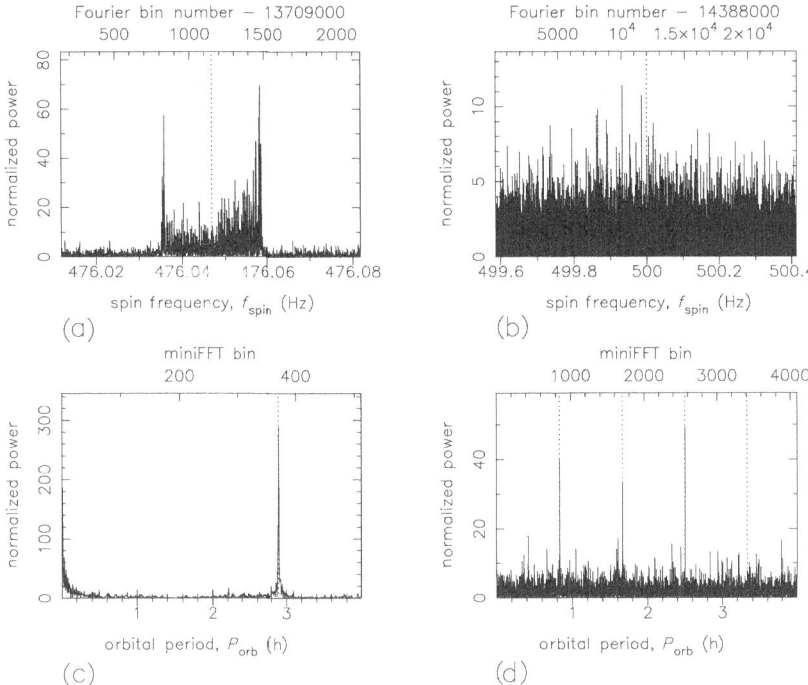

Fig. 6.10. (a) Power spectrum of an 8 h observation of PSR J0023–7203J in the globular cluster 47 Tucanae. (b) Power spectrum of simulated data for a weak 2 ms pulsar in a 50 min orbit about a 0.2 $M_\odot$ companion. (c) FFT of the shaded region in (a) showing the expected modulation at the 2.9 h orbital period of the pulsar. (d) FFT of the shaded region in (b) showing the 50 min orbital period and two harmonics. Figure provided by Scott Ransom.

6.2.2.4 Dynamic power spectrum search

The various acceleration techniques discussed so far perform optimally for distinct ranges of orbital periods relative to integration time, T. For systems with orbital periods greater than a few times T, coherent one-dimensional acceleration searches (either in the time or frequency domain) usually are adequate. When computational demands are large, and some loss of sensitivity can be tolerated, the stack/slide approach also performs well. When the orbital period is significantly less than T, phase modulation techniques provide a very efficient means of detecting ultra-compact systems. In between the two extremes, there is something of a 'sensitivity gap' to pulsars with orbital periods of the order of T.

A promising means of filling this gap is the 'dynamic power-spectrum' search. As in the stack/slide search, the time series is split into a num-

ber of smaller contiguous segments which are Fourier-transformed separately. The individual power/amplitude spectra can be summed harmonically and plotted as a two-dimensional (frequency versus time) image. Orbitally modulated pulsar signals appear as sinusoidal signals in
this plane (as shown in Figure 6.11).

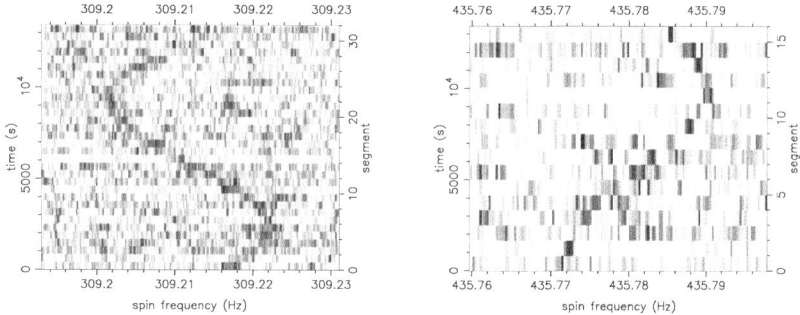

Fig. 6.11. Dynamic power spectra showing two recent pulsar discoveries in
the globular cluster M62 showing fluctuation frequency as a function of time.
Figure and analysis provided by Adam Chandler.

This technique has been used by various groups where spectra are
inspected visually (see, for example, Lyne *et al.* (2000)), or transformed
into a another representation, e.g. using the Hough transform (Aulbert
2004). Recently, Chandler (2003) has developed a hierarchical scheme
for searching these spectra that removes some of the human interven-
tion. This was recently applied to a search of the globular cluster M62
resulting in the discovery of three new pulsars shown in Figure 6.11. One
of the new discoveries – M62F, a faint 2.3 ms pulsar in a 4.8 h orbit,
was detectable only using the dynamic power spectrum technique.

6.3 Searching for pulsars in the time domain

While the frequency domain techniques described above generally are the
most effective and efficient means to find pulsars, there are alternative
approaches based in the time domain. Originally developed in the late
1960s, time-domain techniques (fast-folding and single-pulse algorithms)
are becoming increasingly popular additions in modern analyses that
seek to maximise the volume of phase space covered in pulsar searches.

6.3.1 Fast folding analyses

An essential procedure to check the validity of any good pulsar candidate is to fold[2] the de-dispersed data modulo the candidate period and examine the resulting pulse profile. This suggests that an alternative means of finding pulsars is to fold each de-dispersed time series modulo many different trial periods and look for statistically significant pulse profiles. Such a simple approach would be extremely effective were it not for the fact that the computational power required to fold over all possible periods of interest is enormous. If, however, we can restrict the search to a limited range of periods then folding provides an attractive alternative to the standard Fourier-based algorithms described above.

Shortly after the development of the FFT, Staelin (1969) devised a clever algorithm that avoids the many redundant operations involved in simple folding analyses. The 'fast-folding algorithm' (FFA) works by dividing a time series $\mathcal{T}_j$ of N samples into contiguous groups of n samples chosen such that N/n is an integer power of 2. The simplest folding of these data is at a period P_0 that is n times the data sampling interval, i.e. $P_0 = nt_{\mathrm{samp}}$. The folded profile p then consists of n bins, where the kth bin is simply

$$p_k = \sum_{j=0}^{(N/n)-1} \mathcal{T}_{k+jn}. \qquad (6.19)$$

The FFA works by splitting this summation into $\log_2 n$ stages which can be combined in different ways to fold the data at n slightly different periods. The scheme is best understood with the simple example shown in Figure 6.12 when $N = 16$ and $n = 4$. In this case, there are two stages at which the data are summed to produce four pulse profiles. At each stage, groups of samples are combined with a variable number of time sample shifts between each group. The resulting products represent the accumulated pulse profiles over the appropriate fraction of integration. For example, after the first of the two stages in Figure 6.12, the profiles represent the folded data over half of the integration.

Repeating the FFA procedure over a range of combinations of N and n results in a range of P_0 values that can be searched. The results are best displayed in a periodogram that shows a figure of merit for each folded pulse profile (e.g. reduced χ^2 or S/N; see Chapter 7) versus folding

---

2 The folding procedure is described in detail in Chapter 7.

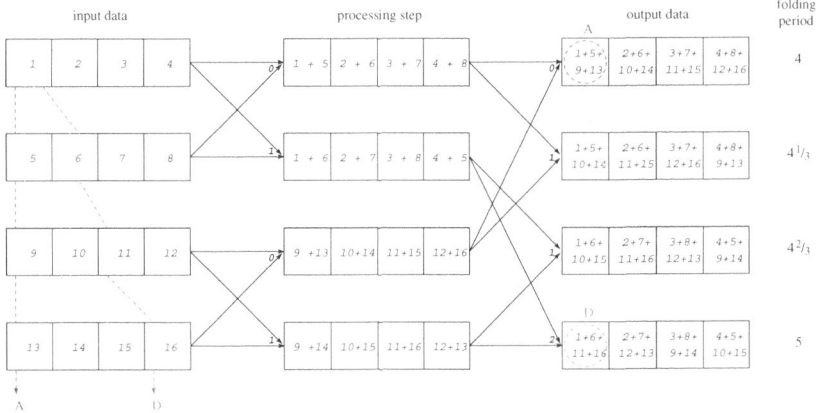

Fig. 6.12. Schematic representation of the fast-folding algorithm applied to a sixteen sample time series that is optimally folded at four different trial periods. Figure provided by Bernd Klein.

period. For a given P_0, the effective folding period of the lth profile

$$P_l = P_0 + \left(\frac{n+1}{N}\right) l\, t_{\text{samp}}, \qquad (6.20)$$

where $0 < l < (n-1)$. As seen in Figure 6.12, the case $l = 0$ reduces to our simple result for P_0 in Equation (6.19). Rather than the $N[(N/n)-1]$ additions required for simple folding, the FFA needs only $N \log_2(N/n)$ additions and therefore becomes very efficient when N is large.

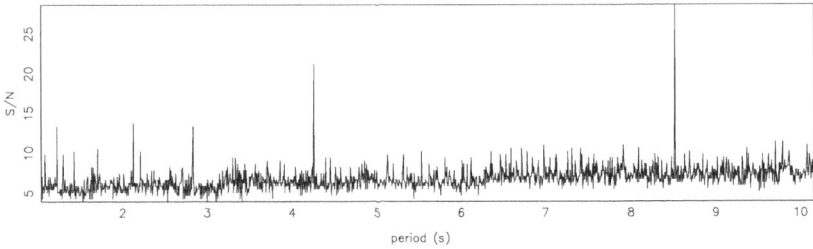

Fig. 6.13. Periodogram output from an FFA showing S/N as a function of trial folding period for an observation of the 8.5 s pulsar J2144–3933. The most significant peak occurs at the pulsar's true period. Most FFT-based codes enforce a long-period cutoff at around 5 s in order to minimise the effects of red noise in the Fourier domain. This results in the detections of sub-harmonics which are less significant, as shown here.

Even with the efficiency of the FFA over simple folding, the high data sampling rates of current surveys mean that searching the entire period range of de-dispersed time series is computationally expensive. The FFA can be used to good effect, however, in the search for long-period ($P > 2$ s) pulsars. The standard FFT-based search means that the signals from such pulsars occupy only a small part of the Fourier spectrum at low frequencies. Due to the physical processes mentioned above, this typically can contain a large amount of red noise, which makes such signals hard to detect. Working in the time domain, the FFA reveals the long-period phase space in much greater detail. This is highlighted by Figure 6.13, which shows an FFA periodogram from an observation of the 8.5 s pulsar J2144−3933, the longest period pulsar so far known.

6.3.2 Single-pulse searches

Throughout this chapter we have concentrated on techniques that find pulsars by virtue of their highly periodic nature. The radiation from some pulsars, however, can vary greatly in amplitude so that *detectable* pulses are not strictly periodic. In such cases, the techniques discussed so far cease to become effective. Indeed, by implicitly assuming an underlying periodicity, we may be selecting against the detection of an important part of the neutron star population. Two classes of pulsars for which this is known to be the case are the giant-pulse emitters, from which pulses with between 100 and 1000 times the mean pulse intensity are occasionally emitted, and the nulling pulsars that emit no pulses for extended periods of time (see Chapter 1). The best known giant-pulse emitter – the Crab pulsar – was first discovered through its giant pulses (Staelin & Reifenstein 1968). Many nulling pulsars that emit only a few pulses during an integration would not be detectable in a periodicity search. This was demonstrated by the discovery of J1918+08, a 2.1 s pulsar discovered by a single-pulse analysis of an Arecibo survey of the Galactic plane (Nice 1999). The pulsar was not detected in the original Fourier analysis of the data (Nice, Fruchter & Taylor 1995).

Searching for single pulses in a time series is basically an exercise in matched filtering. Given a time series of predominantly Gaussian noise of known mean and standard deviation, we seek individual events that deviate by several standard deviations from the mean. Consider a rectangular pulse of amplitude S_{peak} and width W. For the optimal case when W is equal to the sampling time t_{samp}, it can be shown (Cordes

& McLaughlin 2003) that the S/N ratio of the pulse

$$S/N = \frac{S_{\mathrm{peak}}W}{S_{\mathrm{sys}}}\sqrt{\frac{n_{\mathrm{p}}\Delta f}{W}}, \tag{6.21}$$

where S_{sys} is the system equivalent flux density, n_{p} is the number of polarisations summed and Δf is the receiver bandwidth. For a fixed pulse area (i.e. $S_{\mathrm{peak}}W = const.$) it follows that $S/N \propto 1/\sqrt{W}$, i.e. narrow pulses are easier to detect than broader ones. In general, W will not usually be a good match to t_{samp} and the S/N will be less than expected from Equation (6.21). In order to match optimal detection more closely, the time series is 'smoothed' by successively adding groups of neighbouring samples and searching for statistically significant events.

As in the Fourier-based searches, a sensible choice of S/N threshold should be made to avoid recording too many candidate pulses that most likely are caused by random noise fluctuations. For the ideal case of Gaussian noise with zero mean and unit standard deviation, Cordes and McLaughlin (2003) show that the number of events expected to occur by chance above some threshold S/N_{min} is simply

$$n(> S/N_{\mathrm{min}}) \sim 2n_{\mathrm{samples}} \int_{S/N_{\mathrm{min}}}^{\infty} \exp(-x^2)\,\mathrm{d}x, \tag{6.22}$$

where n_{samples} is the number of samples in the time series. Requiring that $n < 3$ usually leads to $S/N_{\mathrm{min}} = 4$. In practice, however, radio frequency interference (Section 6.4) usually increases the number of false detections so that a more practical S/N threshold might be 5–6.

Figure 6.14 shows the single pulse search applied to data from the Parkes multibeam survey. As well as a clear excess of low DM pulses caused by interference, an excess at a DM of 223 cm^{-3} pc is also apparent. No signal was seen in the periodicity search at this DM. Subsequent observations at the same position showed these pulses to have an underlying periodicity of 871 ms and confirmed the existence of a new pulsar J1624–4616. This is one of a number of new pulsars to be confirmed in this way and demonstrates the effectiveness of the single pulse search as a complementary strategy to periodicity searches. Given the simplicity of the algorithm and relatively modest additional computational requirements of a single pulse search, it should soon become a standard part of the pulsar searching procedure.

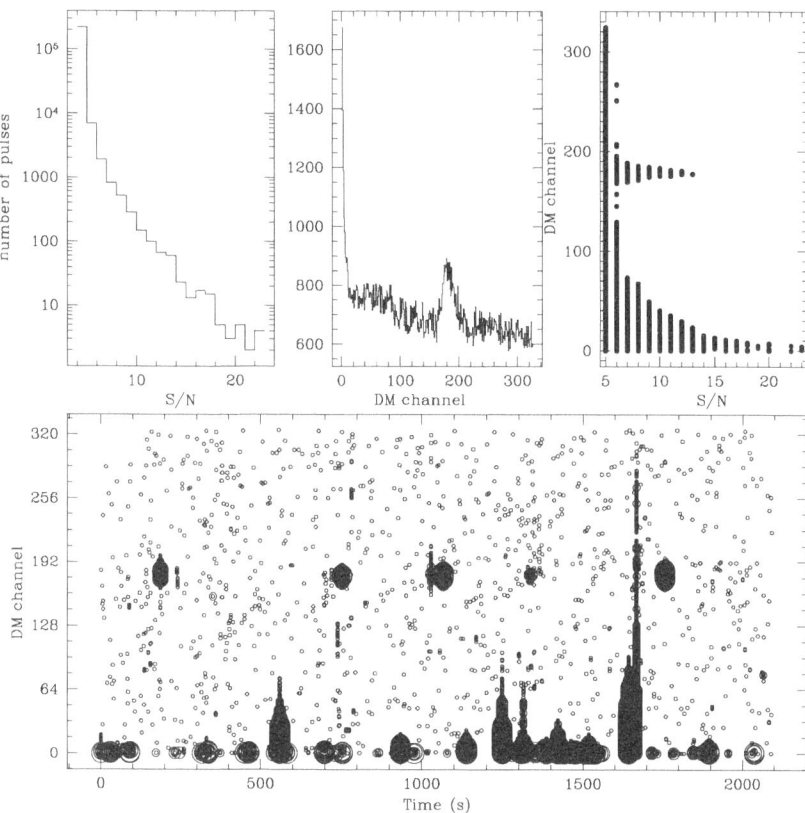

Fig. 6.14. Example output from a single pulse search of data from the Parkes multibeam survey. The top plots show the S/N distribution of the detected pulses, the number of pulses as a function of trial DM as well as S/N as a function of DM. The lower plot shows the individual dispersed pulses as a function of time. The size of the symbols is proportional to the pulse S/N. In addition to the presence of dispersed pulses from, in this case, PSR J1624–4616, persistent interference at low DMs is also apparent. Figure and analysis provided by Maura McLaughlin.

6.4 Radio frequency interference

Even in the remote areas where most radio telescopes are located, terrestrial sources of radio frequency interference (RFI) can have a significant impact on our ability to detect pulsars close to the nominal sensitivity limits of the search. Most search codes have a limit to the number of candidates stored from the analysis. If a particular observation containing a pulsar is affected badly by RFI, the pulsar may not be strong or persistent enough to be saved as a candidate.

Apart from electrical storms (which can saturate the receiver) the main interference problem arises from persistent broadband signals that mimic the periodic and sometimes even dispersed nature of pulsars. Such sources arise predominantly from nearby electrical devices (the mains power line AC frequency usually is detectable) and communications systems such as airport radar systems. Potential sources of RFI at the observatory, such as desktop computers (which now have clock speeds high enough to radiate in the radio band), also need to be monitored. Fortunately, since most sources of RFI are not dispersed they are detectable in the FFT of the zero-DM time series. Two main approaches to excising these unwanted signals are carried out: time domain clipping and frequency domain masking.

6.4.1 *Time domain clipping*

Since most searches are expecting to find weak sources, sporadic bursts of interference represent an unwelcome intrusion on the time series. Such samples can be identified easily by comparing them with the expected mean and standard deviation of the zero-DM time series. As an example, a 1-bit filterbank (see Chapter 5) with n_{chans} channels produces time series with an expected mean and standard deviation of $n_{\mathrm{chans}}/2$ and $\sqrt{n_{\mathrm{chans}}}/2$, respectively. A sample is deemed unsuitable for analysis if it differs from the mean value by more than two standard deviations. In such cases, each of the n_{chans} channels are set to zero ('clipped') to exclude them from influencing all subsequent analyses.

Keeping a count of the number of independent samples provides a measure of how badly effected the data are. If this becomes a significant fraction of the total number of samples (e.g. 20 per cent easily can be reached during a bad electrical storm) the observation should be repeated on a subsequent date. This simple procedure is important for time domain analyses like the FFA, particularly for long-period pulsars, in which the integrated profiles can be significantly biased by sporadic RFI.

6.4.2 *Frequency domain masking*

Most pulsar surveys collect many hundreds or even thousands of observations with the telescope pointing at different parts of the sky. The key signature of RFI is that it will occur frequently and is independent of sky position. In some cases, the signal is seen predominantly when the telescope is pointing toward the source of the source of RFI, e.g. the local airport! Analysing a large number of zero-DM time series from observations of widely spaced sky positions will show persistent RFI sources occurring at the same frequencies in many or even all of the observations. Usually some sort of conservative threshold is placed, e.g. a signal will be considered to be RFI if it occurs in at least ten out of a hundred independent positions with S/N > 7.

This analysis takes place usually on a convenient block of observations, e.g. all those taken in a single observing session or saved to the same magnetic tape. Based on the list of RFI signals that trigger the occurrence threshold, a spectral 'mask' is created. This simply is a list of all Fourier bin numbers corresponding to each RFI frequency. The mask then is applied to all subsequent processing by flagging the relevant Fourier bins so that they are ignored by the analysis software.

The spectral mask is an efficient and effective method of RFI excision. A typical mask usually will require less than 1 per cent of all spectral bins to be ignored. While this does leave a small chance that a pulsar with a frequency coincident with a zero-DM RFI signal could be excised from the analysis, masking usually allows data to be analysed that would be otherwise swamped with candidate signals of RFI origin.

6.5 Pulsar search strategies

Having described the various search techniques in use, we now conclude with a brief discussion of the most profitable search strategies. As we shall see, how a pulsar survey is designed determines to a large extent what type of objects are found. It is important to bear in mind the lessons learnt from previous experiments when planning new ones.

6.5.1 Searches close to the plane of our Galaxy

Young pulsars are most likely to be found near to their place of birth, close to the Galactic plane. This is the target region of one of the Parkes multibeam (PM) surveys and has already resulted in the discovery of over 700 new pulsars, almost half the number currently known. Such a large haul inevitably results in a number of interesting individual objects such as: PSR J1141−6545, a young pulsar in a relativistic 4 h orbit around a white dwarf (Kaspi *et al.* 2000b); PSR J1740−3052, a young pulsar orbiting an $\gtrsim 11$ M$_\odot$ star (Stairs *et al.* 2001); several intermediate mass binary pulsars and two double neutron star systems. For a review on this survey and its major discoveries, see Manchester (2001).

Due to the severe propagation and sky background effects on the sensitivity at low frequencies (< 1 GHz), most Galactic plane surveys are now carried out in the 1–2 GHz band. The centre frequency of the Parkes multibeam system is 1.37 GHz. The generally high pulsar density along the Galactic plane means that deep searches like the PM survey are rewarded by a large yield. One reason for the depth of the PM survey is its relatively long integration time (35 min). While this is an ideal means to maximise the sensitivity to faint isolated pulsars, and those that null, the sensitivity to binary systems is compromised. In order to combat this, the PM survey applies a stack/slide acceleration search similar to that described in Section 6.2.2.2.

6.5.2 All-sky searches for millisecond pulsars

The oldest radio pulsars form a virialised population of stars oscillating in the Galactic gravitational potential. The scale height for such a population is at least 500 pc, about 10 times that of the massive stars that populate the Galactic plane. Since the typical ages of millisecond pulsars are several Gyr or more, we expect, from our vantage point in the Galaxy, to be in the middle of an essentially isotropic population of nearby old and low-luminosity neutron stars.

All-sky searches for millisecond pulsars at high Galactic latitudes have been very effective in probing this population. Much of the initial interest and excitement in this area was started at Arecibo when Wolszczan discovered two exciting pulsars at high latitudes: the double neutron star binary B1534+12 (Wolszczan 1991) and PSR B1257+12 (Wolszczan & Frail 1992), a millisecond pulsar with three orbiting planets. Surveys carried out at Arecibo, Parkes, Jodrell Bank and Green Bank by other groups in the 1990s found many other millisecond pulsars in this way.

Since the interstellar propagation effects (scattering and dispersion) are much less severe away from the Galactic plane, the optimal frequency for these surveys is less than 1 GHz to take advantage of the generally higher flux densities of pulsars at these frequencies (see Section 1.1.3) and the larger telescope beam widths (see Appendix 1). In addition, the short integration times necessary to cover a reasonable area of sky mean that the effects of binary acceleration are far less problematic than deep searches of the Galactic plane.

6.5.3 Searches at intermediate Galactic latitudes

In order to probe more deeply into the population of millisecond and recycled pulsars than possible at high Galactic latitudes, Edwards *et al.* (2001) used the PM system to survey intermediate latitudes in the range $5° < |b| < 15°$. Among the fifty-eight new pulsars discovered, eight are relatively distant recycled objects, including two mildly relativistic neutron star-white dwarf binaries (Edwards & Bailes 2001). The success of this survey has lead to an extension of the search area by two groups. One of these surveys lead to the discovery of the double-pulsar binary system J0737–3039 (Burgay *et al.* 2003; Lyne *et al.* 2004).

6.5.4 Targeted searches of supernova remnants

Ever since the discovery of pulsars, numerous deep surveys of supernova remnants for pulsations from young neutron stars have been carried out. With a few exceptions, these have been surprisingly unsuccessful. The main problem faced by these surveys has been the uncertain position of the putative pulsar, which could lie anywhere within, or around, the vicinity of its associated remnant. This positional uncertainty was quite often much larger than an individual telescope beam width (see Appendix 1) that therefore required a grid of pointings to cover properly the target area. Given that telescope time is always hard-fought, this ultimately leads to a compromise in sensitivity.

This situation has, however, changed quite dramatically with the new generation of X-ray telescopes such as the *Chandra* observatory revealing point-like X-ray emission from within a number of Galactic supernova remnants. These observations suggest strongly the presence of the young neutron star associated with the supernova explosion and, for the first time in most cases, localise the position of the neutron star candidate to sub-arcsecond precision. A number of searches by Camilo and collaborators (see Camilo (2003) for a review) have utilised these positions to carry out very deep searches for pulsations. So far, four young, < 3000 yr old neutron stars have been discovered in these searches. Perhaps not surprisingly, these are all extremely faint objects that are well below the detection threshold of current large-scale surveys. Currently a significant effort is being made to systematically search the known neutron star point sources as deeply as possible.

6.5.5 Targeted searches of globular clusters

Globular clusters have long been known as breeding grounds for millisecond and binary pulsars. The main reason for this is the high stellar density in globular clusters relative to most of the rest of the Galaxy. As a result, low-mass X-ray binaries are almost 10 times more abundant in clusters than in the Galactic disk. In addition, exchange interactions between binary and multiple systems in the cluster can result in the formation of exotic binary systems.

Since a single globular cluster usually fits well within a single telescope beam, deep targeted searches can be made without the positional uncertainty that plagued earlier supernova remnant searches. In addition, once the DM of a pulsar is known in a globular cluster, the DM parameter space for subsequent searches essentially is fixed. This allows computation power to be invested in acceleration searches for short-period binary systems. Multiple observations of clusters also benefit from the occasional boosting of otherwise weak pulsars due to scintillation. To date, searches have revealed eighty pulsars in twenty-four globular clusters. Highlights include the double neutron star binary in M15 (Prince *et al.* 1991), and a low-mass binary system with a 96 min orbital period in 47 Tucanae, one of over twenty millisecond pulsars currently known in this cluster alone (Freire *et al.* 2003). On-going surveys of other clusters continue to yield new discoveries (see, for example, Possenti *et al.* (2003)). It is worth pointing out in this regard that the periodicity searches discussed above do not require any special modification to detect multiple signals in the time series, i.e. provided that the signals are above the noise threshold, the pulsars do not 'interfere' with one another.

6.5.6 Searches of the Galactic centre region

A blank spot on the Galactic map of pulsar discoveries has been the Galactic centre region. Not a single radio pulsar has been found in the inner 500 pc around Sgr A*. This is in contrast to the large population of massive stars expected in this region, which is believed to be a site of past and current star formation (see, for example, Mezger *et al.* (1999)). Due to the extreme environment of the Galactic centre (i.e. tidal forces, magnetic field pressure, high gas densities, high temperatures and turbulences) it is believed that the initial mass function should be peaked at a higher mass than in the Galactic disk. This favours more massive stars probably created by externally triggered star formation (e.g. compression of clouds via collisions). Such massive stars are indeed observed (see, for example, Figer *et al.* 1999).

The increase in the relative importance of high-mass star formation toward the Galactic centre als leads naturally to the expectation of a much larger number density of neutron stars and stellar black holes in comparison to the disk. Moreover, the remnants of massive stars have been detected in the form of supernova remnants (Ekers *et al.* 1983; Kassim & Frail 1996) and possibly neutron stars (Muno *et al.* 2003). Therefore we can expect to find a population of pulsars that would be extremely useful in probing the Galactic centre and its conditions (Kramer *et al.* 1996a; Cordes & Lazio 1997): their number and age distribution would probe the past star formation history (Hartmann 1995); their period derivatives can constrain the gravitational potential in the Galactic centre (see Chapter 8); pulsar timing would enable us to probe the space-time around the super-massive black hole in the Galactic centre due to a variety of relativistic effects (Wex & Kopeikin 1999). The high stellar density of the Galactic centre makes it, like the globular clusters, a possible site of a millisecond pulsar orbiting a stellar-mass black hole. A recent summary can be found in Pfahl and Loeb (2004).

The potential rewards for finding a pulsar in the Galactic centre are high, and searches have been performed. The large distance of ~ 8 kpc to the Galactic centre means that large-area surveys are usually not sensitive enough. The PM survey did have the sensitivity, however, and was still, despite its tremendous success elsewhere, unsuccessful in this region. The reason for the difficulties of finding pulsars around Sgr A* is given by the large amount of interstellar scattering expected for Galactic centre pulsars (Cordes & Lazio 1997). The scattering is so severe that it renders all periodicity searches useless at frequencies below a few GHz. The only way to combat the effects of scattering – which has approximately an f^{-4} dependence (see Chapter 4) – is to search at higher frequencies. Such surveys have been attempted by Kramer *et al.* (1996a; 2000) and Klein (2004) at 4.85 GHz and for a number of selected point sources at 8.5 GHz by Klein (2004). However, increasing system temperatures and the steep spectrum of pulsars (see Chapter 1) worsen the prospects of finding a pulsar at high frequencies. Balancing such effects with the frequency dependence of scattering, Cordes and Lazio (1997) computed an optimal frequency for periodicity searches of Galactic centre pulsars, concluding that the best frequency would lie around 10 GHz. An updated analysis presented by Kramer *et al.* (2000)

is shown in Figure 6.15. At these optimal frequencies the telescope beam usually is too small to cover a large area, in particular for the large telescopes that are required to achieve a sufficient sensitivity. Cordes and Lazio (1997) therefore suggested to find pulsars in targeted searches of steep-spectrum polarised point-sources identified from imaging observations. Ultimately, the best prospects of finding Galactic centre pulsars will be with the Square Kilometre Array (SKA).

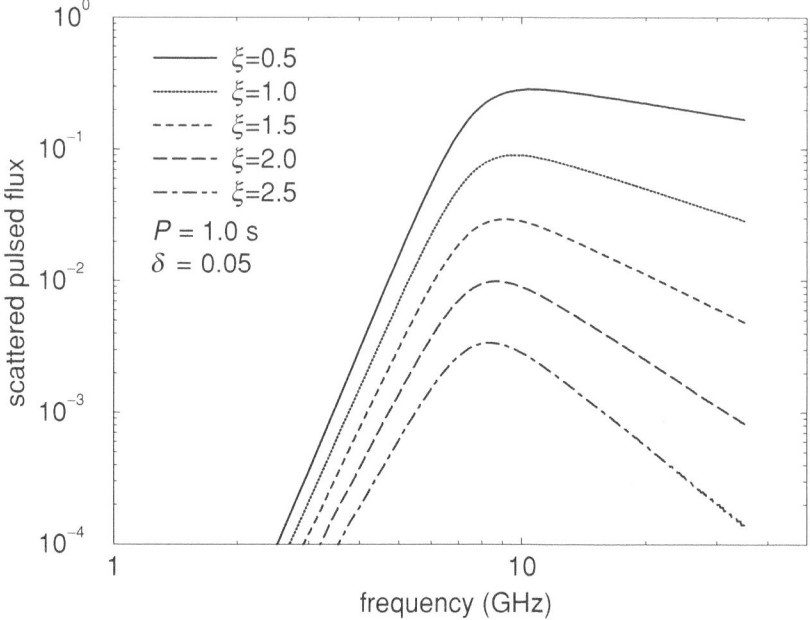

Fig. 6.15. Relative scattered pulsed flux that would be picked up by a periodicity search for a 1 s pulsar in the Galactic centre. Scattering is so severe that it renders the 50 ms wide pulse undetectable at low frequencies. At high frequencies the pulsar spectrum, here shown for various spectral indices ξ, dominates. The optimal search frequencies lies at about 10 GHz.

6.5.7 Searches with the Square Kilometre Array

All the surveys that have been conducted in the past or will be conducted in the next few years appear to be only the prelude to what will be possible with the future SKA. As for other areas in modern astronomy, the SKA will revolutionise the field of pulsar astrophysics. Not only will new science be made possible by the sheer number of pulsars discovered, but also by the unique timing precision achievable with the SKA. The special property of the SKA will be its unique sensitivity. Current design figures indicate that a 1.5 μJy source will be detectable with S/N = 8 in a 1 min integration. This will enable not only the discovery of most pulsars in the Milky Way but also allow present-day survey sensitivities to pulsars in the closest galaxies. With the single-pulse search techniques described in Section 6.3.2, it should be possible to detect giant pulses from pulsars as distant as the Virgo cluster.

Pulsar surveys with the SKA essentially could discover all pulsars in our Galaxy that are beaming toward the Earth. From a simulation of a hypothetical all-sky SKA survey we estimate that between 10 000 and 20 000 pulsars, including over 1 000 millisecond pulsars, could be discovered. This impressive yield also samples effectively every possible outcome of the evolution of massive binary stars, thereby guaranteeing the discovery of very exciting systems. Since the integration times of these surveys will be short (5 min or less), compact binaries should be relatively simple to detect. As a result, we expect at least a hundred compact relativistic binaries, including the elusive pulsar–black hole systems. The search, discovery and study of such systems is one of the main science drivers of the SKA.

6.6 Further reading

Although a large number of articles on pulsar searching have been written over the years, we have attempted in this chapter to bring together the various techniques in a single resource. For the reader wishing to gain a deeper insight into the underlying mathematical and signal processing issues discussed throughout this chapter, the excellent text books *Numerical Recipes: The Art of Scientific Computing* (Press *et al.* 1992) and *The Fourier Transform and its Applications* (Bracewell 1998) are highly recommended.

Nowadays, a large number of reliable FFT routines are freely available. Press *et al.* (1992) provide an excellent discussion of why the FFT is so much faster than a DFT, and provide some excellent subroutines in a number of languages. Perhaps the best of the publicly available set of FFT tools is the `fftw` library (`www.fftw.org`), which is not restricted by the usual power of 2 length requirements.

Early work on pulsar search methodologies was summarised by Burns and Clark (1969) and Hankins and Rickett (1975). An excellent introduction to pulsar searching also can be found in the review by Bhattacharya (1998). Further discussions of the two-dimensional Fourier transform search can be found in Lyne and Smith (2005). Two seminal papers (Ransom *et al.* 2002; 2003a) contain excellent discussions on Fourier domain search techniques. Many of the results quoted in Sections 6.1.3 and 6.2 were developed or discussed in detail for the first time in these papers. Further discussion on the phase-modulation search technique can be found in Jouteux *et al.* (2002). The single-pulse search methods described by Cordes and McLaughlin (2003) and McLaughlin and Cordes (2003) are highly recommended for those wishing to explore these techniques in more detail. A number of aspects concerning pulsar search design and optimisation not discussed here can be found in the excellent review by Cordes (2002). Camilo (1995; 1997; 1999) has written several reviews of pulsar surveys that are good starting points for further reading.

6.7 Available resources

While numerous pulsar search software packages have been developed over the years, only a few are freely available for use. Two currently available packages are PRESTO, developed by Scott Ransom, and SEEK developed by one of us (DRL). Both programs are well tested and have a complementary approach to pulsar searching in the Fourier domain. PRESTO makes full use of the phase information and can be used to carry out sophisticated acceleration and phase modulation searches for pulsars. To date it has been used to great success by Ransom and collaborators (Ransom *et al.* 2003b). SEEK closely follows the standard search approach and has been used successfully in a wide range of different projects. Also incorporated into the package are time domain resampling routines to carry out acceleration searches (used to great effect in searches of 47 Tucanae; see Camilo *et al.* (2000b)) as well as the single-pulse search routines developed by Cordes and McLaughlin (2003). Peter Müller has written an FFA program that has been adapted by one of us (MK) for general use in pulsar searches. Links to all of the above programs and sample data sets that can be used as starting-points for those wishing to develop and test their own software can be found on the book web site, (see Appendix 3).

7

Observing known pulsars

Once a pulsar has been discovered, it is subject to many different types of follow-up observations first in order to characterise its basic properties and later to study it in more detail. In this chapter, we discuss the most commonly used techniques necessary to carry out these observations. A fundamental procedure underpinning most pulsar observations is the synchronous averaging (folding) of the data at the pulse period. Following a description of this process, we discuss the various preliminary observations required to refine pulsar parameters for future observations. These amount basically to an optimisation of the pulse period, dispersion measure and position, as well as determining whether the pulsar is in a binary system. Then we discuss flux density and polarisation calibration procedures, before moving on to review the most commonly used modes of observation and data analyses: single-pulse studies, polarisation and Faraday rotation measurements, scintillation observations and measurements of neutral hydrogen absorption. Finally, some techniques to remove radio-frequency interference from the data are discussed.

7.1 Folding

As pulsars are generally very weak radio sources, the addition (folding) of many pulses so that the signal is visible above the background noise is vital for studying them in detail. Folding can be summarised as follows:

(i) De-disperse the data either coherently (see Section 5.3) or incoherently (see Sections 5.2 and 6.1.1.1). The resulting time series has a sampling time t_{samp}.

(ii) Create an array for storing the folded profile with n_{bins} equally spaced elements (known as 'bins') across the pulse period. Each bin therefore corresponds to a particular phase of the pulse, where the phase centre of bin i is simply $(i - 0.5)/n_{\text{bins}}$.

(iii) Take a sample of the time series and calculate its phase relative to the pulse period. In the simplest case of a constant pulse period, P, the phase of the jth sample is jt_{samp}/P, where t_{samp} is the data sampling interval.

(iv) Find which bin in the profile array has the closest phase to this sample and add the sample to this bin.

(v) Repeat step (iii) for the next sample.

At the end of this process, each phase bin in the folded profile is normalised by the number of samples accumulated per bin. The resulting *integrated pulse profile* represents the average emission from the neutron star as a function of its rotational phase. This algorithm can be extended easily to write out profiles corresponding to short contiguous segments of data (*sub-integrations*), or single pulses (see Section 7.4.2).

The above approach to folding using a constant pulse period usually applies to early observations of a pulsar for which only the discovery period is known. More generally, when high-precision ephemerides are available from timing observations (see Chapter 8), the phase of each sample can be calculated directly. For computational efficiency, an intermediate program is used to interpolate the ephemeris into a set of polynomial coefficients. These then can be used to calculate the phase with sufficient accuracy for the epoch of the observation to avoid any smearing of the pulse due to drifting (see Section 7.2.1). Software packages to carry out these tasks are detailed at the end of this chapter.

7.1.1 Profile significance tests

A quantitative figure of merit for a pulse profile is useful both for calibration issues discussed later and for assessing the significance of weak detections. Representing the profile as an array of n_{bins} phase bins for which the amplitude of ith bin is labelled p_i, in the following we consider two significance measures that are in common use: (a) the profile signal to noise ratio, S/N; (b) the chi-squared statistic, χ^2.

Both measures of significance require the off-pulse mean[1], $\bar{p}$, and standard deviation, σ_{p}. These quantities may be calculated by isolating the on- and off-pulse emission in a process known as pulse *windowing* or *gating*. An alternative method is to calculate the mean and standard devia-

---

1 This is sometimes referred to as the profile baseline, DC level or offset.

tion of a short (~ 1 s) segment of the raw time series which is essentially dominated by noise. This is particularly useful for broad pulse profiles where there is little off-pulse emission in the folded profile. If we denote these 'raw' values by $\bar{r}$ and σ_r, for a folding scheme where the resulting pulse profiles are normalised by the number of samples accumulated per phase bin, it can be shown that $\bar{p} = \bar{r}$ and $\sigma_p = \sigma_r / \sqrt{(n_{\text{samples}}/n_{\text{bins}})}$, where n_{samples} is the total number of samples in the time series.

7.1.1.1 Signal to noise ratio

The most commonly-used measure of profile significance is the signal to noise ratio, S/N, usually defined as follows:

$$ \text{S/N} = \frac{1}{\sigma_p \sqrt{W_{\text{eq}}}} \sum_{i=1}^{n_{\text{bins}}} (p_i - \bar{p}), \tag{7.1} $$

where W_{eq} is the equivalent width (in bins) of a top-hat pulse with the same area and peak height as the observed profile. As defined in Chapter 3, W_{eq} is simply the area under the observed pulse divided by its peak height. Assuming the noise distribution follows Gaussian statistics, the probability of obtaining a profile with a particular S/N by chance

$$ \text{Prob}(> \text{S/N}) = \frac{1}{\sqrt{2\pi}} \int_{\text{S/N}}^{\infty} e^{-x^2/2} \, dx = \frac{1}{2} \left[1 - \text{erf} \left(\frac{\text{S/N}}{\sqrt{2}} \right) \right], \tag{7.2} $$

where the error function $\text{erf}(x) = (2/\sqrt{\pi}) \int_0^x \exp(-x^2) \, dx$ can be solved numerically (see, for example, Press *et al.* (1992)).

7.1.1.2 Chi-squared statistic

An alternative measure of a profile's significance is to consider its deviation from pure Gaussian noise with mean $\bar{p}$ and variance σ_p^2. This test was first used to quantify folding analyses of X-ray data by Leahy *et al.* (1983) who defined a version of the χ^2 statistic

$$ \chi^2 = \frac{1}{\sigma_p^2} \sum_{i=1}^{n_{\text{bins}}} (p_i - \bar{p})^2, \tag{7.3} $$

which, for this simple model, has $n_{\text{bins}} - 1$ degrees of freedom. The significance level for a given χ^2 value can be calculated readily by noting that the probability of exceeding a given χ^2 value by chance

$$ \text{Prob}(> \chi^2) = \frac{1}{\Gamma(a)} \int_{\chi^2/2}^{\infty} e^{-t} t^{a-1} \, dt, \tag{7.4} $$

where $a = (n_{\text{bins}} - 1)/2$ and $\Gamma(a) = \int_0^\infty t^{a-1} e^{-t}\, dt$. Press *et al.* (1992) provide routines to integrate this function numerically.

7.2 Preliminary measurements

Following a pulsar's discovery, its period, dispersion measure and position usually are rather crudely known. Periods and dispersion measures often are limited to the resolution of the search technique (see Chapter 6). Positional uncertainties usually are at the level of the resolution of the radio telescope beam (typically ~ 10 arcmin; see Appendix 1). By reducing the uncertainties in all these parameters through preliminary measurements, we can facilitate greatly the planning and efficiency of subsequent observations. In addition, if a pulsar is a member of a binary system, an initial orbital ephemeris can often be determined.

7.2.1 *Improving the pulse period*

As well as the limited precision from the search, the pulsar's period may have altered from the discovery/confirmation observation due to the presence of an orbiting companion, a high spin-down rate or period glitch. As the grey-scale plot of sub-integration against pulse phase in Figure 7.1 shows, folding the data at a slightly incorrect period will cause the accumulated profile to drift in phase so that the integrated pulse will appear significantly broadened and with reduced S/N.

The period can, however, easily be corrected by measuring the drift rate of the pulse in the incorrectly folded data. To quantify this, consider folding a time series at period P that differs by δP from the true period P_{true}. The drift rate in one period $\dot{P} = \delta P/P$ is small but easily measurable through its accumulation over a longer period of time. If the pulse is observed to drift by an amount Δt over an observation of length t_{obs}, then $\dot{P} = \Delta t/t_{\text{obs}}$ and, hence,

$$P_{\text{true}} = P + \delta P = P + P\dot{P} = P\left(1 + \frac{\Delta t}{t_{\text{obs}}}\right). \tag{7.5}$$

For the drift rate of 111.95 ms measured in Figure 7.1, the data can be re-folded at the correct period to restore the true pulse shape as shown. Any residual drifts when folding the data with period P_{true} are now within a phase bin and the uncertainty of the corrected period $\Delta P_{\text{true}} = P_{\text{true}}^2/(n_{\text{bins}} t_{\text{obs}})$. The period is now sufficiently precise to produce a folded profile for use in timing analyses (see Chapter 8).

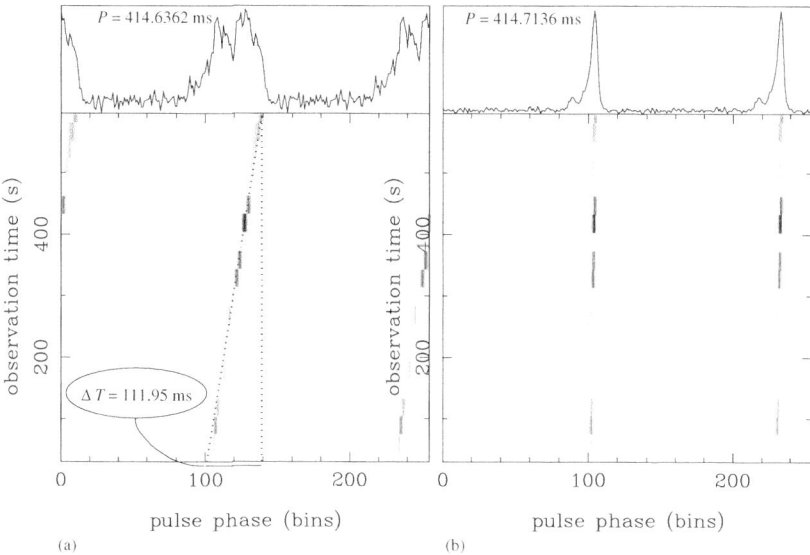

Fig. 7.1. (a) A time series for PSR J0137+1654 folded at a slightly incorrect period. The pulse drifts by ~ 112 ms during the integration. (b) The same time series folded at the period corrected using the measured drift rate.

Looking at this simple analysis another way: for a pulsar of constant period, the fractional folding period uncertainty as a function of time required to keep phase drifts within a single bin is $P/(n_{\mathrm{bins}}t_{\mathrm{obs}})$. This is shown for 512-bin profiles of different pulse periods in Figure 7.2 and gives an idea of the precision required and achievable in folding analyses.

For pulsars in binary systems, as we saw in Chapter 6, the apparent pulse period changes due to time-variable Doppler shifts as the pulsar moves about the centre of mass of the binary system. The simple approximation of a constant folding period is therefore no longer valid. This is noticeable particularly for observations of short orbital period binaries ($P_{\mathrm{b}} \lesssim 24$ h) and long integration times ($t_{\mathrm{obs}} \gtrsim 5$ min). First and higher order period derivatives are often required to avoid drifts due to binary motion when folding the data. i.e.

$$P(t) = P_{\mathrm{true}} + \dot{P}_{\mathrm{bin}}t + \frac{1}{2}\ddot{P}_{\mathrm{bin}}t^2 + \cdots, \qquad (7.6)$$

where the 'bin' subscripts denote period derivatives due to binary motion. For example, the 22 min observation of the 7.75 h binary pulsar B1913+16 discussed in Section 6.2.1 requires an additional $\dot{P}_{\mathrm{bin}} \simeq 3 \times 10^{-9}$ in order to fold the data correctly. These derivatives can be used

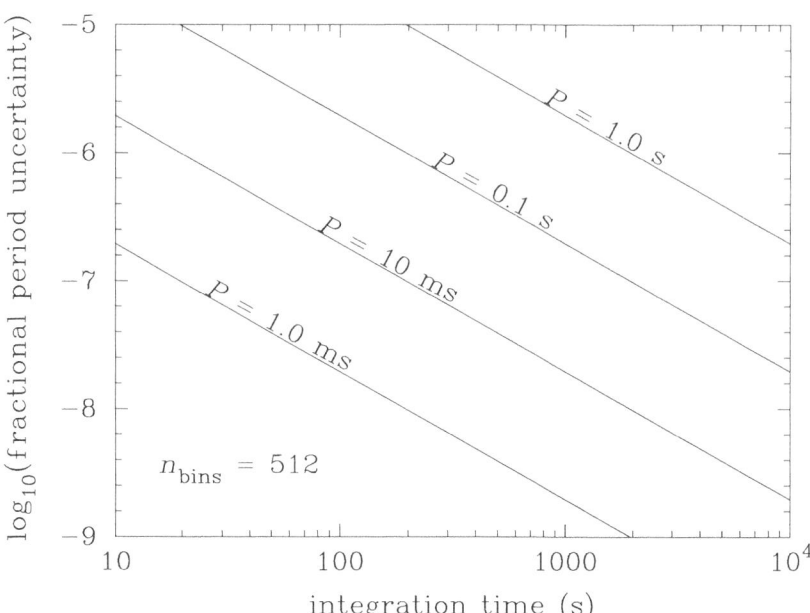

Fig. 7.2. Graph showing the fractional folding period uncertainty required to produce a 512-bin pulse profile for various pulse periods as a function of integration time.

to constrain the orbital parameters of the binary system, in particular $\dot{P}_{\rm bin}$, which we will discuss further in Section 7.2.4.

7.2.2 Improving the dispersion measure

Although the dispersion measure (DM) resolution of a periodicity search usually is chosen to provide a good indication of a pulsar's DM (see Section 6.1.1.2) it is often useful to refine the estimate by de-dispersing the data into a number of sub-bands and folding at the best period obtained from the above analysis. By analogy with the period optimisation, an improved estimate for DM can be found if the pulse profile is delayed across the de-dispersed sub-bands. For a delay Δt (s) over a total bandwidth Δf (MHz), a better DM estimate

$$\mathrm{DM}_{\rm improved} = \mathrm{DM} + 1.21 \times 10^{-4} \left(\frac{\Delta f^3}{f} \right) \Delta t \ \mathrm{cm}^{-3}\mathrm{pc}, \qquad (7.7)$$

where f (MHz) is the centre frequency. This improved estimate usually is quite adequate to de-disperse all subsequent observations. For high precision DM determinations, observations at widely spaced frequencies can be used directly in a timing analysis (see Chapter 8).

At this point it is worth mentioning that when attempting to get the best resolution out of incoherent devices such as filterbanks and correlators (see Sections 5.2.1 and 5.2.2) it is often preferable to fold the individual frequency channels separately. If a single de-dispersed profile is required, the most precise method is to Fourier-transform the resulting profiles and apply the shift theorem in the Fourier domain to rotate the phase of each profile to a common reference frequency. For example, denoting the kth Fourier component of a single profile by

$$\mathcal{P}_k = \sum_{j=0}^{n_{\mathrm{bins}}-1} p_j \exp(-2\pi ijk/n_{\mathrm{bins}}), \tag{7.8}$$

where as usual $i = \sqrt{-1}$, and an arbitrary time shift t_0 is achieved by multiplying each $\mathcal{P}_k$ by the factor $\exp(2\pi ikt_0/P)$. Choosing t_0 appropriately using the dispersion relation (see Equation (4.6)), the profile corresponding to each frequency channel then can be shifted relative to some common reference frequency. The resulting Fourier coefficients of each shifted profile then are summed before applying the inverse Fourier transform to obtain a single de-dispersed profile back in the time domain. For detailed observations, this more exact approach is preferable over a 'search-mode' de-dispersion where integer bin shifts are used (see, for example, Section 6.1.1.1).

7.2.3 Improving the position

Aside from pulsars discovered in targeted searches of radio or high-energy point sources where positions are known to high precision (see, for example, Camilo (2003)), the initial position of most newly discovered pulsars has an uncertainty comparable to the size of the radio telescope beam. Since even a small error in position produces large effects in timing analyses (see Chapter 8), pulsar timing is simplified greatly when the positional uncertainty is small.

Although interferometric determinations of pulsar positions offer one means of improving the position (see Chapter 9), a useful means of reducing the positional uncertainty via single-dish observations is the so-called 'gridding' scheme developed for pulsars discovered in the Parkes multibeam survey (Morris et al. 2002). As part of the confirmation procedure, each pulsar from the survey was observed in five positions:

the nominal search position, and four offset[2] pointings: north, south, east and west of the true position. Folding the data at each position results in several detections clustered around the nominal position. The average position of the detections, weighted by the profile S/N, provides a more accurate determination of the true position.

By comparing these positions with the final values determined from timing observations, Morris *et al.* (2002) demonstrated that the gridding procedure reduced the root mean square scatter between estimated and true positions by a factor of about 4.4. As well as the reduced uncertainties in the timing analysis, the improved positions reduce the pointing offset between the telescope and the pulsar and therefore provide greater S/N in subsequent observations. This reduces the subsequent amount of observing time required to obtain a satisfactory detection and therefore increases the overall observing efficiency.

7.2.4 Orbital determination

In the case of binary pulsars, the orbital parameters need to be taken into account to provide an ephemeris suitable for correctly folding the data. As these parameters initially are unknown, their determination begins as an iterative procedure. Several observations, ideally closely spaced, are used to determine the pulse period at each epoch using the period optimisation procedure described in Section 7.2.1.

Care should be taken at this stage to remove the effects of the motion of the Earth from the data that will Doppler-shift the observed pulse periods with a 1 year orbital period! Unless the time series has already been corrected for this effect prior to folding (see Section 6.1.2), the observed pulse periods and epochs should be converted appropriately to the equivalent values at the Solar System barycentre (SSB). Software tools to perform this conversion are available on the book web site (see Appendix 3).

For a binary orbit, a graph of observed period P_{obs} versus time can be fitted to a simple Keplerian model:

$$P_{\mathrm{obs}} = P_{\mathrm{int}}(1 + V_{\mathrm{l}}(t)/c). \tag{7.9}$$

Here P_{int} is the intrinsic period of the pulsar and $V_{\mathrm{l}}(t)$ is its projected velocity along the line of sight as a function of time t. Full details of the Keplerian orbital prescription for $V_{\mathrm{l}}(t)$ are given in Chapter 8.

2 In this case, based on the 14 arcmin half-power beam width of the Parkes telescope, the offset was chosen to be 9 arcmin.

Figure 7.3(a) shows an example orbital fit to a set of data points for the long-period binary pulsar J0407+1607 (Lorimer *et al.* 2004b).

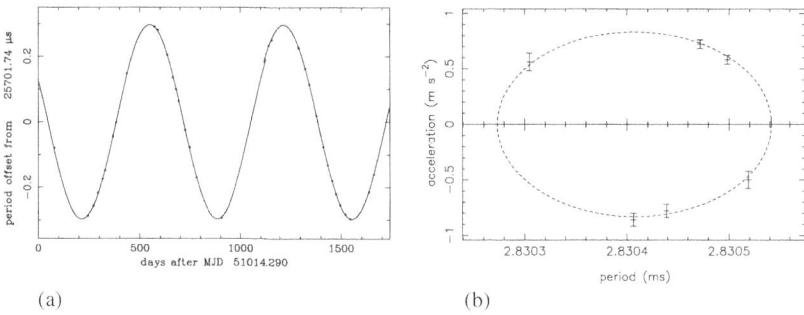

(a)   (b)

Fig. 7.3. (a) Keplerian orbital fit to the 669-day binary pulsar J0407+1607 showing the observed barycentric periods as a function of time. (b) Orbital fit in the period–acceleration plane for the globular cluster pulsar 47 Tuc S.

For pulsars that are less frequently detectable, (e.g. weak pulsars in globular clusters which are prone to scintillation) the density of observations may not be sufficient to resolve orbital period ambiguities using the above technique. In these cases, Freire *et al.* (2001) demonstrated that if, in each observation a significant period derivative due to orbital motion can be detected (as described in Section 7.2.4), the data points for any given orbit will map out a closed loop in $\dot{P}$–P space. Alternatively, since the line of sight component of the orbital acceleration $a_l = \dot{P}c/P$ can be determined through an acceleration search (see Section 6.2.1), the orbit also can be viewed in a_l–P space. For a circular orbit, the loop is an ellipse as shown in Figure 7.3. Fitting a model ellipse to these data points, Freire *et al.* (2001) were able to determine the orbital period and semi-major axis from these sparsely-sampled data points. Without the use of this technique, it would have been practically impossible to obtain an initial timing solution.

7.3 Calibrating pulsar data

As with most areas of astronomy, pulsar data do not automatically come from the telescope in physically meaningful units. In particular, the output of pulsar data acquisition devices is designed to be compact. The data are quantised and, depending on the data acquisition device, recorded with a range of precision ranging between 1 and 32 bits per sample. The post-observation treatment of pulsar data is not truly complete before some sort of calibration from 'machine units', which we will hereafter refer to simply as *counts*, to flux density in Janskys (Jy) measured by the telescope. Although for some applications (e.g. timing and scintillation studies) one can often 'get away'[3] without performing any calibration procedures, for other studies (e.g. polarimetry and spectral measurements) calibration is essential. In general, we recommend that calibration is carried out routinely so that data can be used for many purposes, e.g. flux density determinations from data taken originally for timing observations.

In the following discussion, we begin in a fairly simple manner by considering various schemes to calibrate the flux density scale (see Sections 7.3.1 and 7.3.2). Although these simple estimates are instructive, we recommend that observers make use of a noise diode signal, as we describe in Section 7.3.3, to calibrate more reliably their data wherever possible. Polarisation calibration then will be discussed relative to this flux density scale in Section 7.3.4. Those unfamiliar with the radio-astronomical terminology used below should consult Appendix 1.

7.3.1 Flux density estimates from the signal to noise ratio

Two particular quantities are of interest: (a) the peak flux density (i.e. the maximum point on the folded profile); (b) the mean flux density (i.e. the area under the pulse divided by the period). The mean flux density represents the flux density of a pulsar if it were a continuum source, and is sometimes known as the *equivalent continuum flux density* and is the standard quantity used to interpret the strength of the signal.

The simplest way to estimate the flux density of a folded profile is from its S/N ratio and the (presumed known) local observing system parameters. In Appendix 1 we show that if a pulsar of period P (s) and equivalent width W (s) is observed with a given S/N, then its mean flux density

$$S_{\mathrm{mean}} = \frac{(\mathrm{S/N})\,\beta\,T_{\mathrm{sys}}}{G\sqrt{n_{\mathrm{p}}t_{\mathrm{obs}}\Delta f}}\sqrt{\frac{W}{P-W}}. \qquad (7.10)$$

Here, β is a correction factor accounting for imperfections in the system due to finite digitisation (Chapter 5), T_{sys} is the system noise temperature (K), G is the gain of the telescope (K Jy^{-1}), n_{p} is the number of polarisations summed, Δf is the observing bandwidth (MHz) and t_{obs} is the observation length (s). With this choice of units, we find S_{mean} in mJy from this expression.

3 The authors are as guilty of this crime as others!

As well as measuring S/N and equivalent width (see Section 7.1.1.1), care should be taken to obtain the most appropriate values for the other parameters in Equation (7.10). As explained in Appendix 1, the system noise is the sum of the receiver noise temperature T_{rec} (typically > 20 K), sky background temperature T_{sky} and the 'spillover' noise from the ground T_{spill}. The sky background temperature, a strong function of observing frequency and position, can be estimated from the Haslam *et al.* (1982) all-sky survey. The spillover noise T_{spill} and telescope gain G usually depend on the telescope elevation.

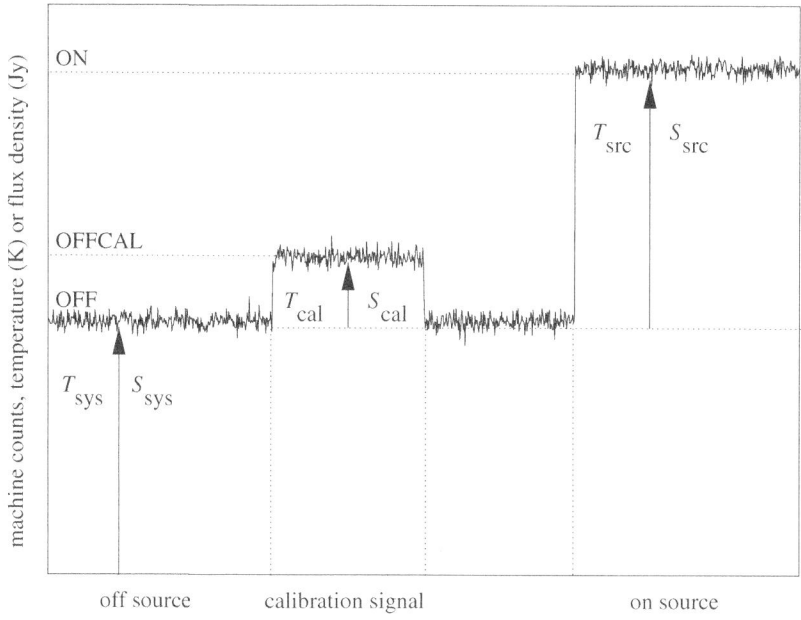

Fig. 7.4. Schematic showing the relationship between machine counts, antenna temperature and flux density during a typical system calibration measurement.

A straightforward means of determining T_{sys} for the purposes of this calculation is to observe a stable radio continuum source with a well-known flux density S_{src} and record the mean number of machine counts received in an *off-source* position, i.e. a nearby patch of sky containing no strong radio sources. As shown in Figure 7.4, the on-source counts (ON) are proportional to the system noise plus the source, while the off-source counts (OFF) represent just the system noise. This means that

$$\frac{ON - OFF}{OFF} = \frac{T_{\mathrm{src}}}{T_{\mathrm{sys}}}, \tag{7.11}$$

where $T_{\mathrm{src}} = GS_{\mathrm{src}}$ is the antenna temperature due to the source.

7.3.2 Flux density estimates from the system noise

Strictly speaking, Equation (7.10) is valid for top-hat pulses. A more general approach is to note that the DC level (off-pulse mean) of a pulse profile, $\bar{p}$, corresponds in counts to the system equivalent flux density of the observing system $S_{\mathrm{sys}} = T_{\mathrm{sys}}/G$ in Jy (see Appendix 1). If T_{sys} and G of the telescope are known, a Jy scale for the profile can be established readily by subtracting the DC offset from the uncalibrated profile and multiplying the result by the factor $G\bar{p}/T_{\mathrm{sys}}$. The peak flux density can be read off the scale directly. The mean flux density is simply the area under the pulse divided by the number of bins across the profile.

A similar procedure makes use of the standard deviation of the off-pulse part of the profile, σ_{p}. From the radiometer equation (see Appendix 1), we expect the root mean square noise fluctuations in Jy

$$\Delta S_{\mathrm{sys}} = \frac{T_{\mathrm{sys}}}{G\sqrt{n_{\mathrm{p}}t_{\mathrm{obs}}\Delta f}} = C\sigma_{\mathrm{p}}, \tag{7.12}$$

where C is the required scaling factor in units of Jy per count. The profile can be calibrated into Jy by simply subtracting $\bar{p}$ and multiplying by C.

7.3.3 Accurate flux density determination

In all three of the above procedures, we assume that the contribution to T_{sys} from the sky background and spillover terms is the same when observing the calibrator source and the pulsar. In practice this may not be correct, even approximately, and the resulting flux densities may contain systematic errors. A more robust approach is to use a semiconductor diode to inject white noise into the receiver system at the waveguide. If the corresponding antenna temperature of the noise diode (T_{cal}) can be measured, then it can be used to establish the flux density scale rather than T_{sys}. Returning to the above measurement of T_{sys} shown in Figure 7.4, we can establish T_{cal} by making one additional measurement of the counts when the noise diode is added during the off-source pointing (OFFCAL). By the same logic as before,

$$\frac{T_{cal}}{T_{sys}} = \frac{OFFCAL - OFF}{OFF}. \tag{7.13}$$

Combining this with Equation (7.11) to eliminate T_{sys}, we find

$$\frac{S_{cal}}{S_{src}} = \frac{T_{cal}}{T_{src}} = \left(\frac{OFFCAL - OFF}{ON - OFF} \right), \tag{7.14}$$

where S_{cal} and S_{src} are the (corresponding) equivalent flux densities of the calibration signal and source, respectively. Based on the known value for S_{src}, S_{cal} can be determined readily from these simple measurements that are independent of the gain and system temperature.

In practice, there are a number of ways the calibration signals can be used to establish the flux density scale for pulsar observations. Three commonly used variants are:

(i) *Carry out multiple observations with the diode switched alternately on and off.* The counts/Jy scaling factor can be readily determined from the difference in the DC levels of pulse profiles when the calibration signal is on and off. This technique is used routinely at Jodrell Bank (see, for example, Gould and Lyne (1998)).

(ii) *Pulse the noise diode at a fixed frequency and fold the signal as if it were a pulsar.* This produces a square-wave profile where the required counts/Jy scaling factor again is simply the difference between the on- and off-pulse levels. These 'pulsed cals' are in fairly common use at telescopes such as Arecibo and Parkes.

(iii) *Pulse the noise diode synchronously at the pulsar period.* Normally this is arranged so that the calibration signal occupies a part of the profile at which no pulsed emission is expected. The area under the pulsar's waveform can be calculated directly from the known area under the calibration signal. This system is used at the Effelsberg telescope (Seiradakis *et al.* 1995).

7.3.4 Polarisation calibration

As reviewed in Chapter 1, pulsars are highly polarised radio sources. Polarisation measurements can yield a wealth of information, not only about the emission process itself, but also about the medium through which it propagates. One example is the structure of the Galactic magnetic field through Faraday rotation measurements. Before discussing such observations, we outline first the considerations required to calibrate the data properly.

As summarised in Appendix 1, the polarisation state of the signal can be described fully by the four Stokes parameters: I, Q, U and V. These may be expressed conveniently as a column matrix known as the Stokes vector:

$$\mathcal{S} = \begin{bmatrix} I \\ Q \\ U \\ V \end{bmatrix}. \tag{7.15}$$

In practice, the measured Stokes vector $\mathcal{S}_{\mathrm{measured}}$ will be different from the intrinsic vector $\mathcal{S}_{\mathrm{int}}$ for a number of reasons related to the telescope and observing system. Before discussing specifics, we can summarise the combination of these effects by the Mueller matrix $\mathcal{M}$ (Mueller 1948) which acts as a transfer function between the measured and true Stokes parameters:

$$\mathcal{S}_{\mathrm{measured}} = \mathcal{M} \times \mathcal{S}_{\mathrm{int}}. \tag{7.16}$$

By calibrating the observing system, we can determine $\mathcal{M}$ and, hence, solve this equation for the true Stokes vector $\mathcal{S}_{\mathrm{int}}$. Following a number of authors (see, for example, Heiles *et al.* (2001)), we can express the Mueller matrix as a set of independent matrices:

$$\mathcal{M} = \mathcal{M}_{\mathrm{Amp}} \times \mathcal{M}_{\mathrm{CC}} \times \mathcal{M}_{\mathrm{Feed}} \times \mathcal{M}_{\mathrm{PA}}, \tag{7.17}$$

where the various terms will be outlined next. Note that, due to the non-commutative nature of matrix algebra, the ordering of the terms here is important! Therefore we proceed from right to left in this expression to follow the order in which the various effects take place.

The first effect applies to any 'alt-azimuth' radio telescope tracking a source across the sky and occurs because the feed rotates with respect to the plane of polarisation by the *parallactic angle*

$$\mathrm{PA} = \arctan\left(\frac{\sin \mathrm{HA} \cos \phi}{\sin \phi \cos \delta - \cos \phi \sin \delta \cos \mathrm{HA}} \right), \tag{7.18}$$

where HA is the hour angle of the source of declination δ and ϕ is the latitude of the observatory. The effect of a change in PA during the observation is a phase rotation of the Stokes vector by 2PA. The associated Mueller matrix

$$\mathcal{M}_{\mathrm{PA}} = \begin{bmatrix} 1 & 0 & 0 & 0 \\ 0 & \cos 2\mathrm{PA} & \sin 2\mathrm{PA} & 0 \\ 0 & -\sin 2\mathrm{PA} & \cos 2\mathrm{PA} & 0 \\ 0 & 0 & 0 & 1 \end{bmatrix}. \tag{7.19}$$

Although elliptical feeds are possible, the incoming radiation is in general sampled by either linear or circular probes. In general, we can write the Mueller matrix for the feed as follows

$$\mathcal{M}_{\mathrm{Feed}} = \begin{bmatrix} 1 & 0 & 0 & 0 \\ 0 & \cos 2\gamma & 0 & \sin 2\gamma \\ 0 & 0 & 1 & 0 \\ 0 & -\sin 2\gamma & 0 & \cos 2\gamma \end{bmatrix}. \tag{7.20}$$

where $\gamma = 0°$ is the case for a dual linear feed and $\gamma = 45°$ implies a dual circular feed. In general, however, the feeds are not perfect, so that the output from the two supposedly orthogonal probes is coupled. We can express this *cross-coupling* effect by the following Mueller matrix

$$\mathcal{M}_{\mathrm{CC}} = \begin{bmatrix} 1 & 0 & A & B \\ 0 & 1 & C & D \\ A & -C & 1 & 0 \\ B & -D & 0 & 1 \end{bmatrix}, \tag{7.21}$$

where the constants are defined in the magnitude and phase (ϵ and ϕ) of the cross coupling from probe 1 to probe 2, i.e.

$$A = \epsilon_1 \cos \phi_1 + \epsilon_2 \cos \phi_2 \tag{7.22}$$
$$B = \epsilon_1 \sin \phi_1 + \epsilon_2 \sin \phi_2 \tag{7.23}$$
$$C = \epsilon_1 \cos \phi_1 - \epsilon_2 \cos \phi_2 \tag{7.24}$$
$$D = \epsilon_1 \sin \phi_1 - \epsilon_2 \sin \phi_2. \tag{7.25}$$

Finally, the two signals pass through slightly different amplifier chains that introduce different gains and phases on the signals. Considering just the differential gain and phase, ΔG and $\Delta \psi$, following Heiles *et al.* (2001) we can write to first order:

$$\mathcal{M}_{\mathrm{Amp}} = \begin{bmatrix} 1 & \Delta G/2 & 0 & 0 \\ \Delta G/2 & 1 & 0 & 0 \\ 0 & 0 & \cos \Delta \psi & -\sin \Delta \psi \\ 0 & 0 & \sin \Delta \psi & \cos \Delta \psi \end{bmatrix}. \tag{7.26}$$

With the above set of Mueller matrices, it is possible to describe the variation of measured Stokes parameters as a function of parallactic angle. Determination of the matrix parameters requires expanding Equation (7.16) into four simultaneous equations for the measured Stokes parameters that can then be solved for the system parameters by least-squares fits to data observed over a wide range of parallactic angles.

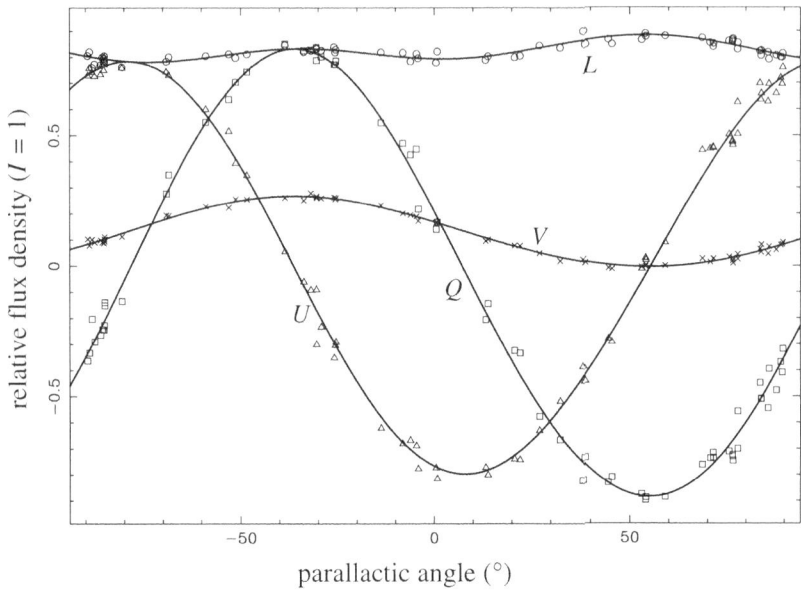

parallactic angle (°)

Fig. 7.5. The data points show measured Stokes parameters Q, U and V as well as the linear polarisation $L = \sqrt{Q^2 + U^2}$ for PSR J1359–6038 as a function of parallactic angle (Johnston 2002). The fitted curves define the Mueller matrix of the system allowing the true Stokes parameters to be inferred. Figure and analysis provided by Simon Johnston.

Early attempts to determine the Mueller matrix parameters (see, for example, Stinebring *et al.* (1984)) were hampered somewhat by the computational power required to perform the fitting and matrix inversion process and a number of simplifications were often made. However, as pointed out by Johnston (2002), modern computers can solve Equation (7.16) easily and without loss of generality. As an example, we show Johnston's fits to calibration data in Figure 7.5. In this case, the highly polarised pulsar J1359–6038 is used effectively to self-calibrate the system relative to the total intensity, I. The resulting Mueller matrix parameters then can be applied to subsequent observations in order to derive reliable polarimetric information for other pulsars. Flux calibration to convert I and, hence, the other three Stokes parameters to a Jy scale is achieved ideally by using a noise diode and a radio source of known flux density as described in Section 7.3.3.

7.4 Various pulsar observing modes

Now we consider more routine pulsar observations, in which it is assumed that we have a well-calibrated system and suitable ephemerides with which the data can be folded to obtain pulse profiles.

7.4.1 Profile stabilisation analyses

The stability of the integrated profile is an important consideration in timing studies in which it is assumed that the integrated pulse profile has a characteristic shape once sufficient pulses have been accumulated. The simplest way to investigate profile stability is to take long integrations (preferably several thousand pulse periods) and compute the correlation coefficient between the integrated profile p and the accumulated profile a formed from the addition of n pulse periods. Deriving a is a simple extension of the folding algorithm discussed in Section 7.1 to write out the accumulated profile after the appropriate number of pulses added. Helfand *et al.* (1975) were the first to perform an analysis of this kind. Forming integrated and accumulated profiles, they examined the behaviour of $1 - \rho_r$ as a function of n, where

$$\rho_r = \frac{\sum_i (a_i - \overline{a})(p_i - \overline{p})}{\sqrt{\sum_i (a_i - \overline{a})^2} \sqrt{\sum_i (p_i - \overline{p})^2}} \tag{7.27}$$

is the familiar linear correlation coefficient (see, for example, Press *et al.* (1992)). In order to minimise the effect of off-pulse noise in this calculation, it is usual to perform this calculation over just the on-pulse bins. As noted by Helfand *et al.* (1975), we expect that $(1 - \rho_r) \propto n^{-0.5}$ for uncorrelated individual pulses. Deviations from this behaviour are common and indicate the presence of non-random behaviour from pulse to pulse, e.g. due to nulling or mode changing. As such, this simple analysis can be useful for the single-pulse studies discussed next.

7.4.2 Single-pulse analyses

Although most pulsars are weak sources, there are a significant number (of the order of 100 out of the 1600 currently known) that are strong enough so that their individual pulses can be studied in detail. Producing single pulses from a de-dispersed time series is a simple extension of the basic folding algorithm in which the accumulated pulse profile is stored after each period. In addition to making excellent illustrations of pulsar emission behaviour (see, for example, Figs. 1.1, 1.5 and 1.6) quantitative studies of the individual pulse properties provide valuable input to theoretical models of the pulsar emission process (see Chapter 3). In the following, we outline some relevant techniques for carrying out single-pulse analyses. For the purposes of this discussion, it is convenient to represent the sampled pulses as a two-dimensional array of pulse intensity $I(j, k)$, where j denotes the pulse phase bin number and k is the pulse number, i.e. $k = 1$ denotes the first pulse, etc. Each profile has n_{bins} phase bins and there are n_{pulses} pulses in total. We shall also assume that each profile has had a DC offset subtracted, i.e. for any k, the off-pulse values of I have a zero mean.

7.4.2.1 Nulling fraction analysis

As discussed in Section 1.1.4.3, it is of interest to determine the fraction of time the pulsar spends in a null state, i.e. its *nulling fraction*, NF. Following earlier studies that used to quantify nulling from histograms of the area under the pulse (i.e. *pulse energy*; see Smith (1973)), Ritchings (1976) proposed a simple procedure that we summarise below and apply to an observation of the nulling pulsar B1944+17. This is shown in Figure 1.6, for which NF ~ 55 per cent (Deich *et al.* 1986). As a control experiment, we also analyse an observation of the pulsar B1933+16 for which very little nulling is observed, i.e. NF< 0.25 per cent.

The first step is to fold the time series and, from the resulting pulse profile, determine suitable on- and off-pulse 'windows', each of width W bins, denoting the starting on-pulse bin number as ON and the starting off-pulse bin number as OFF. Single pulses are then formed in the usual way and the on- and off-pulse energies E_{ON} and E_{OFF} are calculated:

$$E_{\mathrm{ON}}(k) \;\; = \;\; \sum_{j=\mathrm{ON}}^{\mathrm{ON}+W-1} I(j, k) \tag{7.28}$$

$$E_{\mathrm{OFF}}(k) \;\; = \;\; \sum_{j=\mathrm{OFF}}^{\mathrm{OFF}+W-1} I(j, k). \tag{7.29}$$

In order to minimise the effects of short-term variations due to scintillation (see Chapter 4), the resulting pulse energies corresponding to contiguous 2–3 min blocks of data are analysed separately. Within each block, the mean on-pulse energy $\langle E_{\mathrm{ON}} \rangle$ is calculated and used to normalise the individual values of E_{ON} and E_{OFF}.

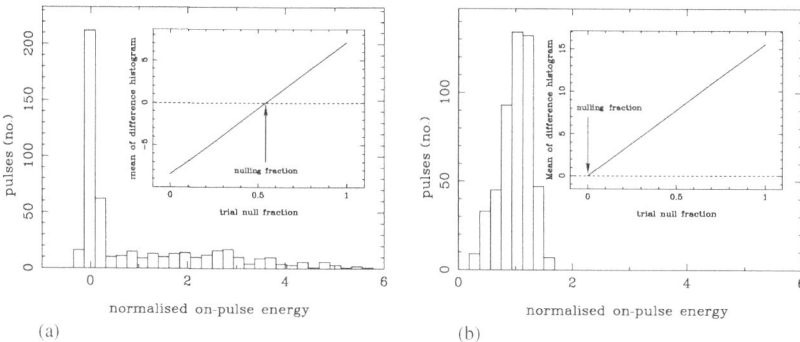

Fig. 7.6. (a) Simple nulling analysis for PSRs B1944+17; (b) B1933+16. The histograms show on-pulse distributions of pulse energy for each pulsar. The mean of the 'residual histogram' between the on- and off-pulse data is shown inset as a function of trial nulling fraction (see text).

Histograms of the resulting normalised on-pulse energies are shown in Figure 7.6. The on-pulse energy distribution for B1944+16 is broad and shows a clear excess at zero, indicating significant nulling. The distribution for B1933+16 is much narrower, showing no significant excess at zero. In order to calculate the nulling fraction, we need to account for the contribution of the system noise in these histograms. The on-pulse distributions represent the distribution of the pulse intensities convolved with the system noise. The off-pulse data represent just the probability distribution of the system noise. Although in principle a Fourier deconvolution technique could be used to recover the true on-pulse distribution, it is sufficient, following Ritchings (1976), simply to subtract the off-pulse histogram from the on-pulse histogram. The resulting 'residual histogram' accounts for the sensitivity limit caused by noise in the observing system.

For an extreme nulling pulsar with NF $\sim$ 100 per cent, the residual histogram would be smooth and continuous with a zero mean. This is not the case for either pulsar, and we need to adjust the fraction of the noise distribution that we subtract from the on-pulse distribution until the mean of the residual histogram is zero. In the case of B1933+16, and as shown in Figure 7.6, the fraction of the noise distribution subtracted indicates that NF $\sim$ 0 as expected. For significant nulling pulsars like B1944+17, we can place a lower limit on NF by reducing the fraction of the subtracted noise distribution until the mean of the residual histogram again deviates from zero. From the plot of shown in Figure 7.6, we find NF ~ 55 per cent, in excellent agreement with Deich *et al.* (1986).

7.4.2.2 Microstructure analyses

As mentioned in Chapter 1, single pulses reveal structure on microsecond and even smaller timescales when studied with sufficient time resolution. In analyses of such microstructure, we are interested primarily in establishing: (a) whether any periodicities are present; (b) a characteristic timescale for the micropulses. Now we outline the techniques to measure these quantities starting from our two-dimensional array $I(j,k)$ defined above, where j is the pulse phase bin and k is the pulse number.

A typical microstructure analysis is shown in Figure 7.7. It is of particular interest to identify and quantify any periodicities and characteristic timescales of micropulses present in the data. From each high-resolution single pulse, the first step (see, for example, Hankins (1972)) is to form the autocorrelation function (ACF). In general, the ACF of the kth pulse

$$\mathrm{ACF}(k,l) = K_k \sum_{j=1}^{n_{\mathrm{bins}}} I(j,k)I(j+l,k), \qquad (7.30)$$

where l is the lag and K_k is a normalising constant, usually chosen such that $\mathrm{ACF}(k,0) = 1$ as shown in Figure 7.7(a).

In order to quantify any periodicities present, the power spectrum of the data (PSD) is usually calculated. Given the ACF, it follows from the autocorrelation theorem (see, for example, Bracewell (1998)) that the PSD of a given pulse is simply the Fourier transform of its ACF. As shown in Figure 7.7(b), the PSD identifies a clear peak in the data with a fluctuation frequency of 2 kHz. Lange *et al.* (1998) introduced a more sensitive means of identifying periodicities by Fourier-transforming the time derivative of the ACF and forming its power spectrum. The advantage of this resulting quantity, known as the ACF derivative power spectrum or ADP, is that low-level features present often can be masked by the overall slope of the ACF. Taking the time derivative before the Fourier transform effectively removes the slope. Figure 7.7(d) shows the ADP as a function of fluctuation frequency and identifies the 2 kHz periodicity with greater significance. As shown by, for example, Cordes *et al.* (1990), these quasi-periodicities are not stable and vary from pulse to pulse. Often it is useful to investigate this behaviour via a histogram of periodicities identified from PSDs or ADPs of individual pulses.

In order to determine the timescale for microstructure, following Hankins (1972), it is standard practice to plot the average ACF over all

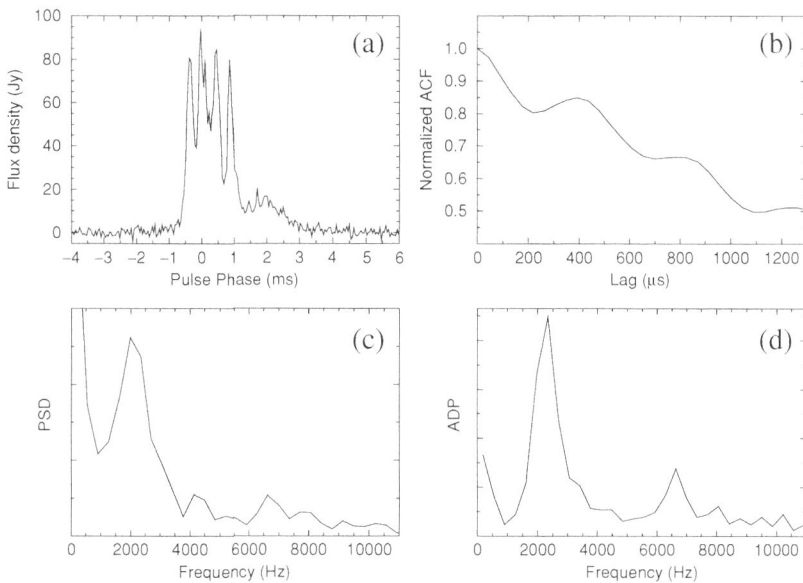

Fig. 7.7. Examples of microstructure analyses for a single pulse from the Vela pulsar (Kramer *et al.* 2002) shown in panel (a). Panels (b), (c) and (d) show respectively the ACF, PSD and ADP of the pulse (see text).

pulses. The average ACF typically is made up of two straight-line components with a characteristic break-point or flattening which represents the typical duration (i.e. width) of the micropulses (Rickett *et al.* 1975) and is defined to be the microstructure timescale. Lange *et al.* (1998) proposed an extension of this simple approach, known as the 'turn-off point method', that searches automatically for changes in the slope due to flattening (see Appendix A of Lange *et al.* (1998) for details of this procedure).

We note in passing that in low-frequency microstructure analyses (see, for example, Kuzmin *et al.* (2003)) multiple breaks are observed in the ACFs due to other effects, e.g. digitisation and sky background variations across the band. In order to identify the intrinsic microstructure timescale in these cases, Kuzmin *et al.* (2003) form the ACFs of on- and off-pulse regions separately and take the difference of the two functions to subtract the instrumental effects common to both regions. The only remaining features are due then to the intrinsic timescale of the microstructure.

7.4.2.3 Drifting subpulse analyses

The highly ordered structure of the drifting subpulses discussed in Chapter 1 (see, for example, Figure 1.6) clearly is a phenomenon that needs to be explained by any viable emission process. As shown schematically in Figure 1.6, in addition to the pulse period $P = P_1$, we identify the characteristic spacing between sub-pulses, P_2, and the period at which a pattern of pulses crosses the pulse window, P_3. Now we discuss briefly ways in which P_2 and P_3 can be determined.

For P_2, Sieber and Oster (1975) devised a method in which successive pulses $I(j, k)$ and $I(j + l, k + 1)$ were cross-correlated as a function of different phase lags l. The resulting correlation coefficients then are averaged over all pulses, and a plot of correlation coefficient versus l reveals two peaks separated in phase by P_2. The classical determination of P_3 was pioneered by Backer (1970b). In this scheme, data are selected in the same phase bin of each of the individual pulses to form a time series at a fixed pulse longitude. A straightforward one-dimensional Fourier transform (see, for example, Section 6.1.3) of this time series can be applied to form the power spectrum as a function of cycles per pulse period (C/P_1). This so-called *fluctuation spectrum* may be averaged over a range of pulse longitudes to increase S/N if desired. The fluctuation spectrum shows three main features: (a) a strong low-frequency 'red-noise' component caused by a combination of interstellar scintillation, receiver fluctuations and intrinsic variations in intensity; (b) a constant 'white noise' component corresponding to independent pulse-to-pulse fluctuations; (c) spectral features at $1/P_3$ and harmonics thereof caused by the drifting sub-pulses.

Advances in computational power now favour two-dimensional analyses to determine P_2 and P_3 simultaneously. Two related techniques have recently been proposed: (a) the harmonic-resolved fluctuation spectrum method of Deshpande and Rankin (1999; 2001); (b) the two-dimensional fluctuation spectrum approach of Edwards and Stappers (2002). As pointed out by the latter authors, both techniques are equivalent. The two-dimensional fluctuation spectrum $\mathcal{S}$ is simply the power spectrum of the two-dimensional Fourier transform, i.e.

$$\mathcal{S}(u, v) = \left| \frac{1}{K} \sum_{j=0}^{n_{\mathrm{bins}}-1} \sum_{k=0}^{n_{\mathrm{pulses}}-1} I(j, k) \exp[-2\pi i(uj + vk)] \right|^2, \qquad (7.31)$$

where the normalisation factor $K = n_{\mathrm{bins}} \times n_{\mathrm{pulses}}$, u is in cycles per radian of pulse longitude and, as for the one-dimensional fluctuation spectrum, v is in cycles per pulse period. The signature of P_2 is detectable in u as harmonics of $P_1/(2\pi P_2)$, while P_3 shows up in v as harmonics spaced by $1/P_3$ as before.

7.4.3 Polarisation profiles

In addition to total intensity, Stokes' I, we have three further profiles: Stokes' Q, U and V. Following the application of a polarisation calibration procedure such as the one outlined in Section 7.3.4, we can dedisperse and fold all four Stokes parameters separately. It is customary usually to display total intensity, I, circular polarisation V, linear polarisation $L = \sqrt{Q^2 + U^2}$ and the associated position angle $\Psi = \frac{1}{2}\arctan(\frac{U}{Q})$.

7.4.3.1 Considerations for linear polarisation

While the individual Stokes parameters are observed to follow Gaussian noise statistics with standard deviations σ_I, σ_U, σ_V and σ_Q, this is not true for the linearly polarised intensity, L. Due the quadrature summation $L = \sqrt{Q^2 + U^2}$, the probability density function of L is non-Gaussian (see below). In addition, there exists a positive bias in the measured values at low L (see, for example, Wardle and Kronberg (1974)). In the extreme case, in which L is zero, we may infer erroneously a non-zero value simply from the quadrature sum of the noise in U and Q! Following Everett and Weisberg (2001), we recommend the following correction scheme to remove the bias and calculate the true value of linear polarisation:

$$L_{\text{true}} = \begin{cases} \sigma_I \left[\left(\frac{L}{\sigma_I}\right)^2 - 1 \right]^{\frac{1}{2}} & \text{if } \left(\frac{L}{\sigma_I}\right) \geq 1.57 \\ 0 & \text{otherwise,} \end{cases} \tag{7.32}$$

where σ_I is the off-pulse standard deviation in Stokes' I.

Similarly, the probability density function of the linear polarisation angle about the true value Ψ_{true} is not a Gaussian distribution. As a consequence, the uncertainty in this angle σ_Ψ does not follow strictly the simple result derived from Gaussian error propagation:

$$\sigma_\Psi = 28.65° \left(\frac{\sigma_I}{L_{\text{true}}}\right). \tag{7.33}$$

This is only useful when $L_{\text{true}}/\sigma_I \gtrsim 10$. For lower values of L, following

Everett and Weisberg (2001), we recommend use of the correct probability density function for Ψ. After Naghizadeh-Khouei and Clarke (1993), this is given by

$$G(\Psi) = \frac{\exp(-L_{\text{true}}^2/2\sigma_I^2)}{\sqrt{\pi}} \left(\frac{1}{\sqrt{\pi}} + \eta \exp(\eta^2)[1 + \text{erf}(\eta)] \right), \quad (7.34)$$

where $\eta = L_{\text{true}} \cos 2(\Psi - \Psi_{\text{true}})/(\sqrt{2}\sigma_I)$ and the error function $\text{erf}(x) = (2/\sqrt{\pi}) \int_0^x \exp(-x^2)\,dx$. To derive a '1-$\sigma$' uncertainty σ_Ψ, we need to integrate numerically $G(\Psi)$ to encompass 68.26 per cent of the total area about the peak value. Everett and Weisberg (2001) recommend that this is carried out for values of Ψ where $L_{\text{true}}/\sigma_I < 10$. For more significant detections, Equation (7.33) may be used. This procedure is particularly useful when fitting the position angle data as described next.

7.4.3.2 Fits to the rotating vector model

As discussed in Chapter 3, the rotating vector model of Radhakrishnan and Cooke (1969) predicts the following relationship between the position angle of linear polarisation Ψ and pulse phase ϕ:

$$\tan(\Psi - \Psi_0) = \frac{\sin\alpha \, \sin(\phi - \phi_0)}{\sin(\alpha + \beta) \, \cos\alpha - \cos(\alpha + \beta)\sin\alpha\cos(\phi - \phi_0)}, \quad (7.35)$$

where α is the angle between the rotation and magnetic axes and the 'impact parameter' β is the angle between the line of sight and the centre of the emission beam (see Section 3.4.4 for further details).

Fits of the position angle swing predicted by this function to the observed values therefore offer a way to determine a pulsar's beam geometry. Again, following Everett and Weisberg (2001), we recommend that care should be taken to weight appropriately the position angles by the uncertainties estimated above. With this in mind, two important additional caveats should be noted: (a) as mentioned in Chapter 1, the position angle data do not always follow the predicted model; (b) even if a statistically good fit is obtained, the often narrow pulses mean that only a small range of pulse longitude is sampled and therefore α and β can be highly covariant. Further details can be found in Section 3.4.4. Software to perform these fits on appropriately calibrated pulse profiles is available on the book web site (see Appendix 3).

7.4.3.3 Faraday rotation measurements

As discussed in detail in Chapter 4, the magnetic field of our Galaxy acts effectively as a Faraday screen that rotates the position angle of linear polarisation of the pulsar signal by an amount proportional to the square of the observing wavelength. The constant of proportionality, known as the *rotation measure* (RM) is the integral of the component of the Galactic magnetic field along the line of sight to the pulsar weighted by electron density (see Equation (4.13)).

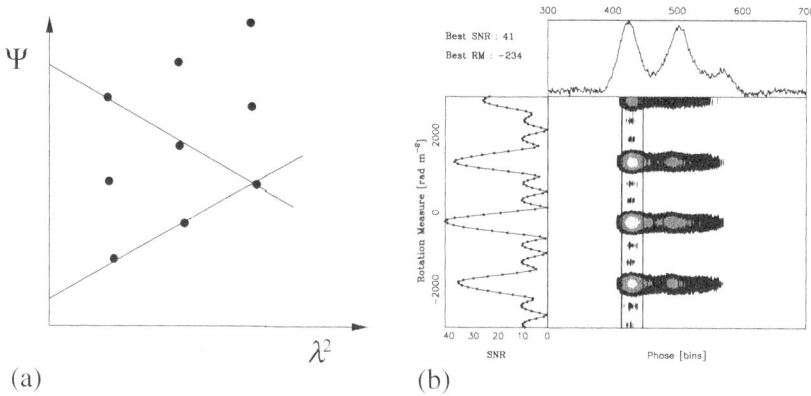

(a) (b)

Fig. 7.8. (a) Schematic showing how position angle ambiguities may lead to an incorrect determination of RM. (b) A search for RM by maximising the linearly polarised profile. Figure provided by Dipanjan Mitra.

Measurements of RM are carried out in two main ways. The simplest approach (see, for example, Rand and Lyne (1994)) is to calculate the linearly polarised flux L in each bin of the pulse profile as described in Section 7.4.3.1. For values of L that exceed the standard deviation of the off-pulse noise σ_L by a certain amount (usually 3–5 times σ_L), the mean value of Ψ is calculated from the vector addition of the linear polarisation across the pulse. Each vector in this sum has components U and Q and is weighted by its S/N in L, i.e. L/σ_L. For bright pulsars, multiple measurements of Ψ as a function of frequency may be possible by splitting the band into several sub-bands and measuring the mean value of Ψ in each band. For fainter pulsars, measurements at multiple frequencies are required. In both cases, RM is derived by fitting to

$$\Psi(\lambda) = \Psi_\infty + \text{RM}\,\lambda^2, \qquad (7.36)$$

where Ψ_∞ is the position angle at infinite frequency. Since, by definition, Ψ is periodic on π rather than 2π, care should be taken to ensure that the frequency spacing is not too large so that position angle ambiguities lead to an incorrect value of RM coming from the fit, as shown in Figure 7.8a.

The second technique, shown in Figure 7.8b, is to perform a search in RM by splitting the band into a number of independent frequency channels that are phase-rotated separately in U and Q before combining to form the average profile in L over the whole band. Rather like the DM search discussed in Chapter 6, a plot of S/N for the resulting L profile versus trial RM should show a maximum at the true value. This method gives consistent results with the simple $\Psi - \lambda$ fit, but with the advantage that it can be applied to weaker pulsars in which S/N prohibits use of the first approach. For further details, see, for example, Mitra *et al.* (2003).

7.4.4 Measuring scintillation parameters

As described in Chapter 4, the observed intensity variations of pulsars tell us much about the structure of the interstellar medium. Here we explain how to measure the basic scintillation parameters.

7.4.4.1 Creating a dynamic spectrum

The characteristic time and frequency scales due to diffractive interstellar scintillation, Δt_{DISS} and Δf_{DISS}, often can be resolved by a single observation with good frequency resolution by forming the *dynamic spectrum* – a two-dimensional image of pulse intensity as a function of observation time and frequency. In order to create a dynamic spectrum, all the individual frequency channels are folded separately and accumulated pulse profiles representing contiguous sub-integrations are saved. The length of each sub-integration depends on the available S/N, but ideally should be ~ 30 s in order to resolve scintillation structure on short timescales. Here, only the relative intensity is of interest, and it is sufficient to calculate the area under the pulse and subtract the mean of the off-pulse emission. The resulting two-dimensional array represents pulse energy as a function of observation frequency f and time t, i.e. $E(f, t)$. An example spectrum is shown in Figure 7.9 for PSR B0834+06.

Although often discernible by eye from the dynamic spectrum, Δf_{DISS} and Δt_{DISS} are more robustly obtained from a two-dimensional autocorrelation analysis. Following Cordes (1986), we first compute the covariance function

$$\mathrm{CF}(\Delta f, \tau) = \sum_{f=1}^{n_{\mathrm{f}}-|\Delta f|} \sum_{t=1}^{n_{\mathrm{t}}-|\tau|} E(f, t) E(f + \Delta f, t + \tau), \qquad (7.37)$$

where n_{f} is the number of frequency channels, n_{t} is the number of subintegrations and f and t are integer values in the range $1 < f < n_{\mathrm{f}}$ and

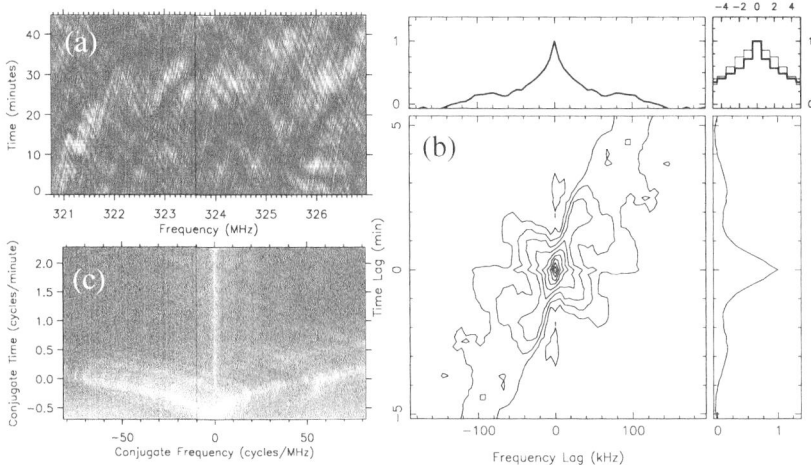

Fig. 7.9. (a) Dynamic spectrum for a 324 MHz observation of PSR B0834+06. (b) Autocorrelation analysis showing the characteristic scintillation timescale and bandwidth. (c) Secondary spectrum of the data showing symmetric arc-like features. Data and analysis provided by Maura McLaughlin.

$1 < t < n_t$. Likewise, the frequency and time lags are integers in the range $-n_f/2 < \Delta f < n_f/2$ and $-n_t/2 < \tau < n_t/2$. The autocorrelation function (ACF) then is just the covariance function normalised by the zero-lag in both time and frequency, i.e.

$$\mathrm{ACF}(\Delta f, \tau) = \frac{\mathrm{CF}(\Delta f, \tau)}{\mathrm{CF}(0,0)}. \tag{7.38}$$

In practice, care should be taken to exclude parts of the dynamic spectrum corrupted by occasional bursts of interference, or where no pulses are present due to nulling. The bandwidth and timescale Δf_{DISS} and Δt_{DISS} are then determined by fitting a two-dimensional Gaussian function to the ACF. Following Cordes (1986), the standard practice is to measure Δf_{DISS} as the half width at $\mathrm{ACF}(\Delta f, 0)/2$, and Δt_{DISS} as the half width at $\mathrm{ACF}(0, \tau)/e$. From such an analysis of the data for PSR B0834+06 in Figure 7.9, we infer $\Delta f_{\mathrm{DISS}} = 42$ kHz and $\Delta t_{\mathrm{DISS}} = 87$ s.

Recalling the relationship from Chapter 4, we can infer the scintillation speed V_{ISS} from these measurements

$$V_{\mathrm{ISS}} = A \left(\frac{d}{\mathrm{kpc}}\right)^{1/2} \left(\frac{\Delta f_{\mathrm{DISS}}}{\mathrm{MHz}}\right)^{1/2} \left(\frac{f}{\mathrm{GHz}}\right)^{-1} \left(\frac{\Delta t_{\mathrm{DISS}}}{\mathrm{s}}\right)^{-1}. \tag{7.39}$$

where d is the distance to the pulsar and the constant A depends on the geometry, location of the scattering screen and form of the turbulence spectrum. For this choice of units, $A = 3.85 \times 10^4$ km s^{-1} (Gupta 1995) or 2.53×10^4 km s^{-1} (Cordes & Rickett 1998). Adopting the latter value for the data shown in Figure 7.9, we find $V_{\text{ISS}} = 148$ km s^{-1} for a dispersion measure based distance $d = 0.65$ kpc.

7.4.4.2 Application to binary pulsars

An elegant application of the dynamic spectrum is to track the changing orbital velocity of a binary pulsar viewed on the plane of the sky as a change in the scintillation pattern as a function of orbital phase. This was first devised by Lyne (1984) who applied it to dynamic spectra of the binary pulsar B0655+64 in order to determine its velocity and the inclination angle between the plane of the orbit and the plane of the sky. Now we outline briefly this technique, closely following the more detailed description given by Ord *et al.* (2002a).

We begin by assuming that we have a dynamic spectrum from which we can obtain measurements of V_{ISS} as described above at regular intervals across the orbit. Then we construct a graph of V_{ISS} versus orbital phase. For the purposes of the model, it is convenient to display orbital phase in terms of true orbital anomaly A_{T}, rather than mean anomaly (see Section 8.3.1.1 for details of computing A_{T}). The model scintillation speed $V_{\text{ISS}}^{\text{model}}$ is the sum of two orthogonal components:

$$V_{\text{ISS}}^{\text{model}} = \kappa\sqrt{V_1^2 + V_2^2}, \tag{7.40}$$

where κ is a free parameter reflecting the uncertainties in the scattering geometry and the terms

$$V_1 = (V_{\text{r}}\cos\phi - V_{\text{A}}\sin\phi) + V_{\|} \tag{7.41}$$

$$V_2 = (V_{\text{A}}\cos\phi + V_{\text{r}}\sin\phi) - V_{\perp} \tag{7.42}$$

relate the radial component of the *orbital* velocity V_{r} in the direction of the focus of the orbital ellipse, its perpendicular component V_{A} and the components of the pulsar's *space* velocity along and perpendicular to the line of nodes (see Section 8.3.1), respectively $V_{\|}$ and $V_{\perp}$. The orbital phase $\phi = \omega + A_{\text{T}}$ is measured with respect to the ascending node, where ω is the longitude of periastron. (See Section 8.3.1 and in particular Figure 8.3 for definitions of the various angles.) In this framework, the orbital velocity components are

$$V_{\text{r}} = V_{\text{orb}}e\sin A_{\text{T}} \quad \text{and} \quad V_{\text{A}} = V_{\text{orb}}(1 + e\cos A_{\text{T}}), \tag{7.43}$$

where

$$V_{\mathrm{orb}} = \frac{2\pi x c}{\sin i \sqrt{(1 - e^2)} P_{\mathrm{b}}} \qquad (7.44)$$

is the mean orbital velocity expressed in terms of observables and the unknown inclination angle i to be determined. The observables are orbital eccentricity e, projected semi-major axis of the orbit x in light seconds, and the orbital period P_{b} (see Section 8.3.1.1).

The geometrical model above provides a description of V_{ISS} in terms of five free parameters: κ, i, ω, $V_{\parallel}$ and $V_{\perp}$. In practice, ω should be known already to high precision along with the other Keplerian parameters P_{b}, x and e from timing measurements (Chapter 8). As noted by Lyne (1984), fits of this model to the data cannot distinguish between the case of a fast-moving pulsar in an orbit viewed nearly edge on, and a slow-moving pulsar in a more face-on orbit. As a result, two degenerate solutions for different values of i are obtained.

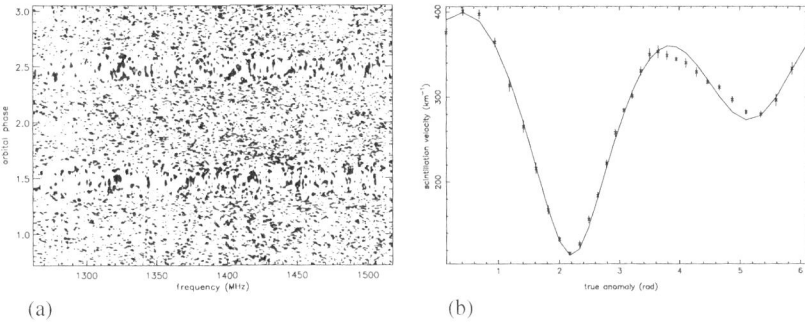

(a)   (b)

Fig. 7.10. (a) Dynamic spectrum for the relativistic binary pulsar J1141−6545 (Ord *et al.* 2002a). The change in relative speed of the pulsar due to its orbital motion stretches and condenses the scintillation timescales. (b) Model fit to the observed scintillation speeds as a function of true anomaly. Figure provided by Steve Ord.

A stunning application of this technique, as applied recently to the relativistic binary J1141−6545 by Ord *et al.* (2002a), is shown in Figure 7.10. In this case, the more edge-on solution ($i = 76 \pm 3°$) is favoured, since the implied pulsar mass (1.29 M$_\odot$) is more in line with neutron star mass measurements (Thorsett & Chakrabarty 1999). The implied transverse component of the binary system's space velocity $V_{\mathrm{T}} = \sqrt{V_{\perp}^2 + V_{\parallel}^2} \sim 115$ km s^{-1} supports the notion that the neutron star received a 'kick' during its formation in a supernova. Subsequent timing observations (Bailes *et al.* 2003) provide an independent confirmation of this more edge-on solution. Similarly, Ransom *et al.* (2004) have applied this technique to PSR J0737–3039A.

7.4.4.3 Secondary spectrum

In some cases, as for the data in Figure 7.9, a 'criss-cross' pattern of fringes can be seen in the dynamic spectrum. In order to investigate these structures in more detail, following Cordes and Wolszczan (1986), we can form the *secondary spectrum* by taking the two-dimensional Fourier transform of the dynamic spectrum. When viewed as a grey-scale image in the conjugate frequency ($\mathcal{F} = 1/f$) and conjugate time ($\mathcal{T} = 1/t$) plane, one often sees beautiful parabolic arc-like features as shown in Figure 7.9.

As pointed out by Stinebring *et al.* (2001), the curvature of these *scintillation arcs*, in the simple thin-screen model for scintillation introduced in Chapter 4, is a function of the location of the screen between the observer and the pulsar. When the scintillation pattern is dominated by V_{ISS}, Stinebring *et al.* (2001) find that

$$\mathcal{F} = \frac{\lambda^2 d}{2cV_{\mathrm{ISS}}^2} \left(\frac{s}{1-s} \right) \mathcal{T}^2, \tag{7.45}$$

where s is the distance to the screen expressed as a fraction of d, λ is the observing wavelength and c is the speed of light. For pulsars for which d and V are known or constrained from proper motion and parallax measurements, a measurement of arc curvature can be used to infer the effective distance to the scattering screen.

7.4.5 Neutral hydrogen measurements of pulsar distances

As mentioned in Chapter 1, one way of constraining the distances to pulsars is by measuring absorption and emission features caused by the neutral hydrogen (HI) distribution along the line of sight. This technique was pioneered by Guelin *et al.* (1969) and is shown schematically in Figure 7.11. The aim of the observation is to detect *absorption by a region of HI in front of the pulsar* and *emission from HI behind the pulsar with no corresponding absorption*. As with all spectral-line astronomy, the detected frequencies are Doppler-shifted from their rest-frame values by differential motion between the source and the observer. The relative velocity therefore can be inferred directly from the observed frequency of the line. The frequency shift, for these observations, is dominated by differential Galactic rotation. Hence, using a model Galactic rotation curve, the inferred velocities of the absorption and emission features can be converted to distances along the line of sight that correspond to lower and upper bounds on the pulsar distance.

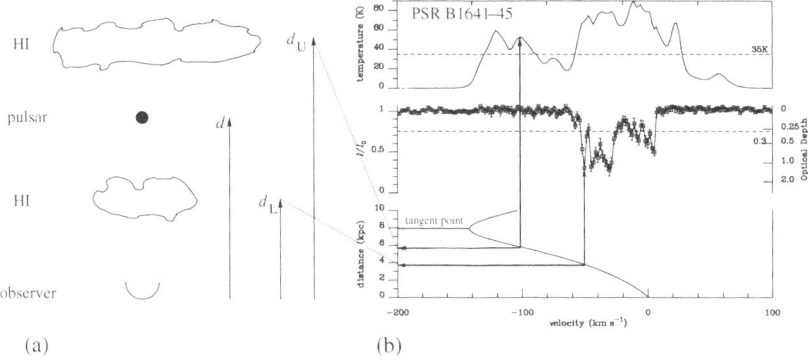

Fig. 7.11. (a) Schematic showing a pulsar bounded by regions of HI. (b) HI emission (top) and absorption (middle) spectra towards the pulsar B1641–45 (Ord *et al.* 2002b). The lower panel shows the expected velocity assuming the Galactic rotation curve of Fich, Blitz and Stark (1989). Figure adapted from the original spectrum provided by Steve Ord.

From Figure 7.11, we note that the sizes of typical spectral features are a few km s^{-1} while the range of velocities is 250 km s^{-1}. For the HI rest frequency (1420.406 MHz), we note that 1 km s$^{-1} \equiv 4.74$ kHz. Hence, this velocity range corresponds to a total bandwidth of just under 1.2 MHz which needs to be covered with good spectral resolution (10 kHz or better). This type of observation therefore is ideal for a narrow-band correlator or software filterbank (see Chapter 5).

The very nature of pulsars makes them particularly well suited to spectral line observations, since their on- and off-pulse regions can be used to derive on- and off-source spectra. In practice, the data are folded modulo the predicted pulse period and spectra are accumulated into a number of phase bins. The width of these bins is chosen usually so that the on-pulse emission occupies several phase bins. Each bin is weighted according to its S/N ratio (see, for example, Koribalski *et al.* (1995)).

The resulting spectra in the off-pulse bins normally are averaged to form a single HI emission spectrum as shown in the upper panel of Figure 7.11. The integral under this curve gives the total HI column density along the line of sight, $N_{\rm HI}$. In the usual optically thin approximation (see, for example, Rohlfs and Wilson (2000)):

$$N_{\mathrm{HI}} = 1.8 \times 10^{18} \int_{-\infty}^{+\infty} T_{\mathrm{B}}(v) \, dv \ \mathrm{cm}^{-2}, \tag{7.46}$$

where v (km s^{-1}) is the relative velocity and the brightness temperature T_{B} (K) can be related directly to the spin temperature of the HI emission. Conversion of the initially uncalibrated ordinate axis to a brightness temperature usually is achieved by normalising the area under the spectrum to N_{HI} determined from dedicated Galactic HI surveys for the particular line of sight (see, for example, Weaver and Williams (1973) and Kerr *et al.* (1986)). The error in T_{B} using this approach is about 5 K (Koribalski *et al.* 1995). The middle panel of Figure 7.11 is the difference between the on- and off-pulse spectra normalised by the pulsar's intensity I_0. This represents the HI absorbed by clouds in between the observer and the pulsar. An absorbed feature I relative to I_0 also can be expressed in terms of the optical depth $\tau = -\ln(I/I_0)$.

Some care should be taken when interpreting distance bounds from emission and absorption spectra. Frail and Weisberg (1990) recommend setting the lower distance limit from the centre of the farthest absorption feature with $\tau > 0.3$ and the upper distance limit from the first emission peak with $T_{\mathrm{B}} > 35K$. As shown by Weisberg *et al.* (1979), the reason for this choice is that HI emission features with $T_{\mathrm{B}} > 35$ K rarely result in optical depths less than 0.3 at the corresponding point in the absorption spectrum. Adopting this procedure for the example spectra for PSR B1641–45 shown in Figure 7.11 and applying the rotation curve of Fich, Blitz and Stark (1989), we find $d = 4 \pm 1$ kpc.

While we have considered an almost 'text-book' example for the HI technique, several caveats are worth noting for less clear-cut cases: (a) both emission and absorption features are not always detectable, so that only upper or lower bounds may be possible (for examples, see Koribalski *et al.* (1995)); (b) Galactic rotation curves do not account for random streaming motions of the HI – this results in an uncertainty of about 7 km s^{-1}, which should be taken into account when using the rotation curves to convert to distance bounds; (c) care should be taken when combining spectra from two polarisations to make sure that the spectra are appropriately weighted by their S/N ratio. These issues are discussed further by Frail and Weisberg (1990).

7.5 Interference excision

As for the search observations discussed in Chapter 6, radio-frequency interference (RFI) is an ever-increasing problem for routine pulsar observations. However, with some care, the effects of RFI on the data can be minimised to a large extent. One example of this is shown in Figure 7.12, a 29 min observation of the relativistic binary B1534+12 by Stairs *et al.* (2000a) which shows numerous bursts of RFI. Since the pulse is generally much weaker than the interference, it is clearly affected by the bursts that modulate the baseline of the integrated pulse profile.

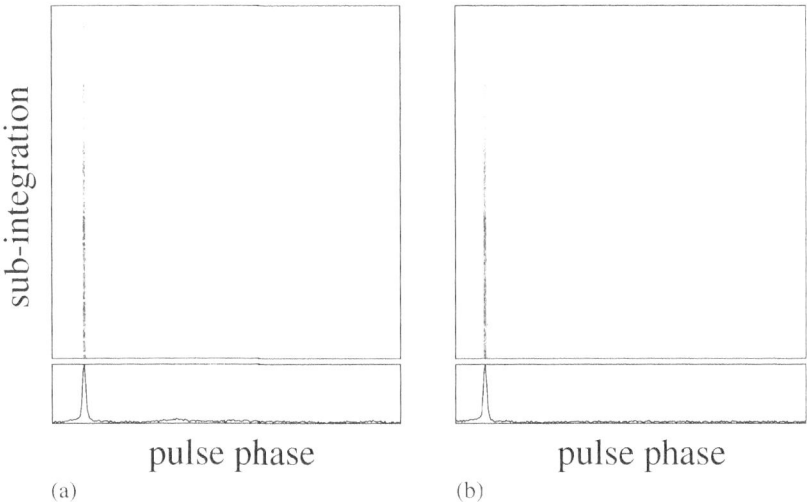

(a)   (b)

Fig. 7.12. (a) Integrated pulse profile and 10 s sub-integrations for an observation of PSR B1534+12. The raw data are clearly affected by intermittent bursts of RFI at random pulse phases. (b) The same data after including a broad-band excision scheme (see text). The baseline of the integrated pulse profile is significantly flatter and the profile S/N ratio is increased. Figure provided by Ingrid Stairs.

This type of interference (strong, short-duration bursts) is relatively easy to excise by keeping track of the statistics of the de-dispersed time series. Useful quantities are the mean, median and standard deviation, σ of short segments of data. Bursts of RFI will show up as significant differences between the mean and median in noisy segments. In the example shown in Figure 7.12, Stairs *et al.* (2000a) employed a simple algorithm which excised short blocks of data if: (a) an individual sample exceeded the median by 30σ; (b) a number of bins exceeded the median by 15σ; (c) the median power of a block exceeded the running median of previous blocks by 10 per cent. The result of excising blocks of data in this way are shown on the right-hand side of Figure 7.12 and show a clear improvement in data quality.

The above example was a broad-band excision scheme. It is often well worth checking individual frequency channels in a similar way to excise narrow-band features. In some cases, rejecting only a small fraction of the total band can result in dramatic improvements in data quality. Since no single RFI excision algorithm is guaranteed to work, experimentation is required to optimise the rejection scheme.

7.6 Further reading

We have attempted in this chapter to review the most commonly used techniques encountered in single-dish observations of pulsars. We have placed, where possible, the emphasis on areas not commonly covered in the literature, e.g. pulse profile statistics, confirmation observations and period optimisation issues. On more conventional areas, some overlap can be found in the reviews by Hankins and Rickett (1975), Bhattacharya (1998) and Cordes (2002). Calibration of pulsar polarimetry is discussed in detail by a number of authors, including Stinebring *et al.* (1984), Xilouris (1991), Britton (2000) and Johnston (2002). Our treatment closely followed the recent work by Heiles *et al.* (2001). An excellent introduction to polarimetry in general can be found in Heiles (2002).

Further discussion on single-pulse techniques can be found in Backer (1973), Ritchings (1976), Deshpande and Rankin (1999; 2001) and Edwards and Stappers (2002). The careful study of Everett and Weisberg (2001) provides a good starting-point for further information about polarimetry analyses. Likewise, the papers by Rand and Lyne (1994) and Mitra *et al.* (2003) should be consulted for further discussions of rotation measure determinations. Much of the techniques concerning scintillation speed measurements were pioneered by Lyne and Smith (1982). A more detailed reference can be found in Cordes (1986), while Stinebring *et al.* (2001) discuss scintillation arcs in more detail. An excellent review of neutron hydrogen measurements can be found in Frail and Weisberg (1990). Further discussion is also given in Koribalski *et al.* (1995).

7.7 Available resources

A number of useful resources are available via the book web site (see Appendix 3) for the analysis of pulsar observations. Most data analysis routines rely on programs to fold and de-disperse the often significant amounts of data. A number of such programs are freely available as part of the SIGPROC, PRESTO and PSRCHIVE packages (Lorimer (2001), Ransom (2001) and Hotan *et al.* (2004), respectively) are now capable of reading raw data in a variety of formats. All these packages require the TEMPO timing program (see Chapter 8) to produce sets of polynomial coefficients (known as *polyco* files) in their period calculations. In order to enhance the functionality of TEMPO, the web site contains a number of useful scripts to generate, read and interpret polyco files. This includes a particularly useful utility to convert topocentric pulse periods and epochs to the equivalent values at the SSB. Software to fit barycentric periods and epochs to a binary model ephemeris are also available, as are packages to perform some of the basic single-pulse and polarisation analyses discussed in Section 7.4. These are part of the European Pulsar Network (EPN) software package designed to work on any pulse profiles in EPN format (Lorimer *et al.* 1998).

8
Pulsar timing

As we saw in Chapter 2, the clock-like stability of pulsars means that through precise monitoring of pulsar rotations we can study a rich variety of phenomena that affect the propagation of their pulses. While the basic spin and astrometric parameters can be derived for essentially all pulsars, millisecond pulsars are the most useful objects for more exotic applications. Their pulse arrival times can be measured much more precisely than for normal pulsars (see Section 2.1) and their rotation is much smoother, making them intrinsically better clocks. Specifically, they usually do not exhibit rotational instabilities such as 'timing noise' (see Section 8.4) and 'glitches' (see Section 2.6) known for normal pulsars. In this Chapter, we discuss the main techniques and methodology used to extract the maximum of information from timing observations.

As we shall see, although conceptually fairly straightforward, the fine details required to carry out pulsar timing analyses successfully (i.e. time transfer considerations, implementations of various models and fitting procedures) are rather involved. Fortunately, many, and in some cases all, of the details described below have been incorporated into three major packages: TEMPO (Australia Telescope National Facility/Princeton University), PSRTIME (Jodrell Bank Observatory) and TIMAPR (Pushchino Observatory/Max-Planck Institut für Radioastronomie).

8.1 Measuring pulse arrival times

Figure 8.1 summarises the basic observational setup required for pulsar timing. As we have seen throughout the other chapters, pulsars are very weak radio sources. In order to obtain a significant detection, the incoming pulses collected by the telescope are amplified by high-sensitivity receivers before being first de-dispersed (see Chapter 5) and then folded (see Chapter 7) to form a mean pulse profile.

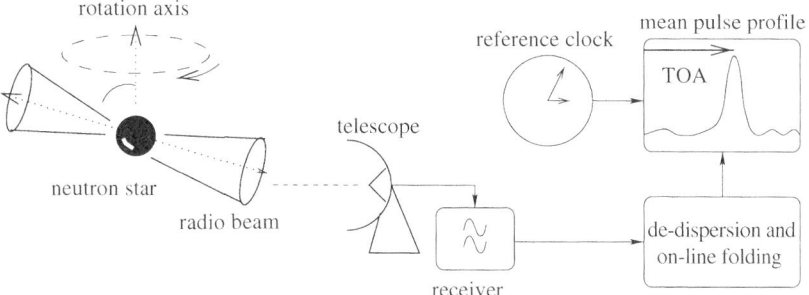

Fig. 8.1. Diagram showing the basic concept of a pulsar timing observation.

The key quantity of interest is the *time of arrival* (TOA) of pulses at the telescope. The TOA is defined usually as the arrival time of the nearest pulse to the mid-point of the observation. As the pulses have a certain width, the TOA refers to some *fiducial point* on the profile. Ideally, this point coincides with the plane defined by the rotation and magnetic axes of the pulsar and the line of sight to the observer.

8.1.1 Template matching

We saw in Section 1.1.2 that the mean profile has a stable form at any given observing frequency. This property means that TOAs can be determined accurately by a cross-correlation of the observed profile with a high signal to noise (S/N) 'template' profile obtained from the addition of many earlier observations at the particular observing frequency. Assuming that the discretely sampled profile, $\mathcal{P}(t)$, is a scaled and shifted version of the template, $\mathcal{T}(t)$, with added noise, $\mathcal{N}(t)$, we may write

$$\mathcal{P}(t) = a + b\mathcal{T}(t - \tau) + \mathcal{N}(t), \tag{8.1}$$

where a is an arbitrary offset and b a scaling factor. The time shift between the profile and the template, τ, yields the TOA relative to the fiducial point of the template and the start time of the observation. The cross-correlation procedure may be carried out in the time domain or, alternatively, in the frequency domain after Fourier-transforming both the template and the profile. While time-domain methods yield a precision of typically one-tenth of the sampling interval, frequency-domain methods using χ^2-minimisation techniques (see, for example, Taylor (1992)) are limited only by the S/N of the pulse profile.

In order to minimise the last term in Equation (8.1), one ideally constructs a noise-free template, for example, by representing the pulse profile as a sum of Gaussian components (Foster *et al.* 1991; Kramer *et al.* 1994). This method is perfectly suited for multi-frequency timing measurements in which one observes profiles that change their shape with frequency (see Chapter 1). In order to produce a template for an additional observing frequency, only the relative amplitudes and widths of the components are adjusted, while their locations are kept fixed. Since the chosen fiducial point therefore is the same at all frequencies, a comparison of the TOAs at different frequencies is possible in an unbiased way (Kramer *et al.* 1999b).

8.1.2 *Arrival-time precision*

Taking the uncertainty of a TOA measurement, σ_{TOA}, to be the ratio of the pulse with, W, to the profile S/N, and using the radiometer equation (see Appendix 1), we expect σ_{TOA} to scale as follows:

$$\sigma_{\text{TOA}} \simeq \frac{W}{\text{S/N}} \propto \frac{S_{\text{sys}}}{\sqrt{t_{\text{obs}}\Delta f}} \times \frac{P\delta^{3/2}}{S_{\text{mean}}}. \tag{8.2}$$

Here, S_{sys} is the system equivalent flux density, Δf is the available observing bandwidth, t_{obs} is the integration time, P is the pulse period, $\delta = W/P$ is the pulse duty cycle and S_{mean} is the mean flux density of the pulsar. Optimal results therefore are obtained with sensitive wide-band systems (low S_{sys} and large Δf) for bright (high S_{mean}) short-period pulsars with narrow pulses (i.e. small duty cycles).

The decrease of TOA uncertainty with S/N is only one reason why *integrated pulse profiles* are used for timing purposes. The arrival time of *indiviual pulses* jitters within the pulse window, so that the timing of those would result in uncertainties scaling roughly with the pulse width, i.e. $\sigma_{\text{TOA}} \propto W$. As each pulsar exhibits a certain profile stabilisation timescale (see Section 1.1.2), a sufficient number of pulses, typically at least a few hundred, need to be added to achieve a stable profile that can be matched to the template. The similarity of the observed pulse profile and the used fixed template and, hence, the timing precision, increases with the number of pulses added, i.e. $\sigma_{\text{TOA}} \propto \sqrt{1/N_{\text{pulses}}}$.

While shorter stabilization timescales result in better timing performance for a given observing time, effects such as mode changing or profile variations due to precession can lead to deviations of the observed pulse profile from the template. In such cases, adaptive template techniques should be employed (Kramer *et al.* 1999c; Stairs *et al.* 2000c). Profile changes that are less obvious, and for which correction is more difficult, can occur due to the effects of the turbulent interstellar medium (see Chapter 4) where a varying pulse scattering and scintillation can broaden the pulse differently for different epochs ('interstellar weather').

For millisecond pulsars, a few thousand pulses can be added easily in a few minutes of observing time. This usually results in extremely stable profiles. In addition to their higher rotational stability, this represents an important factor in explaining the superior timing stability of millisecond pulsars when compared to normal pulsars. Following Backer (1990), we define[1] the timing 'quality factor'

$$\mathcal{Q}_{\mathrm{T}} = S_{\mathrm{peak}} \sqrt{\frac{\int_0^P |\mathcal{T}'(t)|^2 \, \mathrm{d}t}{P}} \tag{8.3}$$

as a measure of the timing potential. Here $\mathcal{T}'(t)$ is the derivative of the normalised template, P is the pulse period and S_{peak} is the peak flux density. A larger value of $\mathcal{Q}_{\mathrm{T}}$ indicates a higher timing potential. Values for selected millisecond pulsars are presented in Table 8.1.

Table 8.1. *Examples for the timing quality factor $\mathcal{Q}_{\mathrm{T}}$ in units of Jy s^{-1} for a selection of millisecond pulsars. A higher value of this factor indicates a better timing potential but does not reflect limiting effects like timing noise (see text). The exact value will depend to some extent on the resolution of the observed profile and, hence, on the capabilities of the data acquisition system (see Chapter 5).*

PSR	J0437–4715	B1937+21	J1713+0747	B1855+09	B1620–26
$\mathcal{Q}_{\mathrm{T}}$	2100	1600	340	170	16

Very often, the mean profiles of the Stokes parameters Q, U and V (see Section 5.3.4) contain sharper features than Stokes' I, especially when the pulsar shows transitions between orthogonal modes (see Section 1.1.5). Although the polarised intensity usually is lower than the total intensity, the sharpness of polarised features with the combined

---

1 We invert Backer's definition to achieve larger values for better performing pulsars.

usage of all Stokes parameters can result in greater timing precision than obtained when using Stokes' *I* (Kramer *et al.* 1999a). Van Straten (2003) presented a detailed treatment and implementation of this procedure. In all cases, whether TOAs are determined from full Stokes information or just total intensity, a proper gain and polarisation calibration (see Section 7.3.4) is required to avoid systematic errors in the TOAs.

8.1.3 Time standards

The time tag of the profile, and hence the TOA is determined from the local time at the observatory, which is maintained typically by hydrogen maser clocks. The observatory time itself is compared to Coordinated Universal Time (UTC), e.g. by using the Global Positioning System (GPS) which provides UTC as measured by the *US National Institute of Standards and Technology* (NIST) and is often referred to as UTC(NIST). UTC is defined to be an integral numbers of seconds from International Atomic Time (TAI) and is never more than 0.9 s away from UT1, a non-uniform timescale based on the rotation of the Earth. Since the Earth is not rotating uniformly, leap seconds are occasionally inserted into UTC, so that the difference between TAI and UTC increases according to TAI = UTC $+\Delta T$, where ΔT is the total algebraic sum of leap seconds. The decision to introduce a leap second in UTC is the responsibility of the *International Earth Rotation Service* (IERS)[2] which monitors the rotation of the Earth. All leap seconds need to be removed from the UTC time standard to obtain a TOA measured in a smoothly running timescale of TAI.

The TAI is maintained as an average of a large number of selected atomic clocks by the *Bureau International des Poids et Mesures* (BIPM), which also publishes a retroactive uniform atomic time standard known as Terrestrial Time (TT) formerly known as Terrestrial Dynamical Time (TDT). The unit of TT is the SI second and may be regarded as the time that would be kept by an ideal atomic clock on the geoid[3] with TT = TAI + 32.184 s, where the offset of about 32 s stems from historic reasons. Intended to represent the idealised geocentric time (Seidelmann *et al.* 1992), this timescale should be used in the final analysis by correcting the initially measured TOAs to TT(BIPM).

We describe in Section 8.2.2 how the arrival time measured at the

2 http://hpiers.obspm.fr
3 The equipotential surface that coincides with the mean sea-level of the Earth.

observatory in TT is transformed to the Solar System barycentre (SSB). The result will be an arrival time in Barycentric Dynamical Time (TDB). Both TT and TDB are consistent with the general theory of relativity (GR).

8.2 Solitary pulsars

In order to study effects on the pulse travel time, first we have to find an expression that describes the pulsar rotation in a reference frame co-moving with the pulsar. We start by expressing the spin frequency of the pulsar in a Taylor expansion:

$$\nu(t) = \nu_0 + \dot{\nu}_0(t - t_0) + \frac{1}{2}\ddot{\nu}_0(t - t_0)^2 + \cdots, \tag{8.4}$$

where $\nu_0 = \nu(t_0)$ is the spin frequency and $\dot{\nu}_0 = \dot{\nu}(t_0)$ and $\ddot{\nu}_0 = \ddot{\nu}(t_0)$ its time derivatives at some reference epoch t_0. The observed parameters ν and $\dot{\nu}$ are associated with the physical processes causing pulsars to spin down (see Section 3.2). Apart from very young pulsars, the value of $\ddot{\nu}$ is usually too small to be measured. However, the presence of timing noise can mimic a significant but typically time varying value of $\ddot{\nu}$ (see Section 8.4). In these cases, braking indices determined from Equation (3.9) do not reflect the intrinsic spin-down behaviour.

Since spin frequency ν is the rate of change of pulse number N, it follows from Equation (8.4) that

$$N = N_0 + \nu_0(t - t_0) + \frac{1}{2}\dot{\nu}(t - t_0)^2 + \frac{1}{6}\ddot{\nu}(t - t_0)^3 + \cdots \tag{8.5}$$

where N_0 is the pulse number at the reference epoch t_0. If t_0 coincides with the arrival of a pulse and the pulsar spin-down is known accurately, the pulses therefore should appear at integer values of N when observed in an inertial reference frame.

Our observing frame is not inertial: we are using telescopes that are located on a rotating Earth orbiting the Sun. Before analysing TOAs measured with the observatory clock ('topocentric arrival times'), we need to transfer them to the centre of mass of the Solar System, i.e. the SSB. To a very good approximation, the SSB is an inertial reference frame. We will encounter deviations from this assumption in Sections 8.2.4 and 8.3.4 in which we discuss how relative accelerations between the SSB and the pulsar affect the observed quantities.

The time transformation also corrects for any relativistic time delay that occurs due to the presence of masses in the Solar System. An

additional advantage of analysing these 'barycentric arrival times' is that
they can easily be combined with other TOAs measured at different
observatories at different times. The transformation from a topocentric
TOA, t_{topo}, to a barycentric TOA, t_{SSB}, can be summarised as follows:

$$t_{\text{SSB}} = t_{\text{topo}} + t_{\text{corr}} - \Delta D/f^2 + \Delta_{\text{R}\odot} + \Delta_{\text{S}\odot} + \Delta_{\text{E}\odot}. \qquad (8.6)$$

Each term is discussed in detail below. Additional terms will be added
when we discuss binary pulsars in Section 8.3.

8.2.1 *Clock and frequency corrections*

The term t_{corr} in Equation (8.6) summarises the various clock corrections
to the topocentric TOAs that have been discussed in Section 8.1.

As described in detail in Chapter 4, the pulses are delayed due to
dispersion in the interstellar medium, so that the arrival time depends
on the observing frequency, f. The TOA therefore is corrected for a pulse
arrival at an infinitely high frequency, thereby removing dispersion from
the data. The $\Delta D/f^2$ term in Equation (8.6) accounts for this correction
in terms of the dispersion measure (DM) and dispersion constant $\mathcal{D}$
both defined in Chapter 4 (see Equations (4.5) and (4.6)). From these
definitions, when the DM is expressed in the usual units of cm^{-3} pc and
f is in MHz, it follows that $\Delta D = \mathcal{D} \times \text{DM}$. For the exact value of the
dispersion constant, $\mathcal{D}$, see Equation (4.6).

If the dispersion measure is changing with time, as observed for a number of pulsars (Hobbs *et al.* 2004, and see Section 4.2), a time-varying,
frequency-dependent drift is introduced into the TOAs. The above term
needs to be modified to include time derivatives of DM, i.e. $\dot{\text{DM}}$, $\ddot{\text{DM}}$
and so on. These can be determined if monitoring observations at two or
more frequencies are available and provide an estimate for the change in
electron density along the line of sight as a result of 'interstellar weather'.
For high-precision timing of millisecond pulsars (e.g. for PSR B1937+21,
and see Lommen and Backer (2001)), multi-frequency observations are
necessary to account for small time delays due to DM changes.

8.2.2 *Barycentric corrections*

The last three terms in Equation (8.6) describe the corrections necessary
to transfer topocentric to barycentric TOAs.

The *Römer delay*, $\Delta_{\text{R}\odot}$ is the classical light-travel time between the phase centre of the telescope and the SSB. Given a unit vector, $\hat{s}$, pointing from the SSB to the position of the pulsar and the vector connecting the SSB to the observatory, $\vec{r}$, we find

$$\Delta_{\mathrm{R}\odot} = -\frac{1}{c}\,\vec{r}\cdot\hat{s} = -\frac{1}{c}\,(\vec{r}_{\mathrm{SSB}} + \vec{r}_{\mathrm{EO}})\cdot\hat{s}. \tag{8.7}$$

Here, c is the speed of light and we have split $\vec{r}$ into two parts. The vector, $\vec{r}_{\mathrm{SSB}}$, points from the SSB to centre of the Earth (geocentre). Computation of this vector requires accurate knowledge of the locations of all major bodies in the Solar System and uses a *Solar System ephemeris* such as the 'DE200' or 'DE405' published by *Jet Propulsion Laboratory* (JPL). It is currently standard practice to use the DE200 ephemeris (Standish 1990) for pulsar timing analyses. The second vector, $\vec{r}_{\mathrm{EO}}$, connects the geocentre with the phase centre of the telescope. In order to compute this vector accurately, the non-uniform rotation of the Earth has to be taken into account, so that the correct relative position of the observatory is derived. This is achieved using the appropriate UT1 corrections published by IERS (see Section 8.1).

The *Shapiro delay*, $\Delta_{\mathrm{S}\odot}$, is a relativistic correction that corrects for extra delays due to the curvature of space-time caused by the presence of masses in the solar system (Shapiro 1964). The delays are largest for a signal passing the limb of the Sun ($\sim 120~\mu s$) while Jupiter can contribute as much as 200 ns. In principle, one has to sum over all bodies in the Solar System, yielding

$$\Delta_{\mathrm{S}\odot} = -2\sum_i \frac{GM_i}{c^3}\ln\left[\frac{\hat{s}\cdot\vec{r}_i^{\,\mathrm{E}} + r_i^{\mathrm{E}}}{\hat{s}\cdot\vec{r}_i^{\,\mathrm{P}} + r_i^{\mathrm{P}}}\right]. \tag{8.8}$$

where G is Newton's gravitational constant, M_i is the mass of body i, $\vec{r}_i^{\,\mathrm{P}}$ is the pulsar position relative to it and $\vec{r}_i^{\,\mathrm{E}}$ is the telescope position relative to that body at the time of closest approach of the photon. In practice, usually only the Sun, and in some cases Jupiter, need to be accounted for in this calculation (Backer & Hellings 1986).

The last term in Equation (8.6), $\Delta_{\mathrm{E}\odot}$, is called the *Einstein delay* and describes the combined effect of time dilation due to the motion of the Earth and gravitational redshift caused by the other bodies in the Solar System. This time varying effect takes into account the variation of an atomic clock on Earth in the changing gravitational potential as the Earth follows its elliptical orbit around the Sun. The delay amounts to an integral of the expression (Backer & Hellings 1986)

$$\frac{d\Delta_{E\odot}}{dt} = \sum_i \frac{GM_i}{c^2 r_i^E} + \frac{v_E^2}{2c^2} - \text{constant}, \qquad (8.9)$$

where the sum is again over all bodies in the Solar System but this time excluding the Earth. The distance r_i^E is the distance between the Earth and body i, while v_E is the velocity of the Earth relative to the Sun.

8.2.3 Positions

The pulsar position determined from timing observations is derived from the annual variation of the pulse arrival introduced by a variation of the Römer delay with a maximum amplitude

$$\Delta_{R\odot}^{max} = \frac{1AU}{c} \cos\beta \approx 500\,\text{s} \cos\beta, \qquad (8.10)$$

where β is the ecliptic latitude. The precision in the position achievable from timing observations therefore depends strongly on the pulsar's ecliptic coordinates. The worst precision is obtained for pulsars near the ecliptic plane where the position can be determined accurately in ecliptic longitude but only rather poorly in ecliptic latitude. When high positional accuracy for pulsars near the ecliptic is required, one should attempt to obtain an interferometric measurement of position (see Chapter 9).

8.2.4 Relative motion

Equation (8.6) is sufficient to measure the clock rate as produced by the pulsar if no further motion occurs between the pulsar and the SSB. If the pulsar is moving relative to the SSB, the transverse component of the velocity, V_T, gradually will change the vector $\hat{s}$ in Equation (8.7), adding a linear time-dependent term to Equation (8.6). This can be measured as a proper motion of the pulsar in the chosen reference frame. In the equatorial coordinate system, one measures a motion in declination, δ, as $\mu_\delta \equiv \dot{\delta}$ and right ascension, α, as $\mu_\alpha \equiv \dot{\alpha}\cos\delta$. The total proper motion is then given by $\mu_T = \sqrt{\mu_\alpha^2 + \mu_\delta^2}$, which is usually quoted in mas yr^{-1}. For a pulsar at a distance d, the corresponding transverse velocity

$$V_T = 4.74\,\text{km s}^{-1} \left(\frac{\mu_T}{\text{mas yr}^{-1}}\right)\left(\frac{d}{\text{kpc}}\right). \qquad (8.11)$$

A transverse velocity usually is easy to recognise when analysing timing data (see Section 8.2.6).

In principle, the radial component of velocity V_R could also be determined through pulsar timing. This is possible because the three-dimensional spatial motion of the pulsar with respect to the SSB, when projected onto the celestial sphere, gives a motion on the sky that is non-linear in time. The non-linear terms allow the determination of V_R. While this effect is too small to be observable for most pulsars the non-linear terms will eventually become significant for fast, nearby objects. According to Wex (1997a), for an optimistic case of a pulsar at $d = 100$ pc with $V_T = V_R = 1000$ km s^{-1}, the non-linear terms will have an amplitude of about 20–50 μs after about 30 yr of observations.

Unfortunately, such amplitudes are measurable only with millisecond pulsars that have much lower velocities (~ 100 km s^{-1}); (see Section 1.3.1). For the currently most precisely timed pulsar – J0437−4715 – this effect could be measurable in 30 yr when the amplitude has increased to 100 ns (van Straten 2003). It will require the timing precision of the Square Kilometre Array (SKA) (see Chapter 2) for timing to provide access to the full three-dimensional velocity vector of a pulsar. A much easier and already accessible method is to have an optically detectable companion to the pulsar, e.g. a white dwarf for which Doppler shifts can be measured from optical spectra (see, for example, van Kerkwijk (1996)).

The consideration made for the positional accuracy is valid equally for the measurements of proper motions. For proper motion measurements of pulsars close to the ecliptic, $\beta \lesssim 5°$, it is advisable to perform the timing analysis in ecliptic coordinates. This minimises covariances between the determined parameters, and allows the determination of at least the proper motion in ecliptic longitude to high precision. In contrast, for fits in equatorial coordinates, the precision of both right ascension and declination are usually degraded for pulsars near the ecliptic. Proper motions measured using this method are currently available for about 300 pulsars (Hobbs *et al.* 2004), with accuracies of a few mas yr^{-1} for normal pulsars and typically < 1 mas yr^{-1} precision for millisecond pulsars.

Another effect arising from a transverse motion is the *Shklovskii effect* (Shklovskii 1970), also known in classical astronomy as 'secular acceleration'. With the pulsar motion, the projected distance of the pulsar to the SSB is increasing, leading to a quadratic correction

$$\Delta t_S = \frac{V_T^2}{2dc} t^2 \qquad (8.12)$$

(Backer & Hellings 1986). Since this delay is inversely proportional to d, the correction usually is too small to be considered for timing. However, it has also the effect that any observed change in a periodicity (i.e. change in pulse or orbital period) is increased over the intrinsic value by

$$\frac{\dot{P}}{P} = \frac{1}{c} \frac{V_T^2}{d} = 2.43 \times 10^{-21} \mathrm{s}^{-1} \left(\frac{d}{\mathrm{kpc}} \right) \left(\frac{\mu_T}{\mathrm{mas\ yr}^{-1}} \right)^2 . \qquad (8.13)$$

For nearby millisecond pulsars for which d and $\dot{P}$ are small, a significant fraction of the observed change in period can be due to the Shklovskii effect. This effect also needs to be considered, when studying the decay of an orbital period due to gravitational wave emission, in which the observed value is increased by the Shklovskii term.

Similarly, any line of sight acceleration a_l of the pulsar due to an external gravitational field produces an apparent period derivative $\dot{P} = a_l P / c$. This effect is observed commonly for pulsars in globular clusters for which the acceleration through the cluster's gravitational field towards our line of sight, a_l often can be large enough to reverse the sign of $\dot{P}$. As a result, the pulsars appear to be spinning up rather than down! Such pulsars place useful constraints on the cluster mass distributions (Phinney 1992) and the intracluster medium (see Chapter 2).

8.2.5 Parallax

Timing residuals of nearby pulsars may demonstrate an annual parallax

$$\Delta t_\pi = -\frac{1}{2cd} (\vec{r} \times \hat{s})^2 = \frac{1}{2cd} \left((\vec{r} \cdot \hat{s})^2 - |\vec{r}|^2 \right). \qquad (8.14)$$

(Backer & Hellings 1986). In comparison to the more familiar positional parallax known from optical astronomy (see also Section 9.2), this *timing parallax* is, in effect, a measurement of the curvature of the emitted wavefronts at different positions of the orbit of the Earth. This effect imposes a variation in the pulse arrival time with an amplitude of $l^2 \cos \beta / (2cd)$ where l is the Earth–Sun distance and β is the ecliptic latitude of the pulsar. This has the interesting consequence that parallax is only measurable for pulsars near the ecliptic plane that is opposite to the parallax measurements known for optical stars. For a pulsar at $d = 1$ kpc, this delay amounts to only $\lesssim 1.2 \ \mu$s, and, hence, is only measurable for a few millisecond pulsars for which it provides a precise distance estimate.

8.2.6 The timing procedure

Equation (8.6) contains a number of parameters that are not known a priori (or only with limited precision after the discovery of a pulsar) and need to be determined precisely in a least squares fit analysis of the measured TOAs. These parameters can be categorised into three groups: (a) *astrometric parameters* (i.e. position, proper motion, parallax contained in the Römer and Shapiro delay, respectively); (b) *spin parameters* (i.e. rotation frequency, ν, and higher derivatives, see Equation (8.4)); (c) *binary parameters* (see Section 8.3).

Given a minimal set of starting parameters, a least squares fit is needed to match the measured arrival times to pulse numbers according to Equation (8.5). We minimise the expression

$$\chi^2 = \sum_i \left(\frac{N(t_i) - n_i}{\sigma_i} \right)^2 \qquad (8.15)$$

where n_i is the nearest integer to $N(t_i)$ and σ_i is the TOA uncertainty in units of pulse period (turns).

The aim is to obtain a phase-coherent solution that accounts for every single rotation of the pulsar between two observations. One starts off with a small set of TOAs that were obtained sufficiently close in time so that the accumulated uncertainties in the starting parameters do not exceed one pulse period. Gradually, the data set is expanded, maintaining coherence in phase. When successful, post-fit residuals expressed in pulse phase show a Gaussian distribution around zero with a root mean square that is comparable to the TOA uncertainties (see Fig 8.2). A good test for the quality of the TOAs and their fit is provided by creating a new set of mean residuals, each formed by averaging $n_{\rm avg}$ consecutive post-fit residuals. The root mean square calculated from the new set should decrease with $\sqrt{n_{\rm avg}}$ if no systematics are present.

After starting with fits for only period and pulse reference phase over some hours and days, longer time spans slowly require fits for parameters like spin frequency derivative(s) and position. Incorrect or incomplete timing models cause systematic structures in the post-fit residuals identifying the parameter that needs to be included or adjusted (see Figure 8.2). The precision of the parameters improves with length of

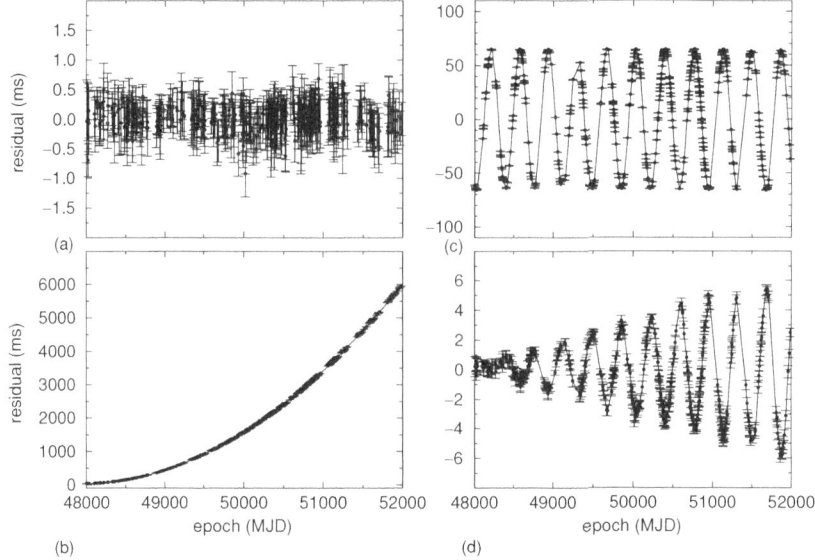

Fig. 8.2. (a) Timing residuals for the 1.19 s pulsar B1133+16. A fit of a perfect timing model should result in randomly distributed residuals. (b) A parabolic increase in the residuals is obtained if $\dot{P}$ is underestimated, here by 4 per cent. (c) An offset in position (in this case a declination error of 1 arcmin) produces sinusoidal residuals with a period of 1 yr. (d) The effect of neglecting the pulsar's proper motion, in this case $\mu_{T} = 380$ mas yr^{-1}. In all plots we have set the reference epoch for period and position to the first TOA at MJD 48000 to show the development of the amplitude of the various effects. Note the different scales on each of the vertical axes.

the data span and the frequency of observation, but also with orbital coverage in the case of binary pulsars.

Uncertainties for the fitted timing parameters usually are estimated from the covariance matrices that are computed by the applied fitting algorithms (see, for example, Press *et al.* (1992)). In cases in which parameters are highly correlated, a more reliable error estimate is obtained by mapping and studying the χ^2 hypersphere.

8.3 Binary pulsars

Observations of pulsars in binary orbits show a periodic variation in pulse arrival time. The timing model therefore needs to be extended to incorporate the additional motion of the pulsar as it orbits the common centre of mass of the binary system. Equation (8.6) is extended by four additional terms to

$$t_{\mathrm{SSB}} = \quad t_{\mathrm{topo}} + t_{\mathrm{corr}} - \Delta D / f^2 + \Delta_{\mathrm{R}\odot} + \Delta_{\mathrm{S}\odot} + \Delta_{\mathrm{E}\odot}$$
$$+ \quad \Delta_{\mathrm{RB}} + \Delta_{\mathrm{SB}} + \Delta_{\mathrm{EB}} + \Delta_{\mathrm{AB}}. \tag{8.16}$$

The new terms describe an additional Römer delay due to the binary orbit, a Shapiro and Einstein delay due to the gravitational field of the companion, and effects due to changing aberration caused by the orbital motion.

For non-relativistic binary systems, the orbit can be described using Kepler's laws, which we describe in detail in Section 8.3.1. For a number of binary systems however, the Keplerian description of the orbit is not sufficient and relativistic corrections need to be applied. Following the description by Damour and Deruelle (1985; 1986), (see also Damour and Taylor (1992)), we discuss these in details in Section 8.3.2.

8.3.1 Keplerian description

Kepler's laws can be used to describe a binary system in terms of six so-called 'Keplerian parameters', shown schematically in Figure 8.3. The five parameters required to refer the TOAs to the binary barycentre are: (a) orbital period, P_{b}; (b) projected semi-major orbital axis, $a_{\mathrm{p}} \sin i$ (see below); (c) orbital eccentricity, e; (d) longitude of periastron, ω; (e) the epoch of periastron passage, T_0. The sixth parameter, Ω_{asc}, describes the position angle of the ascending node and is only measurable in specialised cases (see Section 8.3.4.2 and, for example, van Straten *et al.* (2001)).

The initial fit for the orbital parameters usually is derived from inspecting the variation of pulse period with time (see Section 7.2.4). A subsequent phase-connected timing solution improves greatly the accuracy of the orbital parameters.

8.3.1.1 Kepler's equations

The Keplerian parameters are related to *eccentric anomaly*, E, and *true anomaly*, $A_{\mathrm{T}}(E)$ (see, for example, Roy (1988) and Damour and Taylor (1992)) as follows:

$$E - e \sin E \quad = \quad \Omega_{\mathrm{b}} \left[(t - T_0) - \frac{1}{2} \frac{\dot{P}_{\mathrm{b}}}{P_{\mathrm{b}}} (t - T_0)^2 \right], \tag{8.17}$$

$$A_{\mathrm{T}}(E) \quad = \quad 2 \arctan \left[\sqrt{\frac{1+e}{1-e}} \, \tan \frac{E}{2} \right], \tag{8.18}$$

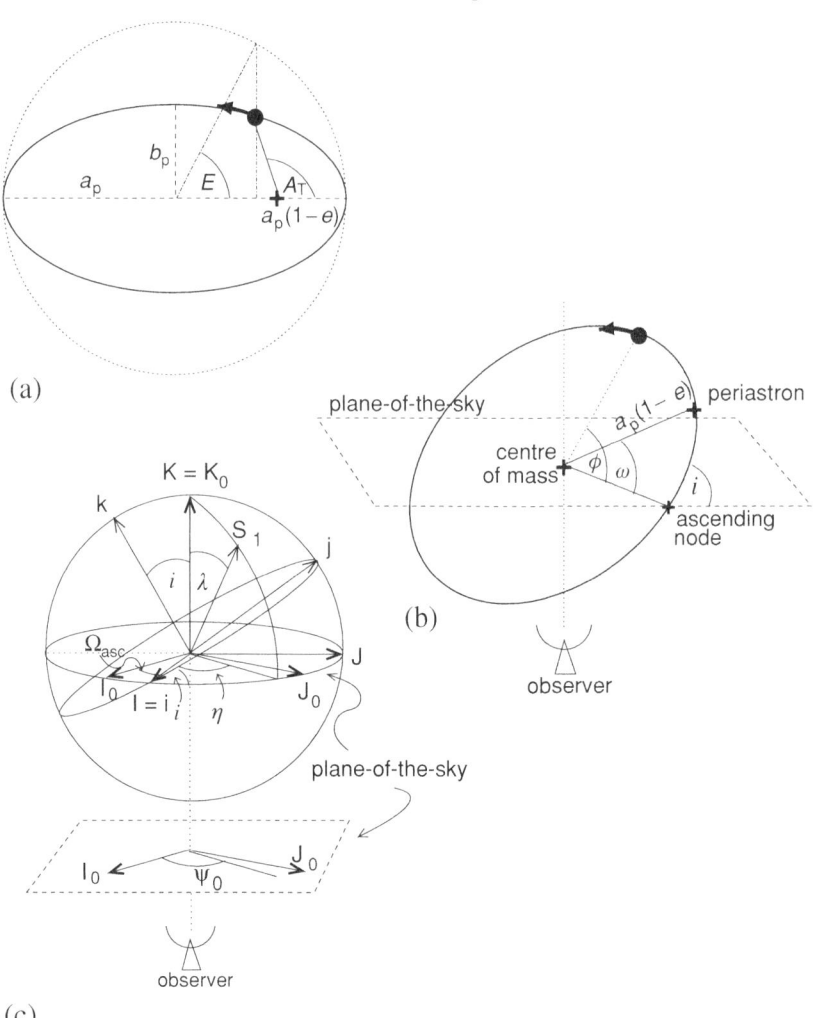

Fig. 8.3. Definition of the orbital elements in a Keplerian orbit and the angles relating both the orbit and the pulsar to the observer's coordinate system and line of sight. (a) The closest approach of the pulsar to the centre of mass of the binary system marks periastron, given by the longitude ω and a chosen epoch T_0 of its passage. The distance between centre of mass and periastron is given by $a_\mathrm{p}(1-e)$ where a_p is the semi-major axis of the orbital ellipse and e its eccentricity. (b) Usually, only the projection on the plane of the sky, $a_\mathrm{p}\sin i$, is measurable, where i is the orbital inclination defined as the angle between the orbital plane and the plane of the sky. The *true anomaly*, A_T, and *eccentric anomaly*, E, are related to the *mean anomaly* as described in the text. The orbital phase of the pulsar Φ is measured relative to the ascending node. (c) The spatial orientation of the pulsar's spin-vector, $\mathbf{S}_1$, is given by the angles λ and η in the coordinate system shown as defined by Damour and Taylor (1992). The angle Ω_asc gives the longitude of ascending node in the plane of the sky. Note that the polarisation angle Ψ_0 as defined by Damour and Taylor and shown here swings in the opposite sense to the normally used observer's convention used in Chapter 3.

$$\omega = \omega_0 + \frac{\dot{\omega}}{\Omega_{\rm b}} A_{\rm T}(E), \qquad (8.19)$$

where we defined the mean angular velocity $\Omega_{\rm b} \equiv 2\pi/P_{\rm b}$. The quantity $M \equiv (2\pi/P_{\rm b})(t - T_0) = \Omega_{\rm b}(t - T_0)$ is known as the *mean anomaly*.

For completeness, we have introduced two post-Keplerian parameters into the above definitions that will be discussed in the next section. These describe a change in longitude of periastron, $\dot{\omega}$, and orbital period, $\dot{P}_{\rm b}$. A secular change in these parameters may be caused by classical or relativistic effects. We discuss relativistic effects such as gravitational wave damping in Section 8.3.2, and discuss some of the classical effects in Section 8.3.4. Without such effects, both post-Keplerian parameters are expected to be zero for point masses, so that the above equations adopt their familiar forms. In particular, the relationship between mean and eccentric anomaly is given by

$$E - e \sin E = M. \qquad (8.20)$$

Software for solving this equation iteratively for E given a value of M (e.g. for use in scintillation velocity calculations of binary pulsars (see Section 7.4.4.2)) is available on the book web site (see Appendix 3).

Kepler's third law relates the size of the semi-major axis of the *relative* orbit, $a_{\rm R} = a_{\rm p} + a_{\rm c}$, to the mean angular velocity, $\Omega_{\rm b}$, and the total mass of the system, as follows

$$\Omega_{\rm b}^2 \left(\frac{a_{\rm R}}{c}\right)^3 = T_\odot(m_{\rm p} + m_{\rm c}), \qquad (8.21)$$

where $m_{\rm p}$ and $m_{\rm c}$ are the masses of the pulsar and orbiting companion, respectively. The constant $T_\odot = GM_\odot/c^3 = 4.925490947\mu$s is used to express these masses in solar units. The size of the pulsar orbit around the common centre of mass, $a_{\rm p}$, can be computed from $a_{\rm R}$ as follows

$$a_{\rm p} = a_{\rm R} \frac{m_{\rm c}}{m_{\rm p} + m_{\rm c}}. \qquad (8.22)$$

Similarly, the semi-major axis of the orbit of the companion

$$a_{\rm c} = a_{\rm R} \frac{m_{\rm p}}{m_{\rm p} + m_{\rm c}}. \qquad (8.23)$$

8.3.1.2 Line of sight motion

As discussed in Chapter 6, the Keplerian parameters of a newly discovered binary pulsar are not known initially. In order to obtain an estimate of their values, one studies the variation of the observed pulse period which changes due to the Doppler effect.

If the velocity along the line of sight

$$V_{\mathrm{l}}(A_{\mathrm{T}}) = \Omega_{\mathrm{b}} \frac{a_{\mathrm{p}} \sin i}{\sqrt{1-e^2}} \left[\cos(\omega + A_{\mathrm{T}}) + e \cos \omega\right] \tag{8.24}$$

is small compared to c, the observed Doppler-shifted pulse period can be written as

$$P(A_{\mathrm{T}}) \simeq P_0 \left(1 + \frac{V_{\mathrm{l}}(A_{\mathrm{T}})}{c}\right), \tag{8.25}$$

where P_0 is the pulse period in the pulsar's rest frame.

Differentiating Equation (8.24) gives the corresponding acceleration along the line of sight

$$a_{\mathrm{l}}(A_{\mathrm{T}}) = -\Omega_{\mathrm{b}}^2 \frac{a_{\mathrm{p}} \sin i}{1-e^2} \left(1 + e \cos A_{\mathrm{T}}\right)^2 \sin(\omega + A_{\mathrm{T}}). \tag{8.26}$$

Equations (8.25) and (8.26) can be used to visualise the orbit on the P–a_{l} plane which traces out a characteristic size and shape based on the Keplerian parameters. Search techniques discussed in Chapter 6 can be used to obtain a_{l}. Alternatively, a_{l} can be measured by the effective $\dot{P} = a_{\mathrm{l}} P/c$ through a timing analysis. For pulsars in which only a few measurements of P and a_{l} are available due to sparse observations/detections – as described in Section 7.2.4 – an analysis of the P–a_{l} plane can be used to constrain the orbit.

8.3.1.3 Römer delay

The Römer delay caused by the orbital motion of the pulsar (see Equation (8.16)) is used to derive the Keplerian parameters of the orbit. It is given by, for example, Blandford and Teukolsky (1976):

$$\Delta_{\mathrm{RB}} = x \left(\cos E - e\right) \sin \omega + x \sin E \sqrt{1-e^2} \cos \omega, \tag{8.27}$$

in which we have expressed the projected semi-major axis as $x \equiv a_{\mathrm{p}} \sin i/c$.

For small-eccentricity binary pulsars, the location of periastron is not well defined. Timing models for such pulsars based on Equation (8.27) lead to very high correlations between ω and T_0 and unacceptably large uncertainties in a χ^2 estimation of these parameters[4] (see Section 8.2.6).

---

4 As a result, ω and T_0 must be specified with many decimal places in a timing solution describing the system in order to maintain phase coherence, no matter how imprecise their determination really is.

This can be avoided by an alternative description of the Keplerian motion implemented in the 'ELL1' timing model (Wex 1999; Lange *et al.* 2001). To first order in e we may write

$$\Delta_{\mathrm{RB}} \simeq x \left(\sin \Phi + \frac{\epsilon_2}{2} \sin 2\Phi - \frac{\epsilon_1}{2} \cos 2\Phi \right), \qquad (8.28)$$

where Φ is measured from the ascending node and terms constant in time are omitted. In the above expression, the three Keplerian parameters T_0, e, and ω are replaced by the epoch of ascending node:

$$T_{\mathrm{asc}} = T_0 - \omega / \Omega_{\mathrm{b}}, \qquad (8.29)$$

and the first and second Laplace–Lagrange parameters

$$\epsilon_1 = e \, \sin \omega \qquad \text{and} \qquad \epsilon_2 = e \, \cos \omega. \qquad (8.30)$$

If necessary, the usual Keplerian parameters can be computed from

$$e = \sqrt{\epsilon_1^2 + \epsilon_2^2}, \qquad (8.31)$$

$$\omega = \arctan(\epsilon_1 / \epsilon_2), \qquad (8.32)$$

$$T_0 = T_{\mathrm{asc}} + \frac{P_{\mathrm{b}}}{2\pi} \arctan(\epsilon_1 / \epsilon_2). \qquad (8.33)$$

Note that the actual time at which the pulsar passes through the ascending node and time of conjunction are given by $T_{\mathrm{asc}} + 2\epsilon_1 / \Omega_{\mathrm{b}}$ and $T_{\mathrm{asc}} + P_{\mathrm{b}} / 4 - 2\epsilon_2 / \Omega_{\mathrm{b}}$, respectively. While this description breaks the covariances, it accounts only for first-order corrections in e. Therefore, the difference between the exact expression (see Equation (8.27)) and the approximation (see Equation (8.28)) can grow up to xe^2. For most of the low-eccentricity binary pulsars, the error in the TOA measurements is much larger than xe^2 and thus the linear-in-e model is sufficient.

8.3.1.4 Mass function

We will see in the next section how relativistic effects can be used to determine the masses of the pulsar and its orbiting companion. Constraints on the mass of the companion can be placed already by using the classical Keplerian parameters. We combine the projected semi-major axis and the orbital period to obtain the *mass function*

$$f(m_{\mathrm{p}}, m_{\mathrm{c}}) = \frac{(m_{\mathrm{c}} \sin i)^3}{(m_{\mathrm{p}} + m_{\mathrm{c}})^2} = \frac{4\pi^2}{G} \frac{(a_{\mathrm{p}} \sin i)^3}{P_{\mathrm{b}}^2} = \frac{4\pi^2}{T_\odot} \frac{x^3}{P_{\mathrm{b}}^2}, \qquad (8.34)$$

where we have again used the constant $T_\odot = (GM_\odot / c^3) = 4.925490947 \, \mu\mathrm{s}$ to express the masses and the mass function in solar units. Assuming

$m_p = 1.35$ M$_\odot$, for white dwarf companions with $m_c \sim 0.3M_\odot$ we expect $f \lesssim 0.01$ M$_\odot$. For a more massive white dwarf ($m_c \sim 1$ M$_\odot$), we expect $f \lesssim 0.2$ M$_\odot$, while for a double-neutron star system, $f \lesssim 0.34$ M$_\odot$.

For an orbit viewed edge-on ($i = 90°$) the companion mass can be computed by solving Equation (8.34) for an assumed typical pulsar mass $m_p = 1.35M_\odot$ (see Section 3.1). In practice, the orbital inclination i is unknown, and the value computed for $i = 90°$ provides a lower limit on m_c. To place an upper bound on m_c, we note that for a random distribution of orbital inclination angles, the probability of observing a binary system at an angle *less* than some value i_0 is $p(< i_0) = 1 - \cos(i_0)$. This implies that the chances of observing a binary system inclined at an angle $\lesssim 26°$ is only 10 per cent. Evaluating m_c for this inclination constrains the companion mass at the 90 per cent confidence level.

8.3.2 *Post-Keplerian description*

Our discussion so far has concentrated on orbits in which relativistic effects can be ignored. For pulsars in close binary systems about white dwarfs, other neutron stars, or perhaps eventually black holes, relativistic effects due to strong gravitational fields and high orbital velocities produce observable signatures in the timing residuals. The orbital velocity of the pulsar at periastron

$$\frac{v_p}{c} = T_\odot^{1/3} \left(\frac{2\pi}{P_b} \right)^{1/3} \left(\frac{1+e}{1-e} \right)^{1/2} \frac{m_c}{(m_p + m_c)^{2/3}}. \qquad (8.35)$$

This gives good indications as to whether relativistic effects are likely to be important. The masses again are expressed in solar units.

Even though general relativity (GR) appears to be the best description of the strong-field regime to date (Will 2001), alternative theories of gravity nevertheless should be considered and tested against it. A straightforward means of comparison is to parameterise the timing model in terms of the so-called 'post-Keplerian' (PK) parameters. For point masses with negligible spin contributions, the PK parameters in each theory should only be functions of the a priori unknown pulsar and companion mass, m_p and m_c, and the easily measurable Keplerian parameters. With the two masses as the only free parameters, an observation of two PK parameters will already determine the masses uniquely in the framework of the given theory. The measurement of a third or more PK parameters then provides a consistency check for the assumed theory (see Section 2.3).

8.3.2.1 Relativistic corrections to Römer delay

A relativistic description of a binary orbit requires not only a correction of the Keplerian parameters ω and P_b (see Equations (8.19) and (8.52)), but also leads to the introduction of two new eccentricities, e_r and e_θ, which appear in a modified Römer delay. Equation (8.27) now takes the form:

$$\Delta_{RB} = x\,(\cos E - e_r)\sin\omega + x\sin E\,\sqrt{1 - e_\theta^2}\,\cos\omega, \qquad (8.36)$$

where the new eccentricities

$$e_r \;\equiv\; e\,(1 + \delta_r) \qquad\qquad (8.37)$$
$$e_\theta \;\equiv\; e\,(1 + \delta_\theta) \qquad\qquad (8.38)$$

introduce two new PK parameters δ_r and δ_θ which represent relativistic deformations of the orbit. These cause periodic changes of order $(v_p/c)^2$ (see Section 8.3.2.5). In classical theory, $\delta_r = \delta_\delta = 0$, and only a single eccentricity occurs in the description of the binary system.

8.3.2.2 Einstein delay

The term Δ_{EB} in Equation (8.16) describes the modification in arrival times caused by the varying effects of the gravitational redshift due to the presence of the companion and time dilation (second-order Doppler) as the pulsar moves in its elliptical orbit at varying speeds and distances from the companion. This correction to 'pulsar time' is given by

$$\Delta_{EB} = \gamma \sin E, \qquad\qquad (8.39)$$

where the PK parameter γ denotes the amplitude measured in seconds.

8.3.2.3 Shapiro delay

The Shapiro delay caused by the gravitational field of the companion is only measurable, depending on timing precision, if the orbit is seen nearly edge on. Its observation is extremely useful as it provides the measurement of two PK parameters, r ('range') and s ('shape'), that lead to mass measurements for the pulsar and companion star. For a binary system, the term in Equation (8.16) becomes

$$\Delta_{SB} = -2r\ln\left[1 - e\cos E - s\left(\sin\omega(\cos E - e) + \sqrt{1 - e^2}\cos\omega\sin E\right)\right], \qquad (8.40)$$

where the maximum is expected for superior conjunction, i.e. at an angle $\Phi = \omega + A_T(E) = \pi/2$.

For small-eccentricity binary pulsars, the Shapiro delay simplifies to

$$\Delta_{\mathrm{SB}} = -2r \ln\left[1 - s \sin \Phi\right], \tag{8.41}$$

where Φ is the orbital phase measured from the ascending node. As a Fourier series Equation (8.41) takes the form (Lange *et al.* 2001):

$$\Delta_{\mathrm{SB}} = 2r(a_0 + b_1 \sin \Phi - a_2 \cos 2\Phi + \ldots), \tag{8.42}$$

where

$$a_0 = -\ln\left(\frac{1 + \sqrt{1 - s^2}}{2}\right), \tag{8.43}$$

$$b_1 = 2\left(\frac{1 - \sqrt{1 - s^2}}{s}\right) \quad \text{and} \tag{8.44}$$

$$a_2 = 2\left(\frac{1 - \sqrt{1 - s^2}}{s^2}\right) - 1. \tag{8.45}$$

Only for nearly edge-on orbits (i.e. $\sqrt{1 - s^2} \ll 1$) are higher harmonics (indicated as ... in Equation (8.42)) significant. Otherwise, the Shapiro delay cannot be separated from the Römer delay (see similarity to Equation (8.28)). As a result, the *observed* orbital parameters differ from their *intrinsic values*. The values for the semi-major axis x and the Laplace–Lagrange parameter ϵ_1 (see Equation 8.30) are modified to

$$x^{\mathrm{obs}} = x^{\mathrm{int}} + 2rb_1 \quad \text{and} \quad \epsilon_1^{\mathrm{obs}} = \epsilon_1^{\mathrm{int}} + 4ra_2/x^{\mathrm{int}}. \tag{8.46}$$

Even for orbits with high inclination, any Shapiro delay present may be absorbed, at least in part, in a fit for the Keplerian parameters x, ω and e. This affects severely the derived values for the pulsar and companion mass, leading to unrealistic values, and is particularly problematic when the choice of a classical timing model leads to additional covariances in T_0 and ω (see Section 8.3.1.3). Alternatively, an unmodelled small but non-zero eccentricity may be able to mimic a non-existent Shapiro delay. However, a genuine detection of a Shapiro delay can be claimed if the residual structure of the higher harmonics in Equation (8.42) can be observed in the timing residuals after fits for the Keplerian parameters (see Figure 8.4) when the covariance between the Keplerian and Shapiro parameters is broken.

The equations for the Shapiro delay, as well as the other PK parameters, are expressed here only in its lowest post-Newtonian approximation. Higher-order corrections may become important if timing precision

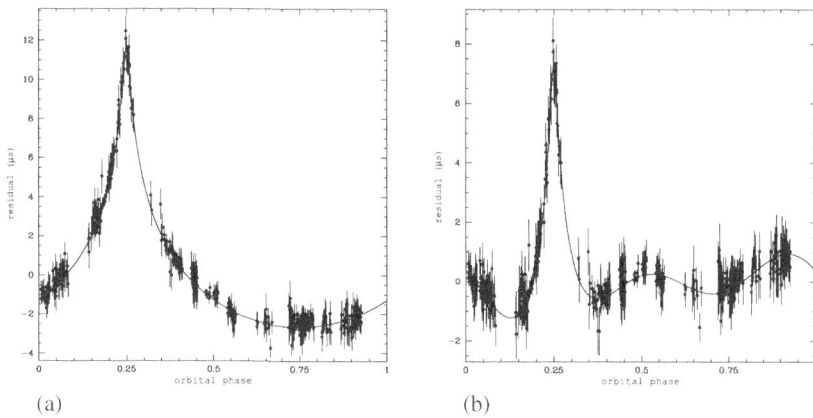

(a)   (b)

Fig. 8.4. Detection of the Shapiro time delay due to space-time curvature near the white dwarf companion to the 2.9 ms pulsar J1909−3744 (Jacoby *et al.* 2003). (a) Timing residuals obtained by subtracting the effects of the measured Shapiro delay from the best-fit solution. As expected, the largest delay occurs at superior conjunction, i.e. when the millisecond pulsar is located behind the companion. (b) Residuals obtained when the Shapiro delay is not included in the timing model. Parts of the delay visible in the left plot are absorbed in the fits for the Keplerian parameters. The clearly visible 'left-over' structure in the timing residuals represents the higher harmonics of Equation (8.42) and confirms the detection of the Shapiro delay. Unpublished data provided by Bryan Jacoby.

becomes sufficiently high. Currently, this is not yet the case, although the double pulsar system may be relativistic enough to allow measurements of these effects in the future (Lyne *et al.* 2004). Corrections would account for the assumptions currently made in the computation of the Shapiro delay that gravitational potentials are static and weak everywhere. Corresponding calculations taking these next higher order corrections into account have been presented by Wex (1995a), Kopeikin and Schäfer (1999) and Kopeikin (2003).

In addition to light propagation effects causing time delays, others exist that are related to light bending and its consequences. Light bending as discussed by Doroshenko and Kopeikin (1995) would be superposed on the Shapiro delay as a typically much weaker signal, which arises due to a modulation of the rotational phase of the pulsar by the effect of gravitational deflection of the light in the field of the companion of the pulsar. This effect would be visible only for very nearly edge-on systems, since it depends crucially on the orientation of the spin axis of the pulsar in space.

8.3.2.4 Aberration delay

The term Δ_{AB} in Equation (8.16) describes the difference between the actual proper time of emission and the corresponding time if the pulsar signal had not been due to a light-house effect, e.g. due to radial oscillation (see, for example, Damour and Taylor (1992)). The difference is due to an aberration effect caused by the motion of the pulsar and is given by

$$\Delta_{\mathrm{AB}} = A\left[\sin\left[\omega + A_{\mathrm{T}}(E)\right] + e\sin\omega\right] + B\left[\cos\left[\omega + A_{\mathrm{T}}(E)\right] + e\cos\omega\right].$$
(8.47)

Aberration occurs already in classical physics, so that A and B are not of relativistic origin, even though they are referred to commonly as PK parameters. In particular, A and B are nearly degenerate with other PK parameters, so that they may not be measurable separately (see Section 8.3.3). However, the situations may change if the system geometry changes on relatively short timescales due to an effect discussed in Section 8.3.2.6.

In addition to the timing effect described by Equation (8.47), aberration also causes the cut through the emission beam to change with orbital phase, so that an *orbital* aberration effect may be detectable as small changes in observed pulse profile. This was achieved recently for the first time for PSR B1534+12 by Stairs *et al.* (private communication).

8.3.2.5 Post-Keplerian parameters in general relativity

In the previous sections, we have discussed some of the most important PK parameters. In GR (after Robertson (1938), Peters (1964), Blandford and Teukolsky (1976) and Damour and Deruelle (1985; 1986)) these are expressed to lowest post-Newtonian order as follows:

$$\dot{\omega} = 3T_\odot^{2/3}\left(\frac{P_{\mathrm{b}}}{2\pi}\right)^{-5/3}\frac{1}{1-e^2}(m_{\mathrm{p}} + m_{\mathrm{c}})^{2/3},$$
(8.48)

$$\gamma = T_\odot^{2/3}\left(\frac{P_{\mathrm{b}}}{2\pi}\right)^{1/3}e\frac{m_{\mathrm{c}}(m_{\mathrm{p}} + 2m_{\mathrm{c}})}{(m_{\mathrm{p}} + m_{\mathrm{c}})^{4/3}},$$
(8.49)

$$r = T_\odot m_{\mathrm{c}},$$
(8.50)

$$s = \sin i = T_\odot^{-1/3}\left(\frac{P_{\mathrm{b}}}{2\pi}\right)^{-2/3}x\frac{(m_{\mathrm{p}} + m_{\mathrm{c}})^{2/3}}{m_{\mathrm{c}}},$$
(8.51)

$$\dot{P}_{\mathrm{b}} = -\frac{192\pi}{5}T_\odot^{5/3}\left(\frac{P_{\mathrm{b}}}{2\pi}\right)^{-5/3}f(e)\frac{m_{\mathrm{p}}m_{\mathrm{c}}}{(m_{\mathrm{p}} + m_{\mathrm{c}})^{1/3}},$$
(8.52)

where all masses are expressed in solar units, G is Newton's gravitational constant, c the speed of light and

$$f(e) = \frac{\left(1 + (73/24)e^2 + (37/96)e^4\right)}{(1 - e^2)^{7/2}}. \tag{8.53}$$

The first PK parameter, $\dot{\omega}$, describes the relativistic advance of periastron in rad s^{-1}. It is the easiest to measure for orbits with non-zero eccentricities (as discussed in Section 8.3.2.3, ω is poorly defined for $e \approx 0$ and so is $\dot{\omega}$). From a measurement of $\dot{\omega}$, we obtain from Equation (8.48) the total mass of the system, $m_p + m_c$.

The orbital decay due to gravitational-wave damping is expressed by the (dimensionless) change in orbital period, $\dot{P}_b$. Any metric theory of gravity that embodies Lorentz-invariance in its field equations predicts gravitational radiation and, hence, $\dot{P}_b$. If a theory satisfies the strong equivalence principle (SEP); (see Chapter 2), like general relativity (GR), gravitational *dipole* radiation is not expected, but *quadrupole* emission will be the lowest multipole term. This arises because the dipole moment (centre of mass) of isolated systems is independent of time due to the conservation of momentum, and the inertial mass that determines the dipole moment is the same as the mass that generates gravitational waves.

In alternative theories, while the *inertial* dipole moment may remain uniform, the *gravitational wave* dipole moment may not. This is because, in the absence of the SEP, the mass generating gravitational waves depends differently on the internal gravitational binding energy of each body than the inertial mass (Will 2001). If dipole radiation is predicted, the magnitude of this effect depends on the difference in gravitational binding energies, expressed by the difference in coupling constants to a scalar gravitational field. While strong limits have been set on the existence of dipole radiation (see, for example, Lange *et al.* (2001)), orbital decay due to gravitational quadrupole radiation was impressively demonstrated for the first time in PSR B1913+16 (Taylor & Weisberg 1989). The corresponding expression for $\dot{P}_b$ is given by Equation (8.52).

In GR, the PK parameters δ_r and δ_θ introduced in the modified Römer delay via the eccentricities, e_r and e_θ (Equations (8.36), (8.37) and (8.38)) take the form (Damour & Deruelle 1986):

$$\delta_r = T_\odot^{2/3} \left(\frac{2\pi}{P_b}\right)^{2/3} \frac{3m_p^2 + 6m_p m_c + 2m_c^2}{(m_p + m_c)^{4/3}} \tag{8.54}$$

$$\delta_\theta = T_\odot^{2/3} \left(\frac{2\pi}{P_b} \right)^{2/3} \frac{\frac{7}{2}m_p^2 + 6m_p m_c + 2m_c^2}{(m_p + m_c)^{4/3}}, \quad (8.55)$$

where, as usual, pulsar and companion mass are expressed in solar units. Note that it is sufficient to use the Keplerian eccentricity e in Equations (8.48)–(8.52) as e_r and e_θ are already of post-Newtonian order. Only δ_θ may be measured separately (see Section 8.3.3).

In GR, the aberration delay depends critically upon the orientation of the pulsar spin axis. The two parameters in Equation (8.47) take the form:

$$A = -\frac{T_\odot^{1/3}}{(2\pi)^{2/3}} \frac{P}{P_b^{1/3}(1-e^2)^{1/2}} \frac{m_c}{(m_p+m_c)^{2/3}} \frac{\sin\eta}{\sin\lambda} \quad (8.56)$$

$$B = -\frac{T_\odot^{1/3}}{(2\pi)^{2/3}} \frac{P}{P_b^{1/3}(1-e^2)^{1/2}} \frac{m_c}{(m_p+m_c)^{2/3}} \frac{\cos i \cos\eta}{\sin\lambda}, \quad (8.57)$$

where λ and η are the polar angles of the pulsar's spin vector (see Figure 8.3) and all masses are expressed in solar units.

8.3.2.6 Geodetic precession

Due to the curvature of space-time near gravitating bodies, the proper reference frame of a freely falling object suffers a precession with respect to a distant observer, called *geodetic precession*. In a binary pulsar system, geodetic precession leads to a relativistic spin–orbit coupling, analogous of spin–orbit coupling in atomic physics. As a consequence, the pulsar spin precesses about the total angular momentum, changing the relative orientation of the pulsar towards Earth (Damour & Ruffini 1974). Since the orbital angular momentum is much larger than the angular momentum of the pulsar, the orbital spin practically represents a fixed direction in space, defined by the orbital plane of the binary system. The rate of geodetic precession of the pulsar spin axis as predicted by GR is given by Barker and O'Connell (1975b):

$$\Omega_{\text{geod}} = \left(\frac{2\pi}{P_b} \right)^{5/3} T_\odot^{2/3} \frac{m_c(4m_p + 3m_c)}{2(m_p + m_c)^{4/3}} \frac{1}{1-e^2} \quad (8.58)$$

where m_p and m_c are again expressed in solar masses. It is useful to note that, in GR,

$$\Omega_{\text{geod}} = \frac{7}{24} \dot\omega \sim 0.3 \, \dot\omega \quad \text{for } m_p = m_c. \quad (8.59)$$

Geodetic precession may have a direct effect on the timing, as it causes

the polar angles of the spin, λ and η, and, hence, the aberration parameters A and B to change with time (see Equation (8.56)). However, indirect consequences are much more apparent. These arise from
changes in the pulse shape and its polarisation properties due to changing cuts through the emission beam as the pulsar spin axis precesses.
Due to these effects, which may complicate the TOA determination with
a template matching procedure (see Section 8.1), geodetic precession is
detected for PSRs B1913+16 (Weisberg *et al.* 1989; Kramer 1998) and
B1534+12 (Arzoumanian 1995; Stairs *et al.* 2000b). For PSR B1913+16
it is used to derive a two-dimensional map of the emission beam (Weisberg & Taylor 2002) and to determine the full system geometry leading
(see Figure 8.5) to the prediction that the pulsar will not be observable
from Earth around 2025 (Kramer 1998).

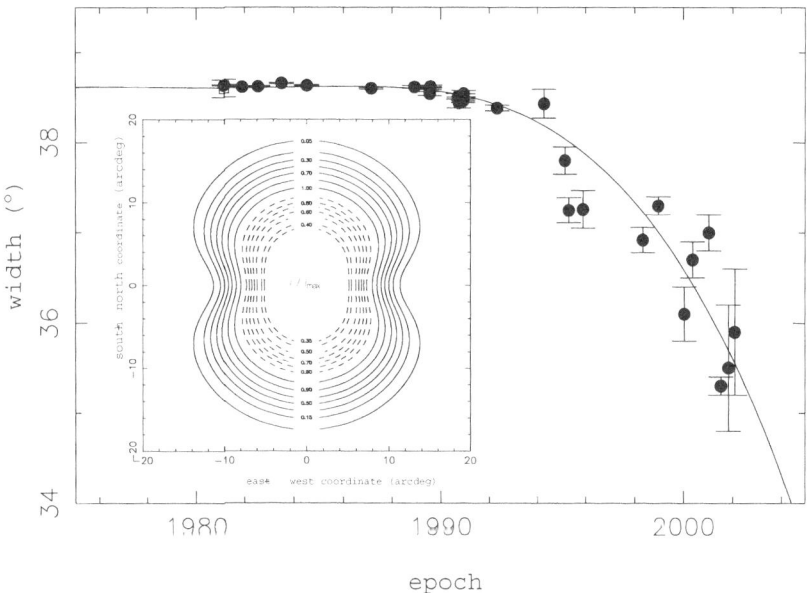

Fig. 8.5. Component separation as measured for PSR B1913+16 (Kramer
1998; Kramer *et al.* 2003d). The component separation is shrinking as the
line of sight changes due to geodetic precession. Using a similar data set,
Weisberg and Taylor (2002) re-assembled the different profiles, representing
different cuts through the emission beam, to construct a two-dimensional beam
map as shown inset.

8.3.3 Measurable or not measurable?

It turns out that not all parameters describing a binary system are separately measurable, i.e. some parameters can be absorbed in a suitable redefinition of other observable parameters (Damour & Deruelle 1986; Damour & Taylor 1992). For example, a radial motion gives rise to a Doppler factor, D, that modifies the intrinsic values of non-dimensionless parameters relative to the observed values as follows:

$$P^{\text{obs}} = D^{-1} P^{\text{int}}, \tag{8.60}$$

$$P_{\text{b}}^{\text{obs}} = D^{-1} P_{\text{b}}^{\text{int}}, \tag{8.61}$$

$$x^{\text{obs}} = D^{-1} x^{\text{int}}, \tag{8.62}$$

$$m_{\text{p}}^{\text{obs}} = D^{-1} m_{\text{p}}^{\text{int}}, \quad \text{etc., while} \tag{8.63}$$

$$e^{\text{obs}} = e^{\text{int}}. \tag{8.64}$$

However, as demonstrated in Section 8.2.4, a radial motion usually is not measurable, so that the intrinsic values cannot be determined. If the Doppler factor has a secular drift, this will contribute to the observed changes in these parameters, leading, for example, to the contributions to $\dot{P}$ and $\dot{P}_{\text{b}}$ as discussed in Section 8.2.4 (see also Section 8.3.4). While also the mass measurements obtained for pulsar and companion are unknown to within the Doppler factor, it is unlikely that the radial velocity exceeds the maximum observed transverse velocity by a large amount, limiting the contribution to $\lesssim 0.5$ per cent.

The aberration parameters A and B (see Equation (8.47)) can be absorbed into redefinitions of T_0, x, e, δ_{r} and δ_θ (Damour & Deruelle 1986; Damour & Taylor 1992), so that we have in addition to the above relations:

$$x^{\text{obs}} = (1 + \epsilon_{\text{A}}) x^{\text{int}}, \tag{8.65}$$

$$e^{\text{obs}} = (1 + \epsilon_{\text{A}}) e^{\text{int}}, \tag{8.66}$$

$$\delta_\theta^{\text{obs}} = \delta_\theta^{\text{int}} - \epsilon_{\text{A}}, \tag{8.67}$$

$$\delta_{\text{r}}^{\text{obs}} = \delta_{\text{r}}^{\text{int}} - 3\epsilon_{\text{A}}, \tag{8.68}$$

$$\left(\frac{2\pi}{P_{\text{b}}} T_0\right)^{\text{obs}} = \left(\frac{2\pi}{P_{\text{b}}} T_0\right)^{\text{int}} - \sqrt{1 - e^2} \, \epsilon_{\text{B}}, \tag{8.69}$$

where we define

$$\epsilon_{\text{A}} \equiv \frac{A}{x} = -\frac{P}{P_{\text{b}}} \frac{1}{\sin i (1 - e^2)^{1/2}} \frac{\sin \eta}{\sin \lambda} \tag{8.70}$$

$$\epsilon_{\text{B}} \equiv \frac{B}{x} = -\frac{P}{P_{\text{b}}} \frac{1}{\sin i (1 - e^2)^{1/2}} \frac{\cos i \cos \eta}{\sin \lambda}. \tag{8.71}$$

The angles η and λ reflect the system geometry and were defined in Section 8.3.2.4. If the pulsar undergoes geodetic precession, this geometry will change, causing secular changes in x^{obs} and e^{obs} (see Section 8.3.4).

The parameters δ_r and δ_θ appearing in the relativistic Römer delay (Equation (8.36)) describe purely periodic relativistic corrections to the Keplerian motion (Damour & Deruelle 1986). It seems impossible to measure δ_r, as it can be absorbed in a redefinition of the rotational phase of the pulsar, leading to small corrections in the spin-down parameters of Equation (8.5). In contrast, the parameter δ_θ could in principle be measured in the long term. As for other PK parameters, the fractional uncertainty decreases with time span of the timing observations, T. Additionally, the orbital period is important, as well as the relative orientation of the orbit to our line of sight which changes with time due to a relativistic periastron advance (see, for example, Damour and Taylor (1992)). The dependence of the fractional uncertainties for some of the PK parameters is listed in Table 8.2.

Table 8.2. *Expected dependence of fractional uncertainty in selected PK parameters on observing time, T, and orbital period, P_{b}. The uncertainties scale according to T^a and P_{b}^b, where a and b are listed (Damour & Taylor 1992).*

	Parameter					
Scaling	$\dot{\omega}$	γ	$\dot{P_{\text{b}}}$	r	s	δ_θ
a	$-3/2$	$-3/2$	$-5/2$	$-1/2$	$-1/2$	$-5/2$
b	1	$4/3$	3	0	0	$10/3$

8.3.4 Apparent variation of binary parameters

A number of effects can lead to changes in the observed orbital elements. In some cases, the properties of the orbit may really change, e.g. due to gravitational wave emission, mass loss or interaction with the companion. Alternatively, the measured values may be affected by acceleration effects, i.e. changing Doppler factors.

8.3.4.1 Changes in orbital period

The most important terms affecting the observed orbital period can be summarised as follows:

$$\left(\frac{\dot{P}_{\rm b}}{P_{\rm b}}\right)^{\rm obs} = \left(\frac{\dot{P}_{\rm b}}{P_{\rm b}}\right)^{\rm GW} - \left(\frac{\dot{D}}{D}\right) + \left(\frac{\dot{P}_{\rm b}}{P_{\rm b}}\right)^{\dot{m}} + \left(\frac{\dot{P}_{\rm b}}{P_{\rm b}}\right)^{\rm T} \qquad (8.72)$$

where the contributions to the observed orbital period derivative are due to the emission of gravitational waves (GW), (see Section 8.3.2.5), acceleration of the binary system, mass loss from the system, $\dot{m}$, and tidal dissipation of the orbit, T.

The second term in Equation (8.72), $\dot{D}/D$, is due to the combined effect of a changing Doppler shift, D, and the proper motion on the plane of the sky (see Sections 8.2.4 and 8.3.3). A change in Doppler shift can be caused by a change in distance between the SSB and the binary pulsar system. This can be due to differential rotation of the Galaxy, acceleration in the gravitational field of the Galaxy or a globular cluster or may be caused by a third massive body in the vicinity of the binary system. In total, we obtain

$$-\frac{\dot{D}}{D} = \frac{1}{c}\,\vec{K}_0 \cdot (\vec{a}_{\rm PSR} - \vec{a}_{\rm SSB}) + \frac{V_{\rm T}^2}{c\,d}\,, \qquad (8.73)$$

where $\vec{K}_0$ is the unit vector from the Earth towards the pulsar (see Figure 8.3), and $\vec{a}_{\rm PSR}$ and $\vec{a}_{\rm SSB}$ are the Galactic accelerations at the location of the binary system and the SSB. The last term including the transverse velocity, $V_{\rm T}$, is the Shklovskii term (see Section 8.2.4).

Interestingly, it is the precision to which the acceleration of PSR B1913+16 in the Galactic gravitational potential is known that limits the precision in its tests of GR (see Section 2.3.1). For PSR B1534+12, the contribution from this term is so large that the observed $\dot{P}_{\rm b}$ produces a line in the $m_{\rm p}$–$m_{\rm c}$ diagram that misses the intersection point of the other PK parameter lines (see, for example, Stairs *et al.* (2002)). The deviation from the intersection allows a high-precision determination of the distance to the pulsar, assuming that GR is the correct theory of gravity.

For pulsars near the Galactic plane, one may follow the description provided by Damour and Taylor (1991) to compute the various terms in Equation (8.73), (see also Nice and Taylor (1995)). In a more general approach, one can use a model for the Galactic potential (see, for example, Carlberg and Innanen (1987)) and determine velocity and acceleration numerically as explained, for example, by Arzoumanian *et al.* (1999) or Wex *et al.* (2000).

As discussed in Section 8.2.4, acceleration affects $\dot{P}_{\rm b}$ and the spin period derivative $\dot{P}$ in the same way. Under the assumption that the observed value of the spin period derivative is due totally to acceleration, the contribution to the binary period derivative can be constrained to $(\dot{P}_{\rm b}/P_{\rm b})^{\rm acc} \leq (\dot{P}/P)$.

The mass loss from the binary system leads to a change in the orbital period as the change in mass, $\dot{m}$, can be related to the a loss in rotational energy via $\dot{E} = \dot{m}c^2$. The rate is given by Damour and Taylor (1991)

$$\left(\frac{\dot{P}_{\rm b}}{P_{\rm b}}\right)^{\dot{m}} = \frac{8\pi^2 G}{T_\odot c^5} \frac{I_j}{m_{\rm p} + m_{\rm c}} \frac{\dot{P}_j}{P_j^3}, \tag{8.74}$$

where I_j denotes the moment of inertia of body j ($j = {\rm p}$ for the pulsar, $j = {\rm c}$ for the companion), P_j and $\dot{P}_j$ are the spin period and period derivative. Assuming $I_{\rm p} \sim 10^{45}$ g cm^2 (see Chapter 3), the change of the orbital period due to mass loss can be estimated.

The remaining terms describe contributions from tidal torques (Applegate (1992) and references therein) and gravitational quadrupole coupling (Applegate & Shaham 1994). For most binary pulsars, these contributions can be neglected unless the companion is deformed or interacting with the pulsar (see, for example, Arzoumanian *et al.* (1994a) and Doroshenko *et al.* (2001)).

8.3.4.2 Changes in projected semi-major axis

An observed change in projected semi-major axis, $x = a_{\rm p} \sin i$, can be caused by a physical change in the intrinsic size of the orbit, $a_{\rm p}$, a change in projection as expressed by the inclination angle, i, or a combination of both. Both periodic and secular changes are observed.

As the Earth orbits the Sun, a binary system is observed under slightly different angles leading to a *periodic* change in apparent inclination angle, i, of the pulsar orbit. For nearby binary systems, this effect, known as the *annual orbital parallax* (Kopeikin 1995), can cause a measurable variation of the projected semi-major axis given by

$$x(t) = x_0 \left[1 - \frac{\cot i}{d} \, \vec{r}_{\rm SSB}(t) \cdot \vec{J}'\right], \tag{8.75}$$

where $\vec{J}' = -\sin\Omega_{\rm asc}\vec{I}_0 + \cos\Omega_{\rm asc}\vec{J}_0$. The vectors $\vec{I}_0$ and $\vec{J}_0$ and the longitude of the ascending node $\Omega_{\rm asc}$ are defined in Figure 8.3. Since

the amplitude of this effect depends on the distance d, a detection of
the annual orbital parallax leads to an accurate pulsar distance measurement. While this periodic change in x could be measured only for
the system of the bright millisecond pulsar J0437−4715 (van Straten *et
al.* 2001) providing the full geometry of the system, *secular* changes in
x are observed much more commonly. They can be the result of various
effects and are summarised as follows:

$$\left(\frac{\dot{x}}{x}\right)^{\mathrm{obs}} = \left(\frac{\dot{x}}{x}\right)^{\mathrm{GW}} + \left(\frac{\dot{x}}{x}\right)^{\mathrm{PM}} + \frac{\mathrm{d}\varepsilon_{\mathrm{A}}}{\mathrm{d}t} - \frac{\dot{D}}{D} + \left(\frac{\dot{x}}{x}\right)^{\dot{m}} + \left(\frac{\dot{x}}{x}\right)^{\mathrm{SO}} + \left(\frac{\dot{x}}{x}\right)^{\mathrm{planet}},$$
$$(8.76)$$

where the contributions to the observed semi-major axis derivative are
due to the emission of gravitational waves (GW), proper motion of the bi-
nary system (PM), varying aberration, $\mathrm{d}\varepsilon_{\mathrm{A}}/\mathrm{d}t$, changing Doppler shift,
$-\dot{D}/D$, spin–orbit coupling (SO) in the binary system and a third com-
panion (planet) around the pulsar.

A shrinking of the pulsar orbit due to gravitational-wave damping
leads also to a change of x as given by Kepler's third law,

$$\left(\frac{\dot{x}}{x}\right)^{\mathrm{GW}} = \frac{2}{3}\left(\frac{\dot{P}_{\mathrm{b}}}{P_{\mathrm{b}}}\right)^{\mathrm{GW}}. \qquad (8.77)$$

With $\dot{P}_{\mathrm{b}}^{\mathrm{GW}}$ calculated from Equation (8.52) for GR, one can compute
the expected contribution. An explicit expression for the change in semi-
major axis of a system of point masses under GR is derived from Peters
(1964) as

$$\dot{a}_{\mathrm{p}} = -\frac{64}{5}\,c\,T_{\odot}^{2}\left(\frac{2\pi}{P_{\mathrm{b}}}\right)^{2} f(e)\,\frac{m_{\mathrm{p}}m_{\mathrm{c}}^{2}}{m_{\mathrm{p}} + m_{\mathrm{c}}} \qquad (8.78)$$

where the masses are expressed in solar units and $f(e)$ is given by Equation (8.53). The corresponding change in the separation of pulsar and
companion, $\dot{a}_{\mathrm{R}}$, can be computed from Equation (8.22). An expression
to compute the time until coalescence of pulsar and companion is given
in Appendix 2.

The next contribution to Equation (8.76) is given by effects caused by
proper motion of the binary system (Arzoumanian *et al.* 1996; Kopeikin
1996):

$$\left(\frac{\dot{x}}{x}\right)^{\mathrm{PM}} = 1.54 \times 10^{-16}\,\cot i\,(-\mu_{\alpha}\sin\Omega_{\mathrm{asc}} + \mu_{\delta}\cos\Omega_{\mathrm{asc}})\,, \qquad (8.79)$$

where Ω_{asc} is the longitude of the ascending node of the orbit, and μ_{α} and

μ_δ are the proper motions in right ascension and declination measured in mas yr^{-1}. In this case, the observed change in x is caused by the secular variation in the inclination angle i due to the proper motion, whereas the intrinsic semi-major axis remains constant. One can estimate the maximal contribution due to proper motion from

$$\left(\frac{\dot{x}}{x}\right)^{\text{PM}} \leq \left(\frac{\dot{x}}{x}\right)^{\text{PM}}_{\text{max}} \simeq 1.54 \times 10^{-16} \cot i \left(\frac{\mu_{\text{T}}}{\text{mas yr}^{-1}}\right) \qquad (8.80)$$

with a total proper motion $\mu_{\text{T}} = \sqrt{\mu_\alpha^2 + \mu_\delta^2}$. In turn, if one has determined an upper limit for $\dot{x}$, we can derive a limit for orbital inclination, i.e.

$$\tan i \lesssim 1.54 \times 10^{-16} \left(\frac{\mu_{\text{T}}}{\text{mas yr}^{-1}}\right) \Big/ \left(\frac{\dot{x}}{x}\right)^{\text{PM}}_{\text{max}} \qquad (8.81)$$

The third term in Equation (8.76), $\text{d}\varepsilon_{\text{A}}/\text{d}t$, is due to varying aberration caused by geodetic precession of the pulsar spin axis (see Section 8.3.3), leading to

$$\frac{\text{d}\varepsilon_{\text{A}}}{\text{d}t} = -\frac{P}{P_{\text{b}}} \frac{\Omega_{\text{geod}}}{(1-e^2)^{1/2}} \frac{(\cot \lambda \sin 2\eta + \cot i \cos \eta)}{\sin \lambda} , \qquad (8.82)$$

where P is the pulsar spin period, and η and λ are again the polar coordinates of the pulsar spin (see Figure 8.3 and Damour and Taylor (1992)).

The fourth term in Equation (8.76), $\dot{D}/D$, is identical to the second term in Equation (8.72), defined in Equation (8.73). The fifth term describes a change in the orbit size due to mass loss. A corresponding expression for the change in P_{b} given by Equation (8.74) can be translated into a change in x using Equation (8.77).

The sixth term in Equation (8.76) describes the effects of a classical spin–orbit coupling, caused by a spin-induced quadrupole moment of the companion star (Smarr & Blandford 1976). This contribution needs to be considered, for example, for main-sequence star companions (see, for example, Kaspi *et al.* (1996) and Wex (1998)) or rapidly rotating white-dwarf companions (see, for example, Kaspi *et al.* (2000b)). Spin–orbit coupling leads a precession of the orbit and therefore to a variation in the inclination angle, a non-zero $\text{d}i/\text{d}t$, while the semi-major axis a remains constant. In this case:

$$\frac{\text{d}i}{\text{d}t} = \left(\frac{\dot{x}}{x}\right)^{\text{SO}} \tan i . \qquad (8.83)$$

Hence, assuming that an observed $\dot{x}$ is totally due to spin–orbit coupling, one can obtain a limit on the change of orbital inclination angle. The change caused by the rotationally induced quadrupole of the companion star is given by Smarr and Blandford (1976), Lai *et al.* (1995) and Wex (1998):

$$\frac{\mathrm{d}i}{\mathrm{d}t} = \frac{1}{2}\,\Omega_{\mathrm{b}}\,\frac{k\,R_{\mathrm{c}}^2\,\hat{\Omega}_{\mathrm{s}}^2}{a_{\mathrm{R}}^2(1-e^2)^2}\,\sin(2\theta)\sin\hat{\Phi}. \tag{8.84}$$

Here, $\hat{\Omega}_{\mathrm{s}} \equiv \Omega_{\mathrm{c}}/(Gm_{\mathrm{c}}/R_{\mathrm{c}}^3)^{1/2}$ is the dimensionless spin of the companion, R_{c} its radius, a_{R} the semi-major axis of the relative orbit, k the apsidal motion constant, θ the angle between the companion spin angular momentum $\mathbf{S}$ and the orbital angular momentum $\mathbf{L}$, and $\hat{\Phi}$ the orbital plane precessional phase (Smarr & Blandford 1976; Wex 1998). Equation (8.84) is valid for the case that $|\mathbf{S}| \ll |\mathbf{L}|$. Note that spin–orbit coupling requires the companion's spin axis to be inclined with respect to the orbital angular momentum vector ($\theta \neq 0$). Measurement of $(\mathrm{d}i/\mathrm{d}t)$ and the classical periastron advance $\dot{\omega}$ due to spin–orbit coupling can, in principle, be used to obtain constraints of the values of θ and Φ (Kaspi *et al.* 1996; Wex 1998). Eventually, all angles can be determined if higher order terms, $\ddot{\omega}$ and $\ddot{x}$, can also be measured (Wex 1998).

The deformation of the orbit due to the presence of an additional companion such as a planet is described by the last term in Equation (8.76). The contribution of this third, outer body can be written as (Rasio 1994; Joshi & Rasio 1997)

$$\left(\frac{\dot{x}}{x}\right)^{\mathrm{planet}} = \frac{3\pi}{2P_{\mathrm{b}}}\,\eta_3\,\cot i\,\sin 2\theta_3\,\cos(\omega+\Phi_3) \tag{8.85}$$

where $\eta_3 = (m_{\mathrm{c}}/m_{\mathrm{p}})\left[(a_{\mathrm{p}}+a_{\mathrm{c}})/r_3\right]^3$ and θ_3, Φ_3 and r_3 are the fixed spherical polar coordinates of the second companion as defined by Rasio (1994); a_{c} is the semi-major axis of the inner binary companion. This description has been applied successfully to the three body system of PSR B1620−26 (Thorsett *et al.* 1999).

8.3.4.3 Changes in longitude of periastron

The annual orbital parallax discussed in Section 8.3.4.2 (Kopeikin 1995) also causes *periodic* variations of the longitude of periastron given by

$$\omega(t) = \omega_0 - \frac{\csc i}{d}\,\vec{r}_{\mathrm{SSB}}(t)\cdot\vec{I'}. \tag{8.86}$$

Here, $\vec{I}' = \cos\Omega_{\mathrm{asc}}\vec{I}_0 + \sin\Omega_{\mathrm{asc}}\vec{J}_0$ where, as before, the vectors $\vec{I}_0$ and $\vec{J}_0$ are defined as in Figure 8.3(c).

Observed *secular* changes in the longitude of periastron can be caused both by relativistic and classical effects. They can be summarised as

$$\dot{\omega}^{\mathrm{obs}} = \dot{\omega}^{\mathrm{Rel}} + \dot{\omega}^{\mathrm{PM}} + \dot{\omega}^{\mathrm{SO}} + \dot{\omega}^{\mathrm{planet}}. \tag{8.87}$$

The first term is a relativistic periastron advance, which for GR is given by Equation (8.48) to the lowest first post-Newtonian order if spin contributions are negligible. The spins of the neutron stars contribute to the observed value of $\dot{\omega}$ via spin–orbit coupling (Barker & O'Connell 1975a) formally at the same level of first post-Newtonian approximation. However, it turns out that for binary pulsars these effects have a magnitude equivalent to second post-Newtonian order (Wex 1995b), so that they only need to be considered if $\dot{\omega}$ is to be studied at this higher level of approximation. An expression for $\dot{\omega}$ at the second post-Newtonian level was derived by Damour & Schäfer (1988).

A proper motion of the pulsar leads, in addition to a change in the projected semi-major axis, also to a change in the measure longitude of periastron. Using the definitions and units used for Equation (8.79), we find (Kopeikin 1996)

$$\dot{\omega}^{\mathrm{PM}} = 2.78 \times 10^{-7} \operatorname{cosec} i \, (\mu_\alpha \cos\Omega_{\mathrm{asc}} + \mu_\delta \sin\Omega_{\mathrm{asc}}) \ \deg \ \mathrm{yr}^{-1}. \tag{8.88}$$

A change in longitude of periastron also can be caused by a classical spin–orbit coupling if the companion star has a spin-induced quadrupole moment (Wex 1998). We have discussed this effect and its impact on the observed projected semi-major axis in Section 8.3.4.2. This contribution can have both positive and negative values, depending on the actual orientation, of pulsar and companion spin relative to the orbital momentum vector. We can relate the change in the longitude of periastron to the observed change in semi-major axis by using the definitions following Equation (8.83) (Wex 1998):

$$\dot{\omega}^{\mathrm{SO}} = \left(\frac{\dot{x}}{x}\right)^{\mathrm{SO}} \frac{1 - \frac{3}{2}\sin^2\theta + \cot i \sin\theta\cos\theta\cos\hat{\Phi}}{\cot i \sin\theta\cos\theta\sin\hat{\Phi}}. \tag{8.89}$$

The existence of a third body around the pulsar contributes also to a change in the longitude of periastron (Rasio 1994; Joshi & Rasio 1997). We can express it in terms of a change in x using Equation (8.85):

$$\dot{\omega}^{\mathrm{planet}} = \left(\frac{\dot{x}}{x}\right)^{\mathrm{planet}} \frac{2\left[\sin^2\theta_3(5\cos^2\Phi_3 - 1) - 1\right]}{\cot i \, \sin 2\theta_3 \, \cos(\omega + \Phi_3)} \tag{8.90}$$

with quantities as defined above.

8.3.4.4 Changes in orbital eccentricity

Observed secular changes in the eccentricity can be caused by a circularisation of the orbit due to gravitational wave emission and a change in aberration due to geodetic precession. They can be summarised as follows:

$$
\left(\frac{\dot{e}}{e}\right)^{\mathrm{obs}} = \left(\frac{\dot{e}}{e}\right)^{\mathrm{GW}} + \frac{\mathrm{d}\varepsilon_{\mathrm{A}}}{\mathrm{d}t} + \left(\frac{\dot{e}}{e}\right)^{\mathrm{planet}}, \tag{8.91}
$$

where the second term is given by Equation (8.82) (Damour & Taylor 1992). In GR, the first term can be computed (Peters 1964) as

$$
\left(\frac{\dot{e}}{e}\right)^{\mathrm{GW}} = -\frac{304}{15} T_{\odot}^{5/3} \left(\frac{2\pi}{P_{\mathrm{b}}}\right)^{8/3} \frac{\left(1 + (121/304)e^2\right)}{(1 - e^2)^{5/2}} \frac{m_{\mathrm{p}} m_{\mathrm{c}}}{(m_{\mathrm{p}} + m_{\mathrm{c}})^{1/3}} \tag{8.92}
$$

indicating that the orbit becomes more circular as it shrinks. As usual, the masses are expressed in solar units.

The presence of a third body affects the eccentricity (Rasio 1994; Joshi & Rasio 1997):

$$
\left(\frac{\dot{e}}{e}\right)^{\mathrm{planet}} = -\frac{15\pi}{2P_{\mathrm{b}}} \eta_3 \sin^2\theta_3 \sin 2\Phi_3 \tag{8.93}
$$

with quantities as defined in Equation (8.85).

8.3.5 Timing models for binary pulsars

Soon after the discovery of PSR B1913+16 (Hulse & Taylor 1975) it became clear that a simple Keplerian timing model is not sufficient to describe the timing observations of this DNS. Blandford and Teukolsky (1976) derived a model that contained the largest short-periodic effect, $\Delta_{\mathrm{Einstein,B}}$, and secular drifts in the orbital parameters ('BT model'). Later, Damour and Deruelle (1985; 1986) derived a simple analytic solution of the post-Newtonian two-body problem and developed a theory-independent relativistic model that includes treatment of short-term periodic terms, Shapiro delay and aberration ('DD model'). Assuming that GR is the correct theory of gravity, the masses, m_{p} and m_{c}, are the only free parameters, so that the 'DDGR model' was developed (Taylor 1987; Taylor & Weisberg 1989). The timing effects caused by the companion's deviation from a mass point source in the case of a pulsar/main-sequence star binary system have been studied by Wex (1998) and incorporated in the 'MSS model'. We have discussed the main features of the 'ELL1 model' for small-eccentricity pulsars which uses the definition of ascending node and Lagrangian parameters to break co-variances in T_0 and ω (Lange *et al.* 2001). A list of these models and their timing parameters as implemented in the TEMPO software package (see Section 8.6) is given in Table 8.3.

8.4 Timing noise

Timing noise manifests itself in a quasi-random walk in one or more of the rotational parameters on timescales of months to years. It appears mostly for young pulsars and scales with some power of the period derivative, $\dot{P}$. The older millisecond pulsars show much less timing noise, although it is observed in some sources, albeit with a much smaller amplitude. It nevertheless affects the timing precision of some millisecond pulsars, with the fastest rotating pulsar B1937+21 as the most prominent example (Kaspi *et al.* 1994).

Usually, timing noise is recognised by the measurement of the second derivative of the pulsar spin frequency, $\ddot{\nu}$, that is hugely inconsistent with that expected for a standard pulsar spin evolution with a braking index of $n = 3$ (see Chapter 3). We can estimate this expected value for $\ddot{\nu}$ by inverting Equation (3.10) to obtain

$$\ddot{\nu} = n\dot{\nu}^2/\nu. \tag{8.94}$$

For a typical pulsar with $\nu = 1$ Hz and $\dot{\nu} = 10^{-15}$ Hz s^{-1}, the expected value for $n = 3$ is $\ddot{\nu} \sim 10^{-30}$ Hz s^{-2} (cf. Johnston and Galloway (1999)). Such values are measurable only for the youngest pulsars. Often, however, the timing analysis solving Equation (8.4) produces significant values for $\ddot{\nu}$, positive and negative, that are much larger than this. Unless other explanations are put forward (e.g. planetary companions for PSR B1937+21 (Lommen 2002)), a measurement like this is a clear indication for timing noise.

A stability parameter was introduced by Arzoumanian *et al.* (1994b) in order to characterise the amount of timing noise observed in pulsars. In contrast to the Allan variance or the σ_z parameter introduced in Chapter 2, this quantity is based simply on the observed values for ν and $\ddot{\nu}$, and it is defined as

$$\Delta_8 = \log \left(\frac{1}{6\nu} |\ddot{\nu}| (10^8 \text{s})^3 \right) \tag{8.95}$$

Table 8.3. *Parameters used by the most commonly used timing models for binary pulsars as implemented in TEMPO. In addition to the parameter symbol as introduced in this chapter, we list the parameter name as used in TEMPO and the corresponding timing models. An 'x' indicates a parameter fully implemented. An 'o' indicates the parameter is used at a fixed value, but cannot be fitted. A '*' indicates the parameter is used only to calculate other binary parameters, and is then ignored.*

Parameter	TEMPO	BT	DD	DDGR	MSS	ELL1
x	A1	x	x	x	x	x
e	E	x	x	x	x	*
T_0	T0	x	x	x	x	*
$P_{\rm b}$	PB	x	x	x	x	x
ω	OM	x	x	x	x	*
$T_{\rm asc}$	TASC	*				x
ϵ_1	EPS1	*				x
ϵ_2	EPS2	*				x
$\dot{\epsilon}_1$	EPS1DOT					x
$\dot{\epsilon}_2$	EPS2DOT					x
$1/P_{\rm b}$	FB	*	*	*	*	*
$(1/\dot{P}_{\rm b})$	FB1	*	*	*	*	*
$\dot{\omega}$	OMDOT	x	x		x	x
$\ddot{\omega}$	OM2DOT				x	
$(\dot{\omega} - \dot{\omega}^{\rm GR})$	XOMDOT			x		
$\dot{P}_{\rm b}$	PBDOT	x	x		x	x
$(\dot{P}_{\rm b} - \dot{P}_{\rm b}^{\rm GR})$	XPBDOT			x		
γ	GAMMA	x	x		x	
$(m_{\rm p} + m_{\rm c})$	MTOT			x		
$m_{\rm c}$	M2		x	x	x	x
$s = \sin i$	SINI		x		x	x
δ_θ	DTHETA		x		x	
$\delta_{\rm r}$	DR		o		o	
$\dot{x}$	XDOT	x	x	x	x	x
$\ddot{x}$	X2DOT				x	
$\dot{e}$	EDOT	x	x	x	x	x
$\sin\eta/\sin\lambda$	AFAC			o		
A	A0		o		o	o
B	B0		o		o	o

where the subscript '8' indicated that this quantity is computed over a (somewhat arbitrarily chosen) time span of 10^8 s. The indicated trend – that millisecond pulsars are more stable rotators than normal pulsars –

is seen easily in Figure 8.6, in which we show Δ_8 as a function of period derivative.

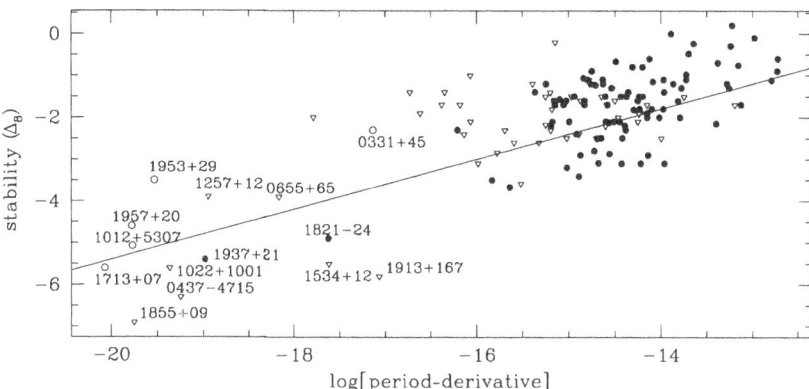

Fig. 8.6. The timing noise parameter, Δ_8, plotted as a function of period derivative. Figure provided by Don Backer.

One way of removing timing noise from the data is to include higher-order derivatives of ν in the timing model. This procedure is called 'polynomial whitening' as sometimes several orders are needed to remove most of the red noise. A superior method models the long-term behaviour as a sum of harmonically related sine waves which is subtracted from the timing residuals before further analysis is done. This method, known as harmonic whitening (Hobbs *et al.* 2004), has the advantage of allowing control of the timescales affected by the whitening. For example, a 1 yr periodicity visible as a sine wave with increasing amplitude caused by proper motion can remain unaffected, so that proper motion can be determined in a subsequent fit.

8.5 Further reading

The general process of pulsar timing has been discussed in the excellent reviews by Backer and Hellings (1986) and Taylor (1992); Taylor also explains the idea of measuring times of arrival with frequency domain template matching in some mathematical detail.

The classic paper by Blandford and Teukolsky (1976) introduces the important aspects in timing binary pulsars. The comprehensive papers by Damour and Deruelle (1985; 1986) develop and present a timing formula that should be used for timing relativistic binaries. Its application for the test of theories of gravity has be presented in great detail by Damour and Taylor (1992). For recent reviews on the use of binary pulsars for tests of strong-field gravity, the interested reader should consult the excellent reviews by Wex (2001) and Stairs (2003). For pulsar–white–dwarf systems these are best performed using the so-called *parameterised post-Newtonian* (PPN) formalism. Further details are given by Will (2001).

As outlined above, several timing models have been developed for timing various types of binary systems. Details can be found in the quoted literature. Other details, like extensions of the Blandford–Teukolsky model for non-linear periastron advance (BT+) or for planetary orbits (BT1P, BT2P) can be found in the TEMPO documentation, accessible from the book web site (see Appendix 3). The aspects of timing a pulsar in a planetary system like PSR B1257+12 in particular are described in a number of papers by Wolszczan (1994), Konacki *et al.* (2000) and Konacki and Wolszczan (2003).

8.6 Available resources

We provide links to TEMPO, PSRTIME and TIMAPR on the book web site (see Appendix 3). By far the most widely used and extensive of these packages is TEMPO. Obtaining timing solutions with these programs is an iterative procedure that is best learnt by experience with real data. In order for the interested reader to gain experience, a set of example TOA files for use with TEMPO are given in the timing area of the web site. Also included is a utility to create fake TOAs for simulating the measurability of various parameters. This is particularly useful when justifying telescope time requests for timing proposals.

One of the shortcomings[5] of TEMPO, particularly to the novice user, is its lack of a graphical user interface (GUI). A number of tools have been developed to provide GUIs to TEMPO and display and analyse the residual files. PLOTRES developed by Norbert Wex is a very useful utility for plotting and analysing timing residuals produced by TEMPO. Two recently developed GUIs to TEMPO are RHYTHM (Hotan *et al.* 2004) and TempoTK developed by one of us (DRL). Links to the source code and documentation on all these packages are available on the web site.

---

5 Although for some TEMPO 'black-belts' GUIs are not considered to be true to the art and the lack of them is considered a bonus rather than a drawback!

A completely new and re-written version of TEMPO designed to address the issues relevant for a pulsar timing array (see Section 2.3.3), and to provide a GUI, is currently under development by George Hobbs (SUPERTEMPO). We plan to provide a link to this package when it becomes available.

9

Beyond single radio dishes

In previous chapters of this book we aimed to provide the reader with the necessary tools and background for single-dish radio observations. Pulsar astronomy, however, is a truly multi-wavelength science, with fantastic results at infrared, optical, X-ray and γ-ray frequencies. In order to carry out or analyse most of these observations, a knowledge of the techniques described in the previous chapters is useful or even required. Moreover, most observations and their interpretation beyond single radio dishes still rely on accurate pulsar ephemerides that are obtained with single-dish radio observations. In this final chapter, we look at different observing modes and frequencies and their relationship to classical radio observations. We outline the basic principles and, in the case of high-energy observations, provide also some background information that is needed to construct the 'big picture' of the pulsar phenomenon.

9.1 Co-ordinated multi-station radio observations

In most single-dish radio observations, the telescope is equipped with only one receiver at any one time and therefore can cover only a single-frequency band (see Appendix 1). However, there are a number of reasons why simultaneous multi-frequency observations are desirable. These include the study of the radio emission by investigating radius-to-frequency mapping or possible propagation effects in the pulsar magnetosphere that are frequency-dependent. Multi-frequency observations of individual pulses, most naturally carried out using coordinated observing time on multiple telescopes, promise to teach us much about the elusive radio emission process (see Chapter 3). They are particularly important by offering opportunity to obtain full polarisation information at widely spaced frequencies. This is otherwise extremely difficult to achieve.

After initial multi-frequency observations by Robinson *et al.* (1968) following the discovery of pulsars, a number of experiments were performed in the 1970s and 1980s (see, for example, Backer and Fisher (1974), Bartel *et al.* (1981), Davies *et al.* (1984) and Kardashev *et al.* (1986)) when efficient reduction of the accumulated data volume was a considerable problem. A simple but important hurdle to overcome in these observations was the lack of a common data format. As discussed briefly in Chapter 5, the multitude of special purpose data-acquisition systems result in a variety of data formats. In 1995, a joint European effort was made to derive a common flexible data storage format that became known as the European Pulsar Network (EPN) data format (Lorimer *et al.* 1998).

In the following years the EPN format was employed successfully for simultaneous multi-frequency observations using telescopes across Europe and the world (see, for example, Karastergiou *et al.* (2001), (2002), and (2003) and Kramer *et al.* (2003b)). An example of these observations is shown in Figure 9.1. The EPN format allows the sharing of pulsar data containing single-pulse information or simply pulsar profiles. Currently, more than 4000 profiles of about 650 pulsars are freely available in the EPN archive (see Appendix 3). Other data formats such as FITS are being used increasingly for similar purposes (Hotan *et al.* 2004).

The observations at each telescope are performed independently, with each system recording data according to the agreed common schedule. Once calibrated data are obtained for each telescope, they are converted into a common data format. The basic procedures are as detailed in Karastergiou *et al.* (2001) for EPN observations:

 (i) For each telescope, each bin of a given time series is transformed into an arrival time at the Solar System barycentre (SSB) at infinite frequency (see Chapter 8).
 (ii) Intervals are determined where the observations from the different telescopes overlap.
 (iii) All data are re-sampled to a common time resolution. This is usually the coarsest time resolution among the original data sets.
 (iv) The overlapping data segments are combined into a single multi-frequency multi-station EPN file that can be analysed conveniently. This file consists of records for each simultaneously recorded pulse, containing all measured polarisations for each telescope.

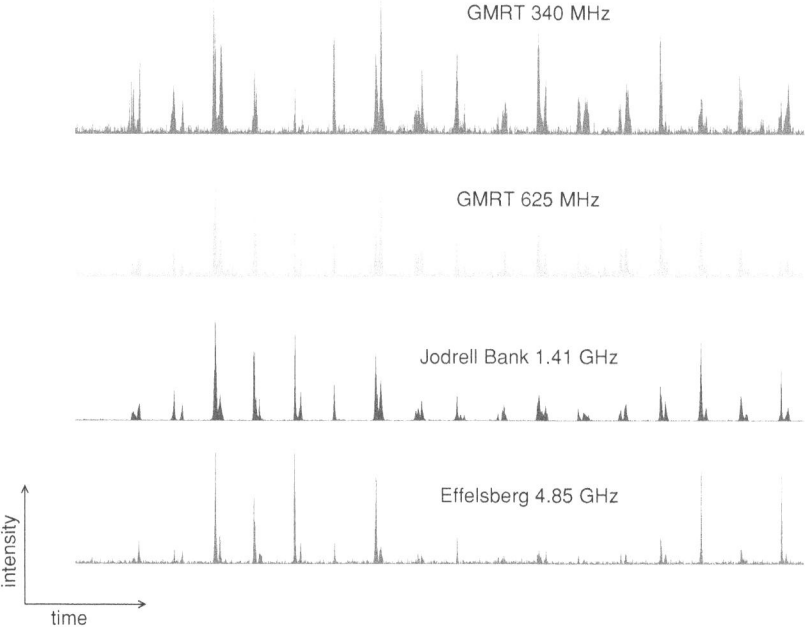

Fig. 9.1. A 22 s data slice obtained during coordinated observations of pulsar B1133+16, simultaneously made with the GMRT, Lovell and Effelsberg telescopes, spanning 4 octaves of frequency. The pulse missing at all frequencies at the beginning of the data set results from the nulling phenomenon (see Section 1.1.4.3). Figure provided by Aris Karastergiou.

Software to perform these steps and to analyse multi-dimensional EPN files is available (see Appendix 3).

9.2 Interferometric observations

Due to their extremely small size, pulsars appear as perfect radio point sources. Consequently, almost all of their studies at radio frequencies require only pointed observations using single-dish telescopes. Radio images, however, are essential to study the environment of a pulsar, e.g. its location in a supernova remnant, or 'bow-shock' interactions between a high-velocity pulsar and the local interstellar medium. The resolution that can be achieved with a single dish of diameter, D, observing at a given wavelength λ is $\sim \lambda/D$. As explained in Appendix 1, the typical resolution of the biggest single telescopes is of the order of a few arc minutes and usually not sufficient for resolving fine structure. Higher

resolution can be obtained only by combining a number of single telescopes into an interferometer – an array of telescopes that synthesise a
much larger telescope with a diameter equivalent to the largest spacing
between individual elements in the array.

In addition to radio images, the resolution provided by an interferometer also can be used to make astrometric measurements of pulsars.
While precise astrometry can be obtained through single-dish pulsar
timing measurements (see Chapter 8), and is better in some cases than
using interferometry, there are a number of situations in which the astrometric precision provided by interferometric observations is extremely
useful:

 (i) If the pulsar is located near the ecliptic plane, its timing position
     and proper motion are measured only poorly in the latitudinal
     component of its ecliptic coordinates (see Section 8.2.3).
 (ii) If a pulsar is young and shows significant timing noise, methods
     to remove it (see Section 8.4) may not be successful, so that high-
     precision timing is not possible.
(iii) The positional parameters and the period derivative of a pul-
     sar are highly correlated in the least squares analysis of timing
     data. Observations spanning a full Earth orbit therefore are nec-
     essary for a reliable determination. If astrometric information
     is required on a timescale significantly shorter than 12 months,
     interferometric observations should be made.
(iv) A parallax measurement provides a precise and reliable distance
     estimate. This has been achieved for twenty pulsars (see, for
     example, Brisken *et al.* (2000)).
 (v) An interferometric position also can help in obtaining an *initial*
     timing solution – in particular for binary pulsars – as the critical
     positional parameters can be held fixed during the fitting process.
(vi) A comparison of interferometric and timing measurements per-
     mits a study of the relationship between different astrometric
     reference frames (see also Chapter 2).

The basic concept of interferometric measurements can be understood
by considering the signals received with two elements of the telescope
array (see Figure 9.2a). As outlined in Appendix 1, a radio telescope
measures the (complex) electric field vector, $\vec{\mathcal{E}}$, of the incident radio
wave. The output of a interferometer is the cross-correlation of the
electric fields measured by each pair of telescopes, i and j, pointing to
the same position in the sky. We call the result of this cross-correlation

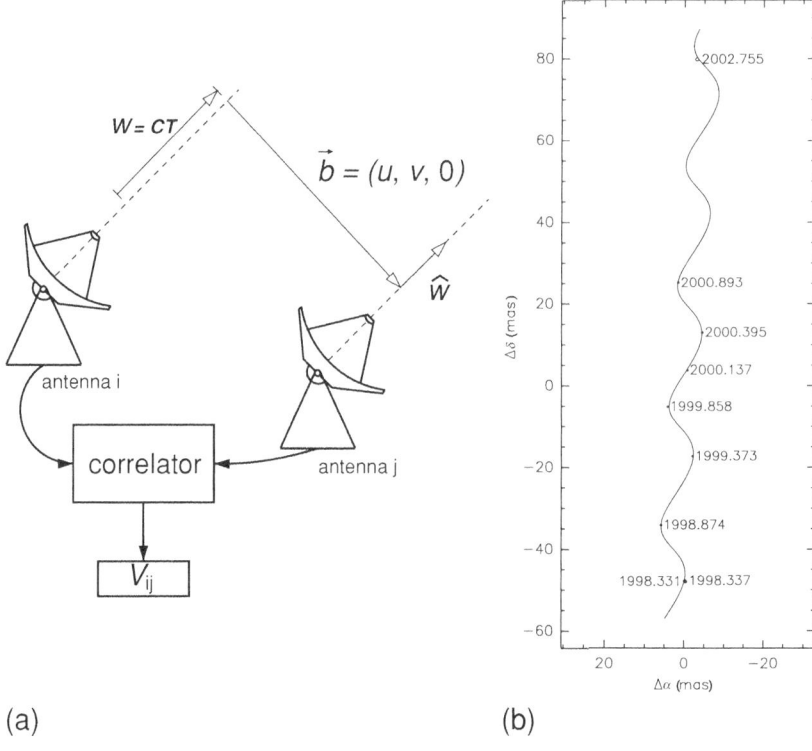

(a)   (b)

Fig. 9.2. (a) Schematic showing a two-element interferometer. Both telescopes are pointed at the same source. Signals arriving at antenna i are delayed by an amount corresponding to the projected light travel distance between the antennae. The baseline vector $\vec{b}$ describes the telescope separation and determines the spatial resolution of the observations. The antenna signals are correlated to produce the visibility function V_{ij}. (b) Positions of PSR B0950+08 measured interferometrically at eight different epochs. The best-fit model for parallax and proper motion model is shown as the solid line. Figures provided by Walter Brisken.

obtained at an observing frequency f the visibility:

$$V_{ij} = \langle \mathcal{E}(\vec{r}_i, t)\mathcal{E}^*(\vec{r}_j, t + \tau)\rangle = \langle \mathcal{E}(\vec{r}_i, t)\mathcal{E}^*(\vec{r}_j, t)e^{i2\pi f\tau}\rangle, \qquad (9.1)$$

which will be non-zero for the correct time delay, τ, that it takes for the electromagnetic wave of a source to travel the distance between telescopes i and j. Signals received from a different direction correlate under a different time delay, so it is clear that a measurement of τ would provide a determination of the source position. In practice, this is achieved by using the rotation of the Earth. Since the visibility depends on the relative position of the telescopes to the source, this orientation changes with time, causing a slow drift in the visibilities. During correlation performed with special purpose hardware ('correlator'), all phases are adjusted to correct for this drift, so that a zero delay is maintained at the image centre (*phase centre*) given by the pointing direction of the telescopes. With the visibility referring to the image centre, any source within the telescope beam but located at an offset position will be left with a non-zero phase that increases linearly with its separation from the phase centre. The non-zero phase of the source changes with time, allowing its location to be determined.

If a sufficient number of baselines can be sampled, one can show that the visibilities represent the Fourier components of an observed radio image that can be reconstructed by an inverse Fourier transform (see, for example, Thompson *et al.* (1986)). In practice, the quality of the image is limited by the number and orientations of the available baselines during the observation. An under-sampling of the Fourier plane causes distortions in the resulting image. While a well-designed interferometer array makes use of the rotation of the Earth to fill the Fourier plane – *Earth-rotation aperture synthesis* (Ryle 1952), the image quality is less important for obtaining astrometric information. However, a high-fidelity image can improve significantly the calibration needed to achieve high-precision astrometric results (see, for example, Brisken (2001)), as the position of a pulsar is best determined relative to well-known calibrator sources in the background observed in the same image.

There are a number of practical issues to be considered when observing with an interferometer, e.g. the influence of the ionosphere on the measured phases (see, for example, Thompson *et al.* (1986)) which limits the achievable precision considerably. While ionospheric effects decrease at higher frequencies (see, for example, Bailes *et al.* (1990)), the typical steep spectrum of pulsars (see Chapter 1) decreases their already low flux densities even further making them harder to detect. The use of 'in-beam' calibrator sources as a phase reference for the pulsar target (see, for example, Fomalont *et al.* (1999) and Chatterjee *et al.* (2001)) is particularly useful in minimising ionospheric effects and, hence, improving positional accuracy. Another issue to contend with is that pulsars are not continuum sources. During a typical observation to determine the visibility lasting perhaps 10 s, the pulsar will switch on and off many times. As a result, for a typical pulse duty cycle (pulse width divided by the period) $\delta = 0.05$, the correlator would try to correlate noise for ~ 95 per cent of the time! In order to recover signal to noise (S/N) ratio, modern correlators can be turned off for the times outside the pulse window (see, for example, McGary *et al.* (2001)). This is achieved by applying a binary gate to the data that is switched synchronously with the pulsar period for the duration of the pulse . This *pulsar gating* process increases the S/N by a factor of approximately $\sqrt{1/\delta}$.

For the gating to be successful, one must have accurate knowledge about the exact expected arrival time of the pulse. Since the correlation usually happens off-line after the actual observation, the most reliable way is to augment the interferometric observations with standard timing observations before and after using a single dish telescope. The resulting timing ephemerides are then used to produce a 'prediction file' that is read in by the correlator software to produce a gate appropriate for the interferometric observation. Finally, the pulsar gate has to take into account that the pulses will be dispersed over the correlator bandwidth, so that the gate width needs to be adjusted for the corresponding pulse smearing. A beautiful example by Brisken *et al.* (2002) of astrometric information obtainable with interferometric observations for PSR B0950+08 is shown in Figure 9.2b.

9.3 Optical observations

Given that a highly magnetised and rotating neutron star had already been suspected at the centre of the Crab nebula even before the discovery of pulsars (Hoyle *et al.* 1964; Pacini 1967), it should not surprise that the Crab pulsar, B0531+21, was also the first pulsar detected at optical wavelengths (Cocke *et al.* 1969). However, it was soon realised that pulsars are generally very weak optical sources, and it took almost a decade before the Vela pulsar, B0833−45, was detected as the second pulsar to show optical pulsations (Wallace *et al.* 1977). Even though the detection rate increased with the arrival of new-generation optical telescopes and the use of charge-coupled detectors (see, for example, Mignani *et al.* (2004)) the sample of pulsars detected at optical wavelengths is still small. We summarise briefly below the detections currently known, where we loosely combine infrared, optical and ultraviolet wavelengths as the 'optical regime', covering frequencies from roughly 10^{12} to 10^{16} Hz.

9.3.1 *Observational status*

In addition to the Crab and Vela pulsars, optical pulsations have been detected from only two further radio pulsars: PSR B0540–69 (Caraveo *et al.* 1992; Shearer *et al.* 1994) and PSR B0656+14 (Caraveo *et al.* 1994a; Shearer *et al.* 1997). Optical pulse profiles are obtained by folding the photons at the period predicted by radio timing ephemerides as described in Section 7.1. Confirmation of an optical detection also can be obtained if the optical counterpart is found to have a proper motion that coincides with that of the radio pulsar measured as described in Chapter 8. Such a confirmation has been achieved for PSR B0656+14 (Mignani *et al.* 2000) and PSR B1929+10 (Pavlov *et al.* 1996; Mignani *et al.* 2002). For the latter, optical pulsations have not been detected yet.

In four further cases, an optical point source is found at the pulsar position but no pulsed emission or proper motion has so far been measured. These sources are PSR B0950+08 (Pavlov *et al.* 1996), PSR B1055−52 (Mignani *et al.* 1997), PSR B1509−58 (Caraveo *et al.* 1994b) and the only millisecond pulsar in the sample of optical pulsars, PSR J0437−4715 (Kargaltzev *et al.* 2004).

9.3.2 Characteristics of optical emission

Although the current sample is not large enough to allow a thorough statistical analysis, a few general statements can be made (Mignani *et al.* 2004). The optical spectra appear in both power law and black body forms. Young and old pulsars show power law spectra while middle aged pulsars also have a black body component produced by thermal emission from the polar cap or the whole surface. The emission of the millisecond pulsar J0437–4715 is consistent with a black body spectrum.

The Crab is the only pulsar bright enough (visual magnitude 16.6) to allow extensive studies of its single pulses and their polarisation properties. Such studies reveal interesting differences between the peak emission of the prominent double-peaked profile and the bridge and non-zero off-pulse emission (see, for example, Romani *et al.* (2001)). The Crab's optical pulse profile (see Figure 9.3) is very similar to that observed at X-ray and γ-ray energies, suggesting that the emission is created by the same incoherent non-thermal radiation process, presumably at the same location. The optical spectrum does not always match the extrapolation of the corresponding components observed in the X-ray domain, and a general optical/X-ray correlation cannot be recognised.

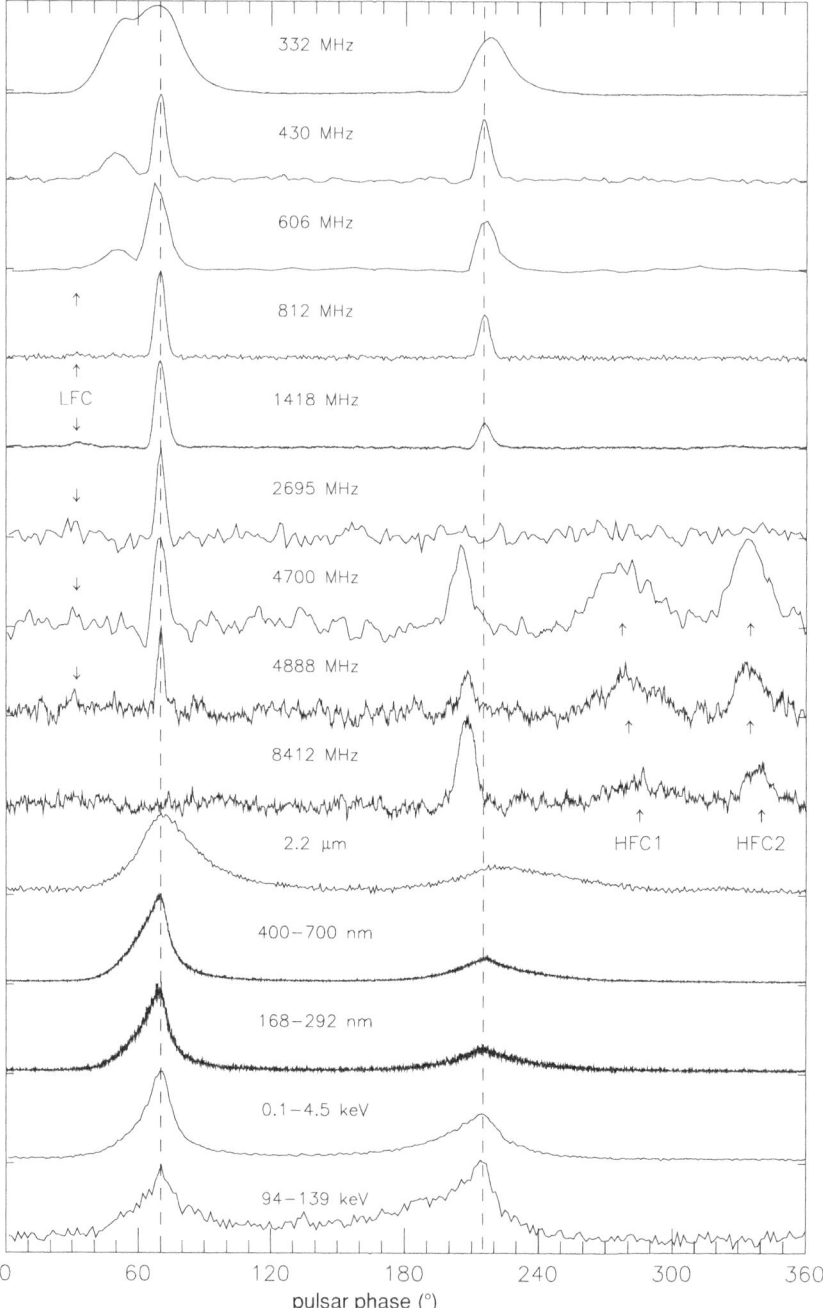

Fig. 9.3. Multi-wavelength profiles for the Crab pulsar showing the evolution of the pulse shape across the electromagnetic spectrum. The two prominent components are aligned from radio to γ-ray frequencies. At 5 GHz a number of additional profile components emerge, indicated by the arrows. Figure provided by Dave Moffett.

9.3.3 Geminga

Another detection of a rotating neutron star seen to emit pulsed optical emission is the enigmatic source *Geminga*[1]. First discovered at γ-ray energies by the SAS-2 satellite (Fichtel *et al.* 1975), Geminga was subsequently shown to be a strong X-ray and γ-ray pulsar by the discovery of 237 ms pulsations with the X-ray satellite *ROSAT* (Halpern & Holt 1992) and is now listed in the pulsar catalogues as PSR J0633+1746. Once the period was known, archival data were used to perform a timing analysis in the same way as described for the radio timing observations (see Chapter 8). An ephemeris from high-energy timing observations now spans almost 30 years (Jackson *et al.* 2002), (see Bignami and Caraveo (1996) for the long history of Geminga observations).

Although Geminga shows all properties of a radio pulsar seen at high energies, no radio pulsations were seen initially, despite extensive searches (see, for example, Seiradakis (1992)) folding at the period predicted by the high-energy ephemeris. While detections of erratic emission from the pulsar have been reported at radio frequencies between 40 and 102 MHz (Kuzmin & Losovskii 1997; Malofeev & Malov 1997; Shitov & Pugachev 1998), attempts to confirm these independently at a variety of similar, but also higher, frequencies have not been successful (Ramachandran *et al.* 1998; Kassim & Lazio 1999; McLaughlin *et al.* 1999).

However, as we shall see below, following the discovery of a number of pulsating X-ray sources that clearly are spinning neutron stars, Geminga now appears much less enigmatic. These cases can be explained either by very weak radio emission or the possibility that a narrow radio beam may not be directed toward Earth, while the wider beam of the high-energy radiation can be observed.

9.4 High-energy observations

The wealth of information that can be obtained outside the classical radio window today justifies some re-definition of the word 'pulsar'. Rather than being characterised solely by appearing as pulsating *radio* sources, a new definition of *pulsar* should perhaps be *a neutron star that emits radiation that is pulsed due to rotation and powered by rotational energy*. This definition also encompasses pulsars like Geminga that are clearly

---

1 As explained by Bignami and Caraveo (1996), Geminga is named after a pun in the Milanese dialect meaning 'it is not there' or 'it does not exist'.

powered by the loss of rotational energy but which have so far not been detected at radio frequencies. This may be due to a misaligned radio beam or a very low radio luminosity. The discovery of an increasing number of very weak radio pulsars coincident with X-ray point sources (see Camilo (2003) for a review) shows clearly that the expression 'radio-quiet pulsar' cannot be used without the discussion of a corresponding flux density limit of the radio searches performed so far. Bearing this in mind, the simplest assumption is that there is no fundamental difference in the physics of radio pulsars and these objects.

In contrast to the above 'rotation-powered pulsars' and so-called 'accretion powered' X-ray pulsars observed as X-ray binaries, another class of rotating neutron stars is observed as high-energy sources that emit short and powerful periodic bursts of X-rays and γ-rays. With typical inferred spin periods in the range 5–12 s, the energy output appears to be too large to be accounted for by the loss of rotational energy. The high-energy emission from these neutron stars is most likely to originate from the their magnetic fields that are inferred to be extremely strong, i.e. $\sim 10^{15}$ G (Duncan & Thompson 1992). These sources are accordingly called *magnetars*, and their class includes both the so-called *soft-gamma repeaters* and *anomalous X-ray pulsars*. A recent review of these sources is given, for example, by Kaspi (2004). For the rest of the discussion we will concentrate on spin-powered pulsars.

9.4.1 X-ray properties

Observations at X-ray energies promise to offer a more direct insight into the magnetospheric processes than the view provided by radio observations. The reason is the confusion by the uncertain physics responsible for the coherence of the radio emission and the requirement to separate radio emission properties from propagation effects in the interstellar medium and the pulsar magnetosphere (see Chapter 3). On the other hand, in many pulsars the observed X-ray emission is, as in the optical regime, due to a mixture of thermal and non-thermal processes that cannot always be discriminated by the available data.

Non-thermal, magnetospheric emission created by an accelerated relativistic plasma should be highly pulsed with a power-law spectrum ranging from optical to γ-ray frequencies (see, for example, Becker and Trümper (1997)). Thermal emission originates from the stellar surface and may still show low-amplitude modulation due to the rotating hot polar cap. In this case, we would expect a black-body spectrum that is modified by a possible atmosphere of the neutron star, ranging from optical to soft X-ray frequencies. Complications may arise due to unresolved emission from pulsar nebulae or winds interacting with dense ambient media.

9.4.1.1 Observational status

At the time of writing, pulsed emission has been observed for fifteen normal radio pulsars. X-ray satellites like *ROSAT*, and now *Chandra* and *XMM-Newton*, have provided increased sensitivity, timing accuracy and temporal resolution so that pulsed X-ray emission is now observed in six of the known millisecond radio pulsars. In order to obtain pulse profiles, as for the optical observations, the photons are folded again with the radio timing ephemerides. In addition, a total of nine normal pulsars and eighteen millisecond pulsars have only been seen as unpulsed X-ray sources (see Becker and Aschenbach (2002) for a recent review).

The X-ray pulsars: Geminga (see above), J0537−6910 (Marshall *et al.* 1998), PKS 1209−51/52 (Zavlin *et al.* 2000), J1811−1925 (Torii *et al.* 1997) and J1846−0258 (Gotthelf *et al.* 2000, and see below) are all clearly spin-powered neutron stars but have so far no corresponding radio detection. In other cases, deep searches of X-ray point sources that appear to be neutron stars have led to the discovery of a number of low-luminosity radio pulsars (Camilo 2003). These include the 65 ms pulsar J0205+6449 in the supernova remnant 3C58, for which X-ray pulsations were first detected by *Chandra* (Murray *et al.* 2002) and later at radio frequencies (Camilo *et al.* 2002). As 3C58 is fairly convincingly associated with the remnant of the historical supernova SN1181 (see, for example, Stephenson and Green (2002)), PSR J0205+6449 now supplants the Crab as the youngest known radio pulsar. Currently, the youngest known spin powered neutron star is PSR J1846−0258, a 325 ms X-ray pulsar in the 700 yr old supernova remnant Kes 75 (Gotthelf *et al.* 2000). This pulsar has no known radio counterpart, yet.

9.4.1.2 Characteristics of X-ray emission

The X-ray detected pulsars are often grouped into Crab-like sources, Vela-like sources, middle-aged pulsars and millisecond pulsars. The Crab-like pulsars show strong features of magnetospheric emission and double-peaked profiles that are, if detected, aligned with similar optical profiles. For Vela-like pulsars, only Vela itself has been detected as a faint optical source. The X-ray spectra of Vela-like pulsars do not represent simple power laws, while the pulse profiles are more complicated and typically misaligned with radio or γ-ray pulses. Middle-aged pulsars show a mixture of thermal and non-thermal emission with energy-dependent profiles.

Among the millisecond pulsars, only the six pulsed sources are bright enough to infer their spectral properties. However, in most cases the data can be equally well described by power law and/or black body fits to their spectra. The similarities of the X-ray profiles with the radio pulse shapes suggest a non-thermal magnetospheric origin of the emission (see, for example, Becker and Aschenbach (2002) for a review) In the case of PSR J0437−4715, for example, the data can be described by a combination of a power law and two thermal black body components, possibly originating from two surface regions of different temperature (Zavlin *et al.* 2002).

9.4.2 Gamma-ray properties

Following early studies of the γ-ray sky by, for example, the SAS-2 and COS-B observatories (see Ramaty and Lingenfelter (1983) for a review), γ-ray observations with the *Compton Gamma-ray Observatory* (CGRO) provided significant advances in γ-ray astrophysics in the 1990s. The three instruments aboard CGRO were: (a) the *Oriented Scintillation Spectrometer Experiment* (OSSE); (b) the *Imaging Compton Telescope* (COMPTEL); (c) the Energetic Gamma Ray Experiment Telescope (EGRET). Now we summarise briefly the main results known. For further details, see Thompson (2001) or Kanbach (2002).

9.4.2.1 Observational status and emission characteristics

The γ-ray emission of pulsars is highly pulsed, beamed and of non-thermal origin. There are seven 'classical' γ-ray detections: Crab, Vela, Geminga and PSRs B1055−52, B1509−58, B1706−44 and B1951+32. These sources are bright enough to obtain pulse profiles by folding the photons using radio timing ephemerides. The profiles are typically double-peaked and change their appearances towards higher energies due to a phase-dependent spectral hardness, suggesting that geometry plays an important role. Similar studies are not possible for other, much weaker, pulsars for which a detection has been suggested based on folding analyses of archival data, e.g. the millisecond pulsar J0218+4232 (Kuiper *et al.* 2000); the young pulsars J1105−6107 (Kaspi *et al.* 1997), B1046−58 (Kaspi *et al.* 2000a) and J1749−2958 (McLaughlin & Cordes

2004). In these cases, the number of photons is limited and significance levels are necessarily much lower.

A survey of the sky with EGRET revealed about 170 sources with no currently known counterparts in other wavebands (Hartmann *et al.* 1999). There are good reasons to suspect that the subset of these unidentified EGRET sources that do not exhibit flux variability (see, for example, McLaughlin *et al.* (1996)) are γ-ray pulsars, some of which may be detectable at other wavebands, including the radio. Currently, steady progress is being made towards resolving the mystery of these enigmatic sources. Two promising new radio pulsar counterparts have been found: PSR J2229+6114 during a systematic multi-wavelength search of the EGRET source 3EG J2227+6122 (Halpern *et al.* 2001); and PSR J2021+3651, the possible counterpart to 3EG J2021+3716, found during a deep survey of selected point X-ray sources within EGRET error boxes (Roberts *et al.* 2002). For the latter source, tentative γ-ray pulsations at the period of J2021+3651 have been proposed recently by McLaughlin and Cordes (2004).

The large number of radio pulsars discovered in the Parkes multibeam survey of the Galactic plane also have produced a number of candidate associations between young pulsars and the EGRET sources (Camilo *et al.* 2001; D'Amico *et al.* 2001; Torres *et al.* 2001; Kramer *et al.* 2003a). Given the relatively large positional error boxes for these sources (often up to a degree or more) such associations naturally are uncertain. The sample of 48 candidate associations recently have been studied in detail by Kramer *et al.* (2003a) who estimate, on statistical grounds, that only about 19±6 are genuine. This number is in good agreement with detailed population syntheses by McLaughlin and Cordes (2000).

We can expect significant progress in the near future as new γ-ray satellite missions are coming online. Already, the *INTEGRAL* satellite is operational and surveying the Galactic plane at the time of writing (Molkov *et al.* 2004). The *AGILE* and *GLAST* telescopes will be launched in the next few years (see Thompson (2004) for a review). The big advantage of these missions over their predecessors will be the greatly reduced positional uncertainties of the detected sources and, in particular for *GLAST*, the increased collecting area. The accompanying sensitivity increase ultimately should result in the discovery of new pulsars in all the unidentified EGRET sources as well as confirming the existing periodicities mentioned above. Indeed, rather than requiring radio observations to detect and time a source at γ-rays, new γ-ray pulsars can be discovered independently and followed up in the radio regime!

9.5 Pulsars across the electromagnetic spectrum

As we have seen in this chapter, pulsars provide not only a wide variety of observational manifestations in the radio regime but also at optical, X-ray and γ-ray frequencies. It is unlikely that a coherent, self-consistent model that can explain *all* the observed properties will be developed any time soon. However, with continued observations across the electromagnetic spectrum we can at least attempt to identify some basic features of the 'big picture'. While it would be naïve to assume that the radiation processes at the various parts of the electromagnetic spectrum are governed by a single mechanism or a few control variables, it is nevertheless instructive to relate some of the radiation properties to observables. In the following we consider the observed spin period P and period derivative $\dot{P}$ that determine the spin-down luminosity $\dot{E} \propto \dot{P}/P^3$, the magnetic field strength at the surface $B_{\mathrm{s}} \propto \sqrt{P\dot{P}}$ and the magnetic field at the light cylinder $B_{\mathrm{LC}} \propto P^{-5/2}\dot{P}^{1/2}$ (see Chapter 3).

9.5.1 Luminosity scaling laws and radiation efficiencies

A comparison of the energy output at radio, optical, X-ray and γ-ray frequencies is complicated by uncertain distance estimates, different spectral shapes, instrumental effects and, in particular, by the unknown beaming fractions. The radio emission appears to be emitted in a cone of radius ρ that scales with period as $\rho \propto 1/\sqrt{P}$ (see Chapter 3). At optical and X-ray frequencies, apparently we observe a mixture of thermal and non-thermal emission, so that one typically assumes isotropic emission in the computation, even though this is obviously wrong for non-thermal emission. At γ-rays a beaming fraction of $1/4\pi$ usually is adopted due to the lack of better knowledge. Bearing these uncertainties in mind, we can try to compare the derived luminosities 'observed' at the different frequencies with the spin-down luminosity, $\dot{E}$.

In their statistical analysis of the radio pulsar population, Arzoumanian *et al.* (2002) favour a model in which the intrinsic population has a radio luminosity $L_{\mathrm{R}} \propto \dot{E}^{1/3}$. This result should be viewed with some caution, since no significant correlation exists between L_{R} and $\dot{E}$ for the observed pulsars. This uncertainty is perhaps not unexpected, given the importance of geometry and the complicated processes that must be present in order to create coherent emission, not to mention uncertainties in the pulsar distance scale.

In the optical range, one can use peak luminosities to mitigate geomet-

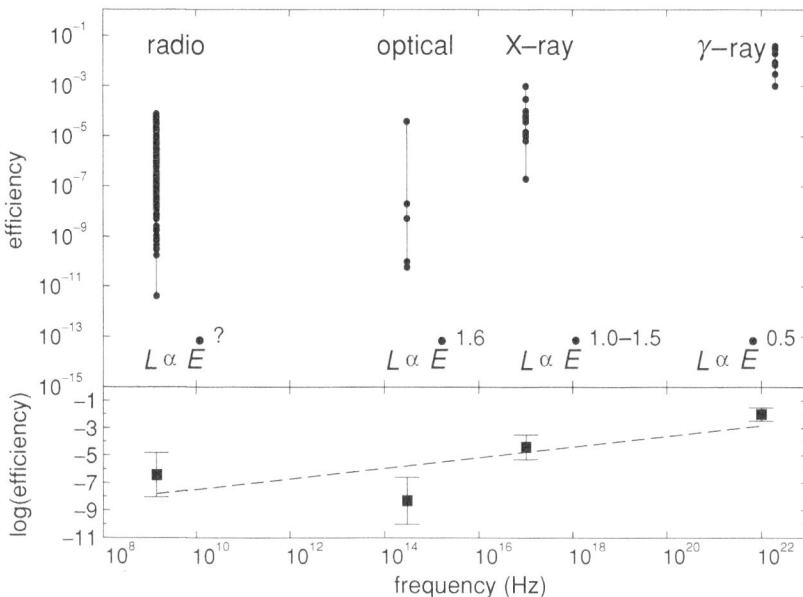

Fig. 9.4. Upper: efficiency, $\eta \equiv L_{\mathrm{f}}/\dot{E}$, as derived for radio, optical, X-ray and γ-ray frequencies. The spread of inferred values differs as indicated for each frequency band. Lower: a fit to the median values shows an increase of efficiency with frequency of $\eta \propto f^{0.17\pm0.10}$.

rical issues or luminosities integrated over the pulse profile. Shearer and Golden (2001) find the peak luminosity $L_{\mathrm{O}} \propto \dot{E}^{1.6\pm0.2}$ (see Figure 9.4). This is consistent with a model proposed by Pacini (1971), explaining optical emission as the result of incoherent synchrotron radiation that is created in the outer magnetosphere and the intensity of which scales with the magnetic field strength at the light cylinder.

The X-ray luminosity, L_{X}, was studied by a number of authors, including Becker and Trümper (1997) who found $L_{\mathrm{X}} \sim \dot{E}$ in the soft X-ray band (0.1–2.4 keV). More recently, Possenti *et al.* (2002) reviewed the harder X-rays (2–10 keV), deriving a relationship remarkably similar to that at optical frequencies, $L_{\mathrm{X}} \propto \dot{E}^{1.5}$. For γ-rays, small-number statistics prevent a detailed analysis, but the values are consistent with the γ-ray luminosity $L_{\gamma} \propto \dot{E}^{0.5}$ (Thompson 2001). Interestingly, this has the same dependence on P and $\dot{P}$ as the accelerating potential above the polar cap (see Chapter 3).

While inferred efficiencies, $\eta_{\mathrm{f}} = L_{\mathrm{f}}/\dot{E}$, show a large variation at each frequency band, f, as a general trend they increase from radio to γ-rays.

A formal fit gives $\eta_f \propto f^{0.17\pm0.10}$ (see Figure 9.4). The least amount of scatter is seen in the soft X-ray band in which it is found that both normal and millisecond pulsars appear to emit only 0.1 per cent of their spin-down luminosity, $\dot{E}$ (Becker & Trümper 1997).

9.5.2 The importance of the magnetic field

In order to explain the lack of radio emission from magnetars, it has been suggested that a surface magnetic field exceeding the quantum critical field, $B_{\mathrm{crit}} = m_e^2 c^3/e\hbar = 4.4 \times 10^{13}$ Gauss (see Equation (3.46)), would quench the radio emission due to the lack of emitting plasma (Zhang & Harding 2000), (see Chapter 3 and Figure 1.13). However, the discovery of pulsars with magnetar-like spin-parameters and derived magnetic fields above the critical field, like the 6.7 s pulsar J1847−0130 with $B \sim 10^{14}$ Gauss (McLaughlin *et al.* 2003), question the importance of the actual field value for the emission process.

A more directly observable effect seems to be caused by the magnetic field strength at the light cylinder, $B_{\mathrm{LC}} \propto P^{-5/2}\dot{P}$. In order to demonstrate this, consider the phenomenon of giant radio pulses discussed in Section 1.1.4.2. All these pulsars with detected giant-pulse emission are detected also at high energies. Another common feature they share is the highest magnetic field at their light cylinders (Cognard *et al.* 1996), (see Figure 9.5). It is unclear whether this is related to the physics of giant pulses, but it is interesting that this property of the outer magnetosphere may connect the giant pulse radio emission to the place of origin that is suggested for high-energy emission (see Section 3.7). Indeed, the giant pulses often occur misplaced from the normal radio profile but appear to be aligned with X-ray and/or γ-ray emission (see Romani and Johnston 2001 and Figure 9.5). This suggests a common origin of the giant pulses and high-energy emission, and indicates that some observed radio emission could be a by-product of the high-energy radiation process. This could explain the highly unusual high-frequency components, seen to emerge at some odd pulse phases in the Crab profile at a few GHz shown in Figure 9.3 (Moffett & Hankins 1996), as the result of such possible by-products and radius-to-frequency mapping effects.

Giant pulses finally may provide the clue to connect the emission across the whole electromagnetic spectrum. Indeed, there appears also to be a direct observational link between the radio giant pulses of the Crab and its optical emission: those optical pulses occurring simultaneously with radio giant pulses appear to be somewhat brighter than

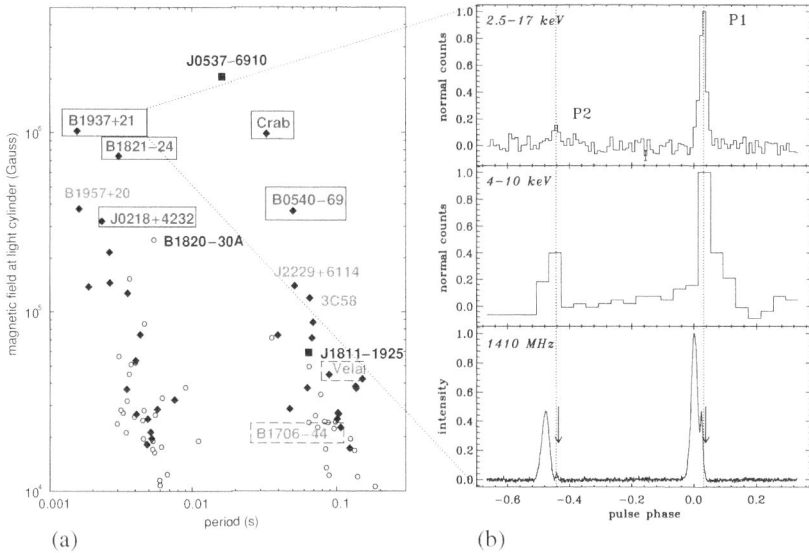

Fig. 9.5. (a) Magnetic field strength at the light cylinder for the pulsars with the highest values. Filled diamonds mark pulsars detected both at radio and high energies. Spin-powered pulsars without known radio detection ('radio-quiet') are shown as filled squares. Pulsars with detected giant-pulse emission are surrounded by a box. The Vela pulsar and PSR B1706−44 show so-called giant 'micro-pulses' (see Section 1.1.4.2). (b) Pulse profile of PSR B1937+21 in the 2–17 keV energy band (top), at 4–10 keV (middle) and at 1410 MHz (bottom). Vertical arrows indicate the phases of observed radio giant pulses that align with the X-ray profile but lag the radio pulse slightly (see Cusumano *et al.* (2003) for details).

other optical pulses (Shearer *et al.* 2003). Therefore, a picture emerges in which the radio emission originates from close to the stellar surface, while the high-energy emission may tend to be created further out. Consequently, one would not expect an alignment of radio and high-energy profiles, as typically observed. Where alignment is observed (e.g. as for the Crab pulsar) the observed radio emission may be of different origin and more related to that at high energies. Only the Crab's pre-cursor component may be considered as the classical radio pulse, while the main and interpulse components are by-products of high-energy processes causing also the high-frequency pulse components at a few GHz. If that is the case, the Crab pulsar should be considered as a much less luminous radio source, similar to PSR B0540−69 when ignoring its giant-pulse emission (Johnston & Romani 2003).

9.6 Further reading

The vast field of pulsar observations beyond single radio dishes is evident from the large body of literature. The state-of-the art of simultaneous multi-telescope radio observations is described in the work by Karastergiou *et al.* (2001; 2002; 2003).

The techniques of very long baseline interferometry (VLBI) are reviewed excellently in the book *Interferometry and Synthesis in Radio Astronomy* by Thompson, Moran and Swenson (1986). Modern techniques for pulsar VLBI observations have been presented by Fomalont *et al.* (1999) and by Chatterjee *et al.* (2001) and Brisken *et al.* (2002) and references therein.

The status of optical pulsar observations has been summarized recently in the reviews by Shearer and Golden (2001) and Mignani *et al.* (2004). Due to the weakness of the optical pulsar signal, progress is relatively slow, in contrast to the rapidly changing field of pulsar X-ray astronomy. A recent overview of the status of this research area is provided by the proceedings of IAU Symposium 218, which concentrated on young, energetic pulsars. A general review is given by Becker and Aschenbach (2002), who also discuss the X-ray detections of millisecond pulsars.

After the success of the CGRO, observational results on pulsars at γ-ray frequencies have been rather sparse. This situation is expected to change dramatically, as soon as the new satellites INTEGRAL, AGILE and finally GLAST will perform extensive pulsar observing progress. Consult Kanbach (2002) for a review about the current observational status and, for example, Thompson (2004) for results anticipated to be obtained with the new observatories.

9.7 Available resources

The software suite developed by the EPN to analyse multi-telescope multi-frequency data includes data format conversion tools, data viewers, and a number of analyse programmes. All software can be downloaded from the book web site (see Appendix 3).

Interferometric observations as well as high-energy observations usually require astrometric and spin-parameter information in order to fold and analyse the data. Such 'ephemerides' are obtained from radio observations. The programme TEMPO (see Chapter 8) uses the determined parameter information to compute polynomial coefficients that enable the observers to determine the expected topocentric pulse period and phase needed for folding the registered photons; TEMPO and software to read the produced prediction files are available from the book web site.

Appendix 1
Radio astronomy fundamentals

In various parts of this book, we assume results well known to most radio astronomers but less so to readers from other backgrounds. The key concepts are summarised in this appendix. Further details can be found in: *An Introduction to Radio Astronomy* (Burke & Smith 2002), *Tools of Radio Astronomy* (Rohlfs & Wilson 2000) and *Single-Dish Radio Astronomy: Techniques and Applications* edited by Stanimirovic *et al.* (2002).

A1.1 Radio wave properties

Radio waves are part of the spectrum of electromagnetic radiation. For observations of astronomical sources, we are always in the *far-field* regime where the received wavefronts can be considered to be parallel. A radio telescope is a device used to sample the plane of the electric field of this radiation which, in the general case, is a time-variable vector quantity, $\vec{\mathcal{E}}$. In addition to the wave amplitude, $\mathcal{E}_0$, the phase or polarisation state of the electromagnetic signal is useful for interpreting the source properties. In general, the electric field vector can be written as

$$\vec{\mathcal{E}}(z,t) = (\hat{a}\mathcal{E}_a + \hat{b}\mathcal{E}_b)\exp(2\pi i(ft - kz)), \qquad (A1.1)$$

where the wave traces out an ellipse in the electric field plane with major and minor axes $\hat{a}$ and $\hat{b}$ and propagates forward along a third orthogonal axis z with time t. As usual, $i = \sqrt{-1}$, the wave frequency $f = c/\lambda$, where λ is the wavelength, c is the speed of light and the wavenumber $k = 2\pi/\lambda$. The wave polarisation is characterised by:

$$\mathcal{E}_a = \mathcal{E}_0\cos\beta; \quad \mathcal{E}_b = -\mathcal{E}_0\sin\beta. \qquad (A1.2)$$

When $\beta = 0°$ or $90°$, the wave is linearly polarised. A purely circularly polarised wave has $\beta = +45°$ for the 'right handed circular polarisation' (RCP) case and $\beta = -45°$ for the 'left handed circular polarisation' (LCP) sense. Other values of β produce an 'elliptically polarised wave'.

For the purposes of actually measuring $\vec{\mathcal{E}}$, the 'feed' of the radio telescope samples two orthogonal axes x and y in the electric field plane. These are, in general, rotated by some angle Ψ with respect to the plane of the ellipse. At this stage, it is useful to introduce the four 'Stokes parameters' I, Q, U and V (Stokes 1852) which are defined in terms of $\mathcal{E}_0$, β and Ψ, and related to the observables $\mathcal{E}_x$ and $\mathcal{E}_y$ as follows:

$$I \equiv \mathcal{E}_0^2 = \mathcal{E}_x^2 + \mathcal{E}_y^2 \tag{A1.3}$$

$$Q \equiv \mathcal{E}_0^2 \cos 2\beta \cos 2\Psi = \mathcal{E}_x^2 - \mathcal{E}_y^2 \tag{A1.4}$$

$$U \equiv \mathcal{E}_0^2 \cos 2\beta \cos 2\Psi = 2\mathcal{E}_x \mathcal{E}_y \cos \phi \tag{A1.5}$$

$$V \equiv \mathcal{E}_0^2 \sin 2\beta = 2\mathcal{E}_x \mathcal{E}_y \sin \phi, \tag{A1.6}$$

where ϕ is the relative phase between the two vectors $\vec{\mathcal{E}}_x$ and $\vec{\mathcal{E}}_y$. We note a few useful identities from these definitions:

$$\mathcal{E}_0^2 = I = \sqrt{Q^2 + U^2 + V^2}, \ \sin 2\beta = V/Q \ \text{and} \ \tan 2\Psi = U/Q. \tag{A1.7}$$

A purely linearly polarised wave has $V = 0$, while $V/Q = 1$ for RCP and $V/Q = -1$ for LCP.

The above discussion implies a 'linear feed' sampling the linearly polarised components of the wave. One can also construct a 'circular feed' sampling the LCP and RCP components. In this case

$$I = \mathcal{E}_{\text{LCP}}^2 + \mathcal{E}_{\text{RCP}}^2 \tag{A1.8}$$

$$Q = -2\mathcal{E}_{\text{LCP}}\mathcal{E}_{\text{RCP}} \sin \phi \tag{A1.9}$$

$$U = 2\mathcal{E}_{\text{LCP}}\mathcal{E}_{\text{RCP}} \cos \phi \tag{A1.10}$$

$$V = \mathcal{E}_{\text{LCP}}^2 - \mathcal{E}_{\text{RCP}}^2. \tag{A1.11}$$

The Stokes parameters can be measured in a number of different ways. Traditionally, a custom-built 'polarimeter' is used to generate Q, U and V in hardware. Nowadays, it is more common to form the necessary cross products using a correlator or baseband recorder (see Chapter 5). Polarisation calibration is described in Chapter 7.

A1.2 Antenna beam width and aperture efficiency

Antennae collect and focus the incoming plane waves so that the radiation arrives at the feed coherently summed in amplitude and phase. The most convenient antenna is a parabolic design that focuses the radiation into a single point. Most large radio telescopes in use (e.g. Effelsberg, the Lovell telescope at Jodrell Bank and Parkes) are paraboloids. Alternative designs are spherical reflectors (e.g. Arecibo), elliptical dishes (e.g. Green Bank) and the elongated transit telescope at Nançay.

For the purpose of planning observations, a key quantity is the field of view or resolution of the telescope. The optics of parabolic and circular antennae can be considered as circular aperture diffraction gratings that are sensitive to radiation away from the focal axis. As a result, the telescope has a characteristic beam shape, or aperture illumination pattern centred around the pointing position. As described in Rohlfs and Wilson (2000), the effective response of a telescope of diameter, D, to incoming radiation of wavelength, λ, as a function of off-axis position (known as the 'power–polar' diagram) is a Bessel function with full-width half maximum (FWHM) $= 1.02(\lambda/D)$ radians and beam width to first null (BWFN) $= 2.44(\lambda/D)$ radians. Note that these expressions apply to the effective diameter (i.e. the illuminated part) of the dish. Practical considerations (mostly a tapering of the feed to minimise spillover radiation, (see below)) mean that only a fraction of the area is illuminated. This fraction η is known as the efficiency of the telescope. Expressing the quantities in more useful units gives

$$\text{FWHM} \simeq 35\,\text{arcmin} \times \left(\frac{\lambda}{\text{cm}}\right) \times \left(\frac{\sqrt{\eta}D}{\text{m}}\right)^{-1} \quad \text{(A1.12)}$$

$$\text{BWFN} \simeq 84\,\text{arcmin} \times \left(\frac{\lambda}{\text{cm}}\right) \times \left(\frac{\sqrt{\eta}D}{\text{m}}\right)^{-1}. \quad \text{(A1.13)}$$

For typical wavelength ranges $\lambda = 10$–100 cm, telescope efficiencies $\eta = 0.5$–0.8 and radio telescope sizes $D = 50$–300 m, beam sizes range from several arcmin up to about $1°$.

A1.3 Antenna gain, system noise and equivalent flux density

In radio astronomy, the unit of flux density is the Jansky (Jy). Expressed in terms of power received per unit collecting area and bandwidth, 1 Jy $\equiv 10^{-26}$ W m^{-2} Hz^{-1}. Consider observing an unpolarised radio source of flux density S with a telescope of effective area $A_\text{e} = \eta A$, where A is the geometric area and, as before, η is the telescope efficiency. Since each polarisation contributes half the total flux, the power received from the source per unit bandwidth and polarisation $P_\text{source} = (SA_\text{e})/2$.

Now consider replacing the antenna by a simple circuit consisting of a resistive load at temperature T. Random thermal motions of electrons induce a time variable voltage in this circuit which, by Nyquist's theorem, produces a power per unit bandwidth $P_{circuit} = k_B T$, where k_B is Boltzmann's constant. If we define the *antenna temperature* T_A as being the value of T such that $P_{source} = P_{circuit}$, it follows that

$$S = \frac{2k_B T_A}{A_e} = \frac{T_A}{G},$$ (A1.14)

where the *gain* $G = A_e/(2k_B)$ is determined by the available collecting area and measures the raw antenna sensitivity. Antenna gains[1] range from 0.64 K Jy^{-1} at Parkes to almost 20 K Jy^{-1} at Arecibo . In practice, the gain often varies as a function of elevation due to the gravitational distortion of fully steerable antennae (although the Effelsberg and Green Bank telescopes do correct for this effect) and the decrease in illuminated area away from the zenith for transit telescopes, e.g. at Arecibo.

From the above discussion, we can think in terms of sources of flux S_{source} having an equivalent *source temperature* $T_{source} = GS_{source}$. Even for a relatively strong pulsar with a flux density of 50 mJy, given the gains of the above telescopes, we expect $T_{source} \lesssim 1$ K. Real observing systems have many sources of noise, so that faint sources have to compete against a background *system temperature* defined by the sum:

$$T_{sys} = T_{rec} + T_{spill} + T_{atm} + T_{sky}.$$ (A1.15)

In this expression, T_{rec} is the receiver noise temperature (typically 20 K for cooled systems), T_{spill} is the 'spillover noise' contribution from the ground (like the antenna gain, this is often a strong function of elevation but usually 10 K or less), T_{atm} is due to emission from the atmosphere of the Earth (usually only important for observing frequencies above 5 GHz) and T_{sky} is the sky background radiation. While there is an ever-present contribution of 3 K to T_{sky} from the cosmic microwave background, the dominant source of noise comes from synchrotron-radiating electrons in the plane of our Galaxy. As a result, the sky background is a strong function of sky position and observing frequency. At 430 MHz, T_{sky} values of several hundred K along the Galactic plane are common, peaking to almost 800 K for observations toward the Galactic centre. At higher Galactic latitudes, typical 430 MHz T_{sky} values are about 30 K. At 1400 MHz, more typical values of T_{sky} on the Galactic plane are 10–20 K, becoming almost negligible at high Galactic latitudes. The book web

---

1 For reference, 1 K Jy^{-1} is equivalent to a collecting area of 2760 m^2 which, coincidentally, is the effective collecting area of the 76 m Lovell telescope.

site (see Appendix 3) provides a program to read the 430 MHz all-sky maps of the radio background (Haslam *et al.* 1982) for T_{sky} calculations.

A useful measure of sensitivity is the 'system-equivalent flux density' (SEFD). Also known as S_{sys}, the SEFD is the strength of a source with a temperature equal to T_{sys}, i.e. SEFD $= S_{\text{sys}} = T_{\text{sys}}/G$. As the ratio of these two key parameters, SEFD is used often to characterise the performance of a system, with low values indicating more sensitive systems. For example, the SEFD of the Parkes 20 cm multibeam system is about 30 Jy (see, for example, Manchester *et al.* (2001)). In contrast, the new multibeam system at Arecibo is expected to have an SEFD of around 3–4 Jy.

A1.4 The radiometer equation applied to pulsar observations

For a pulsar to be detectable, its signal must exceed significantly the noise fluctuations in the receiver system. After Dicke (1946), we may express the root mean square fluctuations in T_{sys} as

$$\Delta T_{\text{sys}} \approx \frac{T_{\text{sys}}}{\sqrt{n_{\text{p}}\, t\, \Delta f}}, \tag{A1.16}$$

where Δf is the observing bandwidth, t is the integration time and $n_{\text{p}} = 1$ for single-polarisation observations, or $n_{\text{p}} = 2$ if two orthogonal polarisations are summed. This so-called *radiometer equation* is at the heart of most radio astronomical calculations of system sensitivity. Following Vivekanand *et al.* (1982) and Bhattacharya (1998), we apply the radiometer equation to derive the sensitivity of a radio receiver to pulsed signals.

Consider a top-hat pulse of period, P, width, W, and peak amplitude, T_{peak}, sitting above a system noise temperature, T_{sys}. If we observe for t_{int} s, the noise fluctuations ΔT_{int} during the integration time t_{int} are obtained from the quadrature sum of the off-pulse ($t_{\text{off}} = (P-W)\, t_{\text{int}}/P$) and on-pulse ($t_{\text{on}} = W\, t_{\text{int}}/P$) fluctuations,

$$\Delta T_{\text{int}} = \sqrt{\Delta T_{\text{sys}}^2\, (t = t_{\text{on}}) + \Delta T_{\text{sys}}^2\, (t = t_{\text{off}})}. \tag{A1.17}$$

Applying the radiometer equation and working through the quadrature sum, making the reasonable approximation that $T_{\text{peak}} \ll T_{\text{sys}}$, we find

$$\Delta T_{\text{int}} = \frac{T_{\text{sys}}}{\sqrt{n_{\text{p}}\, t_{\text{int}}\, \Delta f}} \left(\frac{P}{\sqrt{W(P-W)}} \right). \tag{A1.18}$$

Since, by definition, the pulse signal to noise ratio $S/N = T_{peak}/\Delta T_{int}$, we can use the above result to write

$$S/N = \sqrt{n_p\, t_{int}\, \Delta f}\,\left(\frac{T_{peak}}{T_{sys}}\right)\frac{\sqrt{W(P-W)}}{P}. \tag{A1.19}$$

As a result, for narrow pulses ($W \ll P$), S/N is inversely proportional to the square root of the duty cycle, W/P.

To express S/N in terms of flux density rather than temperature, we use the telescope gain G to convert between temperature and flux density. The peak flux density of the pulsar $S_{peak} = T_{peak}/G$. As defined in Chapter 1, we measure the mean flux density over the whole period

$$S_{mean} = S_{peak}\left(\frac{W}{P}\right) = \frac{T_{peak}W}{GP}. \tag{A1.20}$$

Using this expression to substitute for T_{peak} in Equation A1.19, we obtain

$$S_{mean} = \frac{(S/N)T_{sys}}{G\sqrt{n_p\, t_{int}\, \Delta f}}\sqrt{\frac{W}{P-W}}. \tag{A1.21}$$

This equation can be used for crude calibration purposes (see Section 7.3.1). More usefully, for search applications, we are interested in finding the minimum detectable flux density S_{min} corresponding to a S/N threshold threshold S/N_{min}. Introducing also a 'correction factor' β to account for system imperfections, the result is

$$S_{min} = \beta\frac{(S/N_{min})T_{sys}}{G\sqrt{n_p\, t_{int}\, \Delta f}}\sqrt{\frac{W}{P-W}}. \tag{A1.22}$$

Imperfections arise due to digitisation of the signal and other effects so that $\beta \gtrsim 1$. Two commonly used systems, discussed further in Chapter 5, are one-bit digitisers ($\beta \sim 1.25$) and three-level correlators ($\beta \sim 1.16$)

Appendix 2
Useful formulae

This appendix brings together the most widely used formulae relevant to pulsar observations. Further details are given in the main chapters.

A2.1 Spin-down behaviour and inferred properties

Spin period, P, frequency, ν and their time derivatives are related via:

$$\nu = \frac{1}{P}; \quad \dot{\nu} = -\frac{\dot{P}}{P^2}; \quad \ddot{\nu} = \frac{2\dot{P}^2}{P^3} - \frac{\ddot{P}}{P^2}$$

and

$$P = \frac{1}{\nu}; \quad \dot{P} = -\frac{\dot{\nu}}{\nu^2}; \quad \ddot{P} = \frac{2\dot{\nu}^2}{\nu^3} - \frac{\ddot{\nu}}{\nu^2}.$$

Expressions for the angular velocity $\Omega = 2\pi/P$ are obtained by multiplying the spin frequency by 2π. The spin-down luminosity

$$\dot{E} = 4\pi^2 I \dot{P} P^{-3} \simeq 3.95 \times 10^{31} \text{ erg s}^{-1} \left(\frac{\dot{P}}{10^{-15}}\right) \left(\frac{P}{\text{s}}\right)^{-3}$$

is calculated for the canonical 1.4 $M_\odot$ neutron star of radius 10 km and moment of inertia $I = 10^{45}$ g cm^2. In general, the spin-down equations of motion are of the form $\dot{\nu} \propto \nu^n$, $\dot{\Omega} \propto \Omega^n$ or $\dot{P} \propto P^{2-n}$, where

$$n = \frac{\nu \ddot{\nu}}{\dot{\nu}^2} = \frac{\Omega \ddot{\Omega}}{\dot{\Omega}^2} = 2 - \frac{P \ddot{P}}{\dot{P}^2}$$

is the braking index. Integrating the spin-down law for P yields the age

$$T = \frac{P}{(n-1)\dot{P}} \left[1 - \left(\frac{P_0}{P}\right)^{n-1}\right],$$

where P_0 is the initial spin period. For spin-down due to magnetic dipole radiation ($n = 3$) and $P_0 \ll P$, the 'characteristic age'

$$\tau_c \equiv \frac{P}{2\dot{P}} \simeq 15.8\text{Myr} \left(\frac{P}{\text{s}}\right)\left(\frac{\dot{P}}{10^{-15}}\right)^{-1},$$

and the corresponding surface magnetic field strength

$$B_{\text{S}} = 3.2 \times 10^{19}\text{G} \ \sqrt{P\dot{P}} \simeq 10^{12} \text{ G} \left(\frac{\dot{P}}{10^{-15}}\right)^{1/2}\left(\frac{P}{\text{s}}\right)^{1/2}.$$

In the Goldreich–Julian model of the pulsar magnetosphere

$$n_{\text{GJ}} = \frac{\Omega B_{\text{S}}}{2\pi c e} \simeq 7 \times 10^{10}\text{cm}^{-3} \left(\frac{P}{\text{s}}\right)^{-1/2}\left(\frac{\dot{P}}{10^{-15}}\right)^{1/2}$$

is the plasma density at the polar cap that has a radius

$$R_{\text{p}} = \sqrt{\frac{2\pi R^3}{cP}} \simeq 150 \text{ m} \left(\frac{P}{\text{s}}\right)^{-1/2}.$$

The potential drop between the magnetic pole and the polar cap

$$\Delta\Psi = \frac{B_{\text{S}}\Omega^2 R^3}{2c^2} \simeq 2 \times 10^{13} \text{ V} \left(\frac{P}{\text{s}}\right)^{-3/2}\left(\frac{\dot{P}}{10^{-15}}\right)^{1/2}.$$

The pulsar magnetosphere is bounded by the 'light cylinder' of radius

$$R_{\text{LC}} = \frac{c}{\Omega} = \frac{cP}{2\pi} \simeq 4.77 \times 10^4 \text{ km} \left(\frac{P}{\text{s}}\right),$$

where the corresponding dipole magnetic field

$$B_{\text{LC}} = B_{\text{S}} \left(\frac{\Omega R}{c}\right)^3 \simeq 9.2 \text{ G} \left(\frac{P}{\text{s}}\right)^{-5/2}\left(\frac{\dot{P}}{10^{-15}}\right)^{1/2}.$$

A2.2 Radio beam geometry and polarisation fitting

For the radio beam shown in Figure 3.4, the expression

$$\cos\rho = \cos\alpha\cos(\alpha + \beta) + \sin\alpha\sin(\alpha + \beta)\cos\left(\frac{W}{2}\right)$$

relates the beam radius, ρ, to the magnetic inclination angle, α, the impact angle, β, and the observed pulse width, W, measured in degrees of longitude. When $\rho \lesssim 30°$

$$\rho \simeq 1.24° \left(\frac{r_{\mathrm{em}}}{10\,\mathrm{km}}\right)^{1/2} \left(\frac{P}{\mathrm{s}}\right)^{-1/2} \simeq 86° \left(\frac{r_{\mathrm{em}}}{R_{\mathrm{LC}}}\right)^{1/2},$$

where r_{em} is the height of an emitting region. The expression

$$\tan(\Psi - \Psi_0) = \frac{\sin\alpha\ \sin(\phi - \phi_0)}{\sin(\alpha + \beta)\ \cos\alpha - \cos(\alpha + \beta)\sin\alpha\cos(\phi - \phi_0)}$$

gives the polarisation position angle Ψ as a function of rotational phase ϕ predicted by the rotating vector model (RVM). The maximum gradient

$$\left(\frac{d\Psi}{d\phi}\right)_{\mathrm{max}} = \frac{\sin\alpha}{\sin\beta}.$$

From polarisation measurements (see Chapter 7) we obtain the four Stokes parameters I, Q, U and V. The linear polarisation $L = \sqrt{Q^2 + U^2}$ and the position angle $\Psi = \frac{1}{2}\arctan(\frac{U}{Q})$. Removal of the noise bias in L is recommended (see Section 7.4.3.1) and the true linear polarisation

$$L_{\mathrm{true}} = \begin{cases} \sigma_I \left[\left(\frac{L}{\sigma_I}\right)^2 - 1\right]^{\frac{1}{2}} & \text{if } \left(\frac{L}{\sigma_I}\right) \geq 1.57 \\ 0 & \text{otherwise,} \end{cases}$$

where σ_I is the standard deviation of the total intensity, I. The uncertainty in Ψ required to properly weight fits to the RVM:

$$\sigma_\Psi = 28.65° \left(\frac{\sigma_I}{L_{\mathrm{true}}}\right)$$

is useful only when $L_{\mathrm{true}}/\sigma_I \gtrsim 10$ (see Section 7.4.3.1 for further details).

A2.3 Flux densities, spectra and luminosities

The peak flux density S_{peak} is the maximum intensity of a pulse profile. The more commonly quoted *mean flux density* S_{mean} is the integrated intensity under the pulse averaged over the full period. Defining W_{eq} as the equivalent width of a top-hat pulse having the same area and peak flux as the true profile, it follows that $S_{\mathrm{mean}} = (W_{\mathrm{eq}}S_{\mathrm{peak}})/P$. The inferred brightness temperature for a pulse of width W_{eq}

$$T_{\mathrm{B}} \simeq 10^{30}\,\mathrm{K} \left(\frac{S_{\mathrm{peak}}}{\mathrm{Jy}}\right) \left(\frac{f}{\mathrm{GHz}}\right)^{-2} \left(\frac{W_{\mathrm{eq}}}{\mu\mathrm{s}}\right)^{-2} \left(\frac{d}{\mathrm{kpc}}\right)^2,$$

where d is the distance to the pulsar and f is the observing frequency. The integrated luminosity between two frequencies f_1 and f_2

$$L = \frac{2\pi d^2}{\delta} \left(1 - \cos\rho\right) S_{\mathrm{mean}}(f_0) \frac{f_0^{-\xi}}{\xi + 1} \left(f_2^{\xi+1} - f_1^{\xi+1}\right),$$

where $\delta = W_{\mathrm{eq}}/P$ is the pulse duty cycle and ξ is the spectral index defined by $S_{\mathrm{mean}} \propto f^{\xi}$. Although this expression properly accounts for the pulsar spectrum and beam shape, uncertainties in these quantities force a more common definition of the 'pseudoluminosity'

$$L_{400} \equiv S_{400} d^2 \quad \text{or} \quad L_{1400} \equiv S_{1400} d^2$$

at two 'standard' observing frequencies of 400 and 1400 MHz.

A2.4 Pulse dispersion and Faraday rotation

The plasma and cyclotron frequencies for a cold, magnetised medium

$$f_{\mathrm{p}} = \sqrt{\frac{e^2 n_e}{\pi m_e}} \simeq 8.5\,\mathrm{kHz} \left(\frac{n_e}{\mathrm{cm}^{-3}}\right)^{0.5} ; f_{\mathrm{B}} = \frac{eB_{\parallel}}{2\pi m_e c} \simeq 3\,\mathrm{MHz} \left(\frac{B_{\parallel}}{\mathrm{G}}\right),$$

where for the Galaxy, the electron number density $n_e \sim 0.03\ \mathrm{cm}^{-3}$ and parallel component of magnetic field $B_{\parallel} \sim 1\mu\mathrm{G}$. The associated dispersion meaure, DM, and rotation measure, RM, are:

$$\mathrm{DM} = \int_0^d n_e\,\mathrm{d}l; \quad \mathrm{RM} = \frac{e^3}{2\pi m_e^2 c^4} \int_0^d n_e B_{\parallel}\,\mathrm{d}l.$$

The DM is measured from an observation of the delay

$$\Delta t = 4.148808\,\mathrm{ms} \times \left[\left(\frac{f_{\mathrm{lo}}}{\mathrm{GHz}}\right)^{-2} - \left(\frac{f_{\mathrm{hi}}}{\mathrm{GHz}}\right)^{-2}\right] \times \left(\frac{\mathrm{DM}}{\mathrm{cm}^{-3}\mathrm{pc}}\right)$$

between a pulse received at a high-frequency f_{hi} relative to a lower one f_{lo}. When observed at a centre frequency $f = (f_{\mathrm{lo}} + f_{\mathrm{hi}})/2$ across a bandwidth $\Delta f = f_{\mathrm{hi}} - f_{\mathrm{lo}} \ll f$, we find

$$\Delta t = 8.297616\,\mu\mathrm{s} \times \left(\frac{\Delta f}{\mathrm{MHz}}\right) \times \left(\frac{f}{\mathrm{GHz}}\right)^{-3} \times \left(\frac{\mathrm{DM}}{\mathrm{cm}^{-3}\mathrm{pc}}\right).$$

Observing the polarisation angle Ψ (rad) change with wavelength λ (m)

$$\Psi(\lambda) = \Psi_{\infty} + \mathrm{RM}\,\lambda^2.$$

where Ψ_∞ is the polarisation angle at infinite frequency gives a means to determine the RM. The ratio of DM to RM gives the mean value of $B_\parallel$ along the line of sight weighted by electron density:

$$\langle B_\parallel \rangle \equiv \frac{\int_0^d n_e B_\parallel \, dl}{\int_0^d n_e \, dl} = 1.23 \, \mu G \left(\frac{RM}{\mathrm{rad \, m^{-2}}} \right) \left(\frac{DM}{\mathrm{cm^{-3} \, pc}} \right)^{-1}.$$

A2.5 Interstellar scattering and scintillation

A *rough* estimate of the scattering timescale τ_s (ms) is the empirical fit

$$\log_{10} \tau_s = -6.46 + 0.154 \log_{10}(DM) + 1.07(\log_{10} DM)^2 - 3.86 \log_{10} f,$$

where the units are $\mathrm{cm^{-3}}$ pc for DM and GHz for observing frequency f. See, however, Figure 4.4 for the large scatter about this relation.

In Chapter 4, we identified three types of scintillation: weak, diffractive (DISS) and refractive (RISS). For observational work, a useful parameter is the scintillation strength $u \equiv l_F/s_0$, where

$$l_F \simeq 1.2 \times 10^9 \; \mathrm{m} \left(\frac{d}{\mathrm{kpc}} \right)^{0.5} \left(\frac{f}{\mathrm{GHz}} \right)^{-0.5}$$

is the Fresnel length and, assuming a Kolmogorov turbulence spectrum,

$$s_0 \simeq 5.2 \times 10^5 \; \mathrm{m} \left(\frac{d}{\mathrm{kpc}} \right)^{0.5} \left(\frac{f}{\mathrm{GHz}} \right)^{-1.0} \left(\frac{\tau_s}{\mathrm{ms}} \right)^{-0.5}$$

is the field coherence scale. Alternatively, in terms of the DISS bandwidth (see below),

$$s_0 \simeq 3.8 \times 10^7 \; \mathrm{m} \left(\frac{d}{\mathrm{kpc}} \right)^{0.5} \left(\frac{f}{\mathrm{GHz}} \right)^{-1} \left(\frac{\Delta f_{\mathrm{DISS}}}{\mathrm{MHz}} \right)^{0.5}.$$

For weak scintillation, we have $u < 1$. In this case

$$\Delta t_{\mathrm{weak}} \simeq 3.35 \, \mathrm{h} \left(\frac{d}{\mathrm{kpc}} \right)^{0.5} \left(\frac{f}{\mathrm{GHz}} \right)^{-0.5} \left(\frac{V_{\mathrm{ISS}}}{100 \; \mathrm{km \, s^{-1}}} \right)^{-1}$$

is the scintillation timescale, and V_{ISS} is the transverse speed of the pulsar. The bandwidth for weak scintillation $\Delta f_{\mathrm{weak}} \sim f$.

Both DISS and RISS occur in the strong regime ($u > 1$). For DISS, noting that $\Delta t_{\mathrm{DISS}} = s_0/V_{\mathrm{ISS}}$, using the above expressions we find

$$\Delta t_{\mathrm{DISS}} \simeq 380 \, \mathrm{s} \left(\frac{d}{\mathrm{kpc}} \right)^{0.5} \left(\frac{f}{\mathrm{GHz}} \right)^{-1} \left(\frac{V_{\mathrm{ISS}}}{100 \; \mathrm{km \, s^{-1}}} \right)^{-1} \left(\frac{\Delta f_{\mathrm{DISS}}}{\mathrm{MHz}} \right)^{0.5}$$

is the timescale. The corresponding bandwidth,

$$\Delta f_{\text{DISS}} = \frac{C_1}{2\pi\tau_s} \simeq 185 \text{ Hz} \left(\frac{\tau_s}{\text{ms}}\right)^{-1},$$

where τ_s is the scattering time and the constant $C_1 = 1.16$ for a Kolmogorov turbulence spectrum. As a rough estimate,

$$\Delta f_{\text{DISS}} \sim 11 \text{ MHz} \left(\frac{f}{\text{GHz}}\right)^{4.4} \left(\frac{d}{\text{kpc}}\right)^{-2.2},$$

Measurements of f_{DISS} and t_{DISS} (see Section 7.4.4) yield the transverse speed

$$V_{\text{ISS}} = A \left(\frac{d}{\text{kpc}}\right)^{0.5} \left(\frac{\Delta f_{\text{DISS}}}{\text{MHz}}\right)^{0.5} \left(\frac{f}{\text{GHz}}\right)^{-1} \left(\frac{\Delta t_{\text{DISS}}}{\text{s}}\right)^{-1},$$

where $A = 2.53 \times 10^4 \text{ km s}^{-1}$ depends on the assumed turbulence spectrum. The timescales for DISS and RISS are related as follows:

$$\Delta t_{\text{RISS}} = u^2 \, \Delta t_{\text{DISS}} = \frac{f}{\Delta f_{\text{DISS}}} \, \Delta t_{\text{DISS}} \propto f^{-2.2} \, d^{1.6}.$$

A2.6 Detection thresholds and significance tests

The minimum detectable flux density of a pulsar search in mJy

$$S_{\text{min}} = \frac{(\text{S/N}_{\text{min}}) \, \beta \, T_{\text{sys}}}{G\sqrt{n_p t_{\text{obs}} \Delta f}} \sqrt{\frac{\delta}{1-\delta}}.$$

where $\beta \gtrsim 1$ is the 'degradation factor' (see Appendix 1), T_{sys} (K) is the system temperature, G (K Jy^{-1}) is the telescope gain, n_p is the number of polarisations summed, Δf (MHz) is the observing bandwidth, t_{obs} (s) is the observation length, δ is the pulse duty cycle and

$$\text{S/N}_{\text{min}} = \frac{\sqrt{\ln[n_{\text{trials}}]} - \sqrt{\pi/4}}{1 - \pi/4} \simeq \frac{\sqrt{\ln[n_{\text{trials}}]} - 0.88}{0.47}$$

is the signal to noise threshold determined by the number of trials in the search, n_{trials}. For a typical Fourier domain search (see Section 6.1) of an n_{samples} time series with summing of n_h harmonics for n_{DM} trial dispersion measures, $n_{\text{trials}} = n_{\text{samples}} \times n_{\text{DM}} \times [\log_2(n_h) + 1]/2$.

For time-domain studies, useful quantities are the signal to noise ratio, S/N, and chi-square statistic, χ^2, of a pulse profile p_i with n_{bins} bins defined by:

$$\text{S/N} = \frac{1}{\sigma_p \sqrt{W_{\text{eq}}}} \sum_{i=1}^{n_{\text{bins}}} (p_i - \bar{p}) \quad \text{and} \quad \chi^2 = \frac{1}{\sigma_p^2} \sum_{i=1}^{n_{\text{bins}}} (p_i - \bar{p})^2,$$

where $\bar{p}$ and $\sigma_{\rm p}$ are the off-pulse mean and standard deviation and $W_{\rm eq}$ is the equivalent width in bins. For further details, see Section 7.1.1.

A2.7 Binary pulsars

The mass function $f(m_{\rm p}, m_{\rm c})$ relates the pulsar and companion masses, $m_{\rm p}$ and $m_{\rm c}$ and unknown orbital inclination angle i to the observable quantities orbital period $P_{\rm b}$ and semi-major axis x as follows:

$$f(m_{\rm p}, m_{\rm c}) = \frac{(m_{\rm c}\sin i)^3}{(m_{\rm p} + m_{\rm c})^2} = \frac{4\pi^2}{T_\odot}\frac{x^3}{P_{\rm b}^2} \simeq 0.618 {\rm M}_\odot \left(\frac{x}{\rm s}\right)^3 \left(\frac{P_{\rm b}}{\rm hr}\right)^{-2},$$

where $T_\odot = GM_\odot/c^3 = 4.925490947\mu{\rm s}$. For a binary orbit of eccentricity e, Kepler's equations relate the eccentric anomaly E, the true anomaly $A_{\rm T}$ to the longitude of periastron, ω and any change in orbital period $\dot{P}_{\rm b}$ as follows:

$$E - e\sin E = \Omega_{\rm b}\left[(t - T_0) - \frac{1}{2}\frac{\dot{P}_{\rm b}}{P_{\rm b}}(t - T_0)^2\right],$$

$$A_{\rm T}(E) = 2\arctan\left[\sqrt{\frac{1 + e}{1 - e}}\tan\frac{E}{2}\right],$$

$$\omega = \omega_0 + \frac{\dot{\omega}}{\Omega_{\rm b}}A_{\rm T}(E),$$

where $\Omega_{\rm b} \equiv 2\pi/P_{\rm b}$ and T_0 is the epoch of periastron. In general relativity, to lowest post-Newtonian order, the five most commonly observed post-Keplerian parameters are the rate of advance of periastron

$$\dot{\omega} = 3T_\odot^{2/3}\left(\frac{P_{\rm b}}{2\pi}\right)^{-5/3}\frac{(m_{\rm p} + m_{\rm c})^{2/3}}{1 - e^2},$$

the Einstein delay

$$\gamma = T_\odot^{2/3}\left(\frac{P_{\rm b}}{2\pi}\right)^{1/3}e\frac{m_{\rm c}(m_{\rm p} + 2m_{\rm c})}{(m_{\rm p} + m_{\rm c})^{4/3}},$$

the 'range' and 'shape' of the Shapiro delay

$$r = T_\odot m_{\rm c}, \quad s = \sin i = T_\odot^{-1/3}\left(\frac{P_{\rm b}}{2\pi}\right)^{-2/3}x\frac{(m_{\rm p} + m_{\rm c})^{2/3}}{m_{\rm c}},$$

and the rate of orbital decay due to gravitational radiation

$$\dot{P}_{\rm b} = -\frac{192\pi}{5}T_\odot^{5/3}\left(\frac{P_{\rm b}}{2\pi}\right)^{-5/3}f(e)\frac{m_{\rm p}m_{\rm c}}{(m_{\rm p} + m_{\rm c})^{1/3}}.$$

In these expressions, P_b and x are in s, all masses are in $M_\odot$ and

$$f(e) = \frac{(1 + (73/24)e^2 + (37/96)e^4)}{(1 - e^2)^{7/2}}.$$

The rates of change of semi-major axis a_p and orbital eccentricity e due to gravitational wave emission are:

$$\dot{a}_p = -\frac{64}{5} c\, T_\odot^2 \left(\frac{2\pi}{P_b}\right)^2 f(e)\, \frac{m_p m_c^2}{m_p + m_c}, \quad \text{and}$$

$$\dot{e} = -\frac{304\, G^{5/3}}{15 c^5} \left(\frac{2\pi}{P_b}\right)^{8/3} \frac{m_p m_c}{(m_p + m_c)^{1/3}} \frac{e\left(1 + (121/304)e^2\right)}{(1 - e^2)^{5/2}}.$$

A useful approximation for the corresponding coalescence time

$$\tau_{GW} \simeq 9.83 \times 10^6\, \text{yr} \left(\frac{P_b}{\text{hr}}\right)^{8/3} \left(\frac{\mu}{M_\odot}\right)^{-1} \left(\frac{m_p + m_c}{M_\odot}\right)^{-2/3} (1 - e^2)^{7/2},$$

where the reduced mass $\mu = m_p m_c / (m_p + m_c)$. The relativistic advance of periastron $\dot{\omega}$ is the most commonly measured of these parameters. For a binary system of total mass $M = m_p + m_c$,

$$\dot{\omega} \simeq 39.73^\circ\, \text{yr}^{-1} \left(\frac{P_b}{\text{hr}}\right)^{-5/3} \left(\frac{1}{1 - e^2}\right) \left(\frac{M}{M_\odot}\right)^{2/3}.$$

In terms of a measured $\dot{\omega}$ from timing observations:

$$M \simeq 0.126\, M_\odot \left(\frac{\dot{\omega}}{10^\circ\, \text{yr}^{-1}}\right)^{3/2} \left(\frac{P_b}{\text{hr}}\right)^{5/2} (1 - e^2)^{3/2}.$$

The geodetic precession rate (rad s^{-1})

$$\Omega_{\text{geod}} = \left(\frac{2\pi}{P_b}\right)^{5/3} T_\odot^{2/3} \frac{m_c(4m_p + 3m_c)}{2(m_p + m_c)^{4/3}} \frac{1}{1 - e^2}.$$

In more convenient units for P_b and Ω_{geod} we find

$$\Omega_{\text{geod}} \simeq 6.62^\circ\, \text{yr}^{-1} \left(\frac{P_b}{\text{hr}}\right)^{-5/3} \frac{m_c(4m_p + 3m_c)}{M^{4/3}} \left(\frac{1}{1 - e^2}\right),$$

where we note the useful simplification $\Omega_{\text{geod}} = \frac{7}{24} \dot{\omega} \sim 0.3\, \dot{\omega}$ valid for double neutron star binaries where $m_p = m_c$.

A2.8 Unit conversion factors

The following list is a useful summary for those wishing to convert non-standard units used in this book into SI quantities: 1 Gauss $\equiv 10^{-4}$ Tesla; 1 erg $\equiv 10^{-7}$ Joule; 1 Jansky $\equiv 10^{-26}$ W m^{-2} Hz^{-1}; 1 solar mass (M$_\odot$) $= 1.989 \times 10^{30}$ kg; 1 parsec $= 3.086 \times 10^{16}$ m. A few useful approximate conversions are also worth noting: 1 yr $\simeq \pi \times 10^7$ s; 1 km s$^{-1} \simeq 1$ pc Myr^{-1}; 10^{-9} rad s$^{-1} \simeq 1.8°$ yr$^{-1} \simeq 6.5 \times 10^6$ mas yr^{-1}.

Appendix 3
Useful resources

Pulsar astronomy, like many other fields, has benefited greatly from the ease of information exchange made possible by the internet over the last decade. There are now a number of excellent web sites hosting useful catalogues, data reduction software and archival data. At the current time, the key sites that we are aware of are:

The on-line pulsar catalogue hosted by the Australia Telescope National Facility (ATNF):

`http://www.atnf.csiro.au/research/pulsar/psrcat`

The pulsar timing package TEMPO hosted by Princeton University and the ATNF:

`http://pulsar.princeton.edu/tempo`
`http://www.atnf.csiro.au/research/pulsar/timing/tempo`

The European Pulsar Network (EPN) pulse profile database hosted at the Max Planck Institut für Radiastronomie (MPIfR):

`http://www.mpifr-bonn.mpg.de/div/pulsar/data`

One of the main aims of this book is to expand the scope of these resources further via our web site:

`http://www.jb.man.ac.uk/~pulsar/handbook`

which provides links to existing facilities as well as new resources and data that we find useful. At the time of writing, a number of groups are putting a lot of effort into making pulsar software more freely available. We plan to keep the web site up to date with links to new software packages as and when they become available. Access to the web site is completely free and, as we plan to continually update it, we welcome all comments and feedback via email: `handbook@jb.man.ac.uk`.

References

Ables, J. G. & Manchester, R. N., 1976. *A&A*, **50**, 177.
Abramovici, A., Althouse, W. E., Drever, R. W. P., Gursel, Y., Kawamura, S., Raab, F. J., Shoemaker, D., Sievers, L., Spero, R. E., Thorne, K. S., Vogt, R. E., Weiss, R., Whitcomb, S. E. & Zucker, M. E., 1992. *Science*, **256**, 325.
Allan, D. W., 1966. *Proc. I.E.E.E.*, **54**.
Anderson, P. W. & Itoh, N., 1975. *Nature*, **256**, 25.
Anderson, S. B., 1992. *PhD thesis*, California Institute of Technology.
Applegate, J. H. & Shaham, J., 1994. *ApJ*, **436**, 312.
Applegate, J. H., 1992. *ApJ*, **385**, 621.
Arendt, P. N. & Eilek, J. A., 2002. *ApJ*, **581**, 451.
Armstrong, J. W., Rickett, B. J. & Spangler, S. R., 1995. *ApJ*, **443**, 209.
Arons, J., 1983. *Nature*, **302**, 301.
Arons, J. & Barnard, J. J., 1986. *ApJ*, **302**, 120.
Arzoumanian, Z., 1995. *PhD thesis*, Princeton University.
Arzoumanian, Z., Chernoff, D. F. & Cordes, J. M., 2002. *ApJ*, **568**, 289.
Arzoumanian, Z., Cordes, J. M. & Wasserman, I., 1999. *ApJ*, **520**, 696.
Arzoumanian, Z., Fruchter, A. S. & Taylor, J. H., 1994a. *ApJ*, **426**, L85.
Arzoumanian, Z., Nice, D. J., Taylor, J. H. & Thorsett, S. E., 1994b. *ApJ*, **422**, 671.
Arzoumanian, Z., Joshi, K., Rasio, F. & Thorsett, S. E., 1996. In: *Pulsars: Problems and Progress, IAU Colloquium 160*, p. 525, eds Johnston, S., Walker, M. A. & Bailes, M., Astronomical Society of the Pacific, San Francisco.
Arzoumanian, Z., Taylor, J. H. & Wolszczan, A., 1999. In: *Pulsar Timing, General Relativity, and the Internal Structure of Neutron Stars*, p. 85, eds Arzoumanian, Z., van der Hooft, F. & van den Heuvel, E. P. J., North Holland, Amsterdam.
Asseo, E. & Khechinashvili, D., 2002. *MNRAS*, **334**, 743.
Asseo, E., 1993. *MNRAS*, **264**, 940.
Aulbert, C., 2004. *ApJ*, submitted.
Backer, D. C., 1970a. *Nature*, **228**, 1297.
Backer, D. C., 1970b. *Nature*, **228**, 42.
Backer, D. C., 1973. *ApJ*, **182**, 245.
Backer, D. C., 1976. *ApJ*, **209**, 895.

Backer, D. C., 1990. In: *Impact of Pulsar Timing on Relativity and Cosmology*, ed. Backer, D., Center for Particle Astrophysics, Berkeley.

Backer, D. C., 1996. In: *Compact Stars in Binaries: IAU Symposium 165*, p. 197, eds van Paradijs, J., van del Heuvel, E. P. J. & Kuulkers, E., Kluwer, Dordrecht.

Backer, D. C., Dexter, M. R., Zepka, A., D., N., Wertheimer, D. J., Ray, P. S. & Foster, R. S., 1997. *PASP*, **109**, 61.

Backer, D. C. & Fisher, J. R., 1974. *ApJ*, **189**, 137.

Backer, D. C., Foster, R. F. & Sallmen, S., 1993a. *Nature*, **365**, 817.

Backer, D. C., Hama, S., Van Hook, S. & Foster, R. S., 1993b. *ApJ*, **404**, 636.

Backer, D. C. & Hellings, R. W., 1986. *Ann. Rev. Astr. Ap.*, **24**, 537.

Backer, D. C., Kulkarni, S. R., Heiles, C., Davis, M. M. & Goss, W. M., 1982. *Nature*, **300**, 615.

Backer, D. C. & Ramachandran, R., 2004. In: *Young Neutron Stars and Their Environments, IAU Symposium 218*, in press, eds Camilo, F. & Gaensler, B. M., Astronomical Society of the Pacific, San Francisco.

Backer, D. C. & Rankin, J. M., 1980. *ApJS*, **42**, 143.

Backer, D. C. & Wong, T., 1999. In: *Pulsar Timing, General Relativity, and the Internal Structure of Neutron Stars*, p. 39, eds Arzoumanian, Z., van der Hooft, F. & van den Heuvel, E. P. J., North Holland, Amsterdam.

Backer, D. C., Wong, T. & Valanju, J., 2000. *ApJ*, **543**, 740.

Bahcall, J. N. & Ostriker, J. P., 1997. *Unsolved problems in astrophysics*, Princeton University Press, 1997.

Bailes, M., Manchester, R. N., Kesteven, M. J., Norris, R. P. & Reynolds, J. E., 1990. *MNRAS*, **247**, 322.

Bailes, M., Ord, S. M., Knight, H. S. & Hotan, A. W., 2003. *ApJ*, **595**, L49.

Baring, M. G. & Harding, A. K., 1998. *ApJ*, **507**, L55.

Baring, M. G. & Harding, A. K., 2001. *ApJ*, **547**, 929.

Barish, B. C., 2000. *Advances in Space Research*, **25**, 1165.

Barker, B. M. & O'Connell, R. F., 1975a. *Phys. Rev. D*, **12**, 329.

Barker, B. M. & O'Connell, R. F., 1975b. *ApJ*, **199**, L25.

Barnard, J. J., 1986. *ApJ*, **303**, 280.

Barnard, J. J. & Arons, J., 1986. *ApJ*, **302**, 138.

Bartel, N., Chandler, J. F., I., R. M., Shapiro, I. I., Pan, R. & Capallo, R. J., 1996. *AJ*, **112**, 1690.

Bartel, N., Kardashev, N. S., Kuzmin, A. D. Nikolaev, N. Y., Popov, M. V., Sieber, W., Smirnova, T. V., Soglasnov, V. A. & Wielebinski, R., 1981. *A&A*, **93**, 85.

Bartel, N. & Sieber, W., 1978. *A&A*, **70**, 260.

Bauch, A., 2003. *Measurement Science and Technology*, **14**, 1159.

Baym, G., Pethick, C., Pines, D. & Ruderman, M., 1969. *Nature*, **224**, 872.

Beck, R., 2002. In: *ASP Conf. Ser. 275: Disks of Galaxies: Kinematics, Dynamics and Peturbations*, p. 331, eds E. Athanassoula, A. B. & Mujica, R.

Becker, W. & Aschenbach, B., 2002. In: *WE-Heraeus Seminar on Neutron Stars, Pulsars and Supernova Remnants. MPE Report 278.*, p. 64, eds Becker, W., Lesch, H. & Truemper, J., Max-Plank-Institut für extraterrestrische Physik.

Becker, W. & Trümper, J., 1997. *A&A*, **326**, 682.

Beskin, V. S., 1999. *Physics Uskpekhi*, **42**, 1071.

Beskin, V. S., Gurevich, A. V. & Istomin, Y. N., 1993. *Physics of the Pulsar Magnetosphere*, Cambridge University Press.

Bhat, N. D. R., Cordes, J. M., Camilo, F., Nice, D. J. & Lorimer, D. R., 2004. *ApJ*, **605**, 759.

Bhat, N. D. R., Cordes, J. M. & Chatterjee, S., 2003. *ApJ*, **584**, 782.

Bhat, N. D. R., Gupta, Y. & Rao, A. P., 1998. *ApJ*, **500**, 262.

Bhattacharya, D., 1998. In: *The Many Faces of Neutron Stars (NATO ASI Series)*, p. 103, eds Bucheri, R. & van Paradijs M. A. Alpar, J., Kluwer, Dordrecht, Boston, London.

Bhattacharya, D. & van den Heuvel, E. P. J., 1991. *Phys. Rep.*, **203**, 1.

Biggs, J. D., 1992. *ApJ*, **394**, 574.

Bignami, G. F. & Caraveo, P. A., 1996. *Ann. Rev. Astr. Ap.*, **34**, 331.

Bignami, G. F., Caraveo, P. A., de Luca, A. & Mereghetti, S., 2003. *Nature*, **423**, 725.

Bisnovatyi-Kogan, G. S. & Komberg, B. V., 1974. *Sov. Astron.*, **18**, 217.

Blandford, R. D., Hewish, A., Lyne, A. G. & Mestel, L., 1992. *Philos. Trans. Roy. Soc. London A*, **341**, 1.

Blandford, R. D. & Teukolsky, S. A., 1976. *ApJ*, **205**, 580.

Blaskiewicz, M., Cordes, J. M. & Wasserman, I., 1991. *ApJ*, **370**, 643.

Boriakoff, V., Ferguson, D. C. & Slater, G., 1981. In: *Pulsars, IAU Symposium 95*, p. 199, eds Sieber, W. & Wielebinski, R., Reidel, Dordrecht.

Bracewell, R., 1998. *The Fourier Transform and its Applications*, McGraw–Hill, New York.

Braje, T. M. & Romani, R. W., 2002. *ApJ*, **580**, 1043.

Brinklow, A., 1989. *PhD thesis*, The University of Manchester.

Brisken, W. F., 2001. *PhD thesis*, Princeton University.

Brisken, W. F., Benson, J. M., Beasley, A. J., Fomalont, E. M., Goss, W. M. & Thorsett, S. E., 2000. *ApJ*, **541**, 959.

Brisken, W. F., Benson, J. M., Goss, W. M. & Thorsett, S. E., 2002. *ApJ*, **571**, 906.

Britton, M. C., 2000. *ApJ*, **532**, 1240.

Burgay, M., D'Amico, N., Possenti, A., Manchester, R. N., Lyne, A. G., Joshi, B. C., McLaughlin, M., Kramer, M., Sarkissian, J. M., Camilo, F., Kalogera, V., Kim, C. & Lorimer, D. R., 2003. *Nature*, **426**, 531.

Burke, B. F. & Smith, F. G., 2002. *An Introduction to Radio Astronomy*, Cambridge University Press, Cambridge.

Burns, W. R. & Clark, B. G., 1969. *A&A*, **2**, 280.

Camilo, F., 1995. In: *The Lives of the Neutron Stars (NATO ASI Series)*, p. 243, eds Alpar, A., Kiziloğlu, Ü. & van Paradis, J., Kluwer, Dordrecht.

Camilo, F., 1997. In: *High Sensitivity Radio Astronomy*, p. 14, eds Jackson, N. & Davis, R. J., Cambridge University Press.

Camilo, F., 1999. In: *Pulsar Timing, General Relativity, and the Internal Structure of Neutron Stars*, p. 115, eds Arzoumanian, Z., van der Hooft, F. & van den Heuvel, E. P. J., North Holland, Amsterdam.

Camilo, F., 2003. In: *Radio Pulsars*, p. 145, eds Bailes, M., Nice, D. J. & Thorsett, S. E., Astronomical Society of the Pacific, San Francisco.

Camilo, F., Bell, J. F., Manchester, R. N., Lyne, A. G., Possenti, A., Kramer, M., Kaspi, V. M., Stairs, I. H., D'Amico, N., Hobbs, G., Gotthelf, E. V. & Gaensler, B. M., 2001. *ApJ*, **557**, L51.

Camilo, F., Kaspi, V. M., Lyne, A. G., Manchester, R. N., Bell, J. F.,
    D'Amico, N., McKay, N. P. F. & Crawford, F., 2000a. *ApJ*, **541**, 367.
Camilo, F., Lorimer, D. R., Freire, P., Lyne, A. G. & Manchester, R. N.,
    2000b. *ApJ*, **535**, 975.
Camilo, F., Nice, D. J., Shrauner, J. A. & Taylor, J. H., 1996. *ApJ*, **469**, 819.
Camilo, F., Stairs, I. H., Lorimer, D. R., Backer, D. C., Ransom, S. M.,
    Klein, B., Wielebinski, R., Kramer, M., McLaughlin, M. A.,
    Arzoumanian, Z. & Müller, P., 2002. *ApJ*, **571**, L41.
Caraveo, P. A., Bignami, G. F. & Mereghetti, S., 1994a. *ApJ*, **422**, L87.
Caraveo, P. A., Mereghetti, S. & Bignami, G. F., 1994b. *ApJ*, **423**, L125.
Caraveo, P. A., Bignami, G. F., Mereghetti, S. & Mombelli, M., 1992. *ApJ*,
    395, L103.
Cardall, C. Y., Prakash, M. & Lattimer, J. M., 2001. *ApJ*, **554**, 322.
Carlberg, R. G. & Innanen, K. A., 1987. *AJ*, **94**, 666.
Caron, B. & et al., 1998. In: *Second Edoardo Amaldi Conference on
    Gravitational Wave Experiments*, p. 73, eds Coccia, E., Pizzella, G. &
    Ronga, F.
Chakrabarty, D. & Morgan, E. H., 1998. *Nature*, **394**, 346.
Champion, D. J., Lorimer, D. R., McLaughlin, M. A., Arzoumanian, Z. A.,
    Cordes, J. M., Weisberg, J. M. & Taylor, J. H., 2004. *MNRAS*, **350**, 61.
Chandler, A. M., 2003. *PhD thesis*, California Institute of Technology.
Chatterjee, S., Cordes, J. M., Lazio, T. J. W., Goss, W. M., Fomalont, E. B.
    & Benson, J. M., 2001. *ApJ*, **550**, 287.
Chen, K. & Ruderman, M., 1993. *ApJ*, **408**, 179.
Cheng, K. S., Ho, C. & Ruderman, M., 1986. *ApJ*, **300**, 500.
Cheng, K. S., Ruderman, M. & Zhang, L., 2000. *ApJ*, **537**, 964.
Cocke, W. J., Disney, M. J. & Taylor, D. J., 1969. *Nature*, **221**, 525.
Cognard, I., Bourgois, G., Lestrade, J.-F., Biraud, F., Aubry, D., Darchy, B.
    & Drouhin, J.-P., 1993. *Nature*, **366**, 320.
Cognard, I., Shrauner, J. A., Taylor, J. H. & Thorsett, S. E., 1996. *ApJ*,
    457, L81.
Contopoulos, I., Kazanas, D. & Fendt, C., 1999. *ApJ*, **511**, 351.
Cordes, J. M., 1978. *ApJ*, **222**, 1006.
Cordes, J. M., 1986. *ApJ*, **311**, 183.
Cordes, J. M., 2002. In: *Single-Dish Radio Astronomy: Techniques &
    Applications*, p. 227, eds Stanimirovic, S., Altschuler, D. R., Goldsmith,
    P. F. & Salter, C. J., Astronomical Society of the Pacific, San Francisco.
Cordes, J. M., Bhat, N. D. R., Hankins, T. H., McLaughlin, M. A. & Kern,
    J., 2004. *ApJ*, in press, (astro-ph/0304495).
Cordes, J. M. & Lazio, T. J. W., 1991. *ApJ*, **376**, 123.
Cordes, J. M. & Lazio, T. J. W., 1997. *ApJ*, **475**, 557.
Cordes, J. M. & Lazio, T. J. W., 2001. *ApJ*, **549**, 997.
Cordes, J. M. & Lazio, T. J. W., 2002a. *ApJ*, submitted (astro-ph/0207156).
Cordes, J. M. & Lazio, T. J. W., 2002b. *ApJ*, submitted (astro-ph/0301598).
Cordes, J. M. & McLaughlin, M. A., 2003. *ApJ*, **596**, 1142.
Cordes, J. M. & Rickett, B. J., 1998. *ApJ*, **507**, 846.
Cordes, J. M. & Stinebring, D. R., 1984. *ApJ*, **277**, L53.
Cordes, J. M. & Wolszczan, A., 1986. *ApJ*, **307**, L27.
Cordes, J. M., Pidwerbetsky, A. & Lovelace, R. V. E., 1986. *ApJ*, **310**, 737.
Cordes, J. M., Weisberg, J. M. & Boriakoff, V., 1985. *ApJ*, **288**, 221.
Cordes, J. M., Weisberg, J. M. & Hankins, T. H., 1990. *AJ*, **100**, 1882.

Craft, H. D. & Comella, J. M. & Drake, F., 1968. *Nature*, **218**, 1122.

Crawford, F., Kaspi, V. M., Manchester, R. N., Lyne, A. G., Camilo, F. & D'Amico, N., 2001. *ApJ*, **553**, 367.

Cusumano, G., Hermsen, W., Kramer, M., Kuiper, L., Löhmer, O., Massaro, E., Mineo, T., Nicastro, L. & Stappers, B. W., 2003. *A&A*, **410**, L9.

D'Amico, N., Kaspi, V. M., Manchester, R. N., Camilo, F., Lyne, A. G., Possenti, A., Stairs, I. H., Kramer, M., Crawford, F., Bell, J. & McKay, N. P. F., 2001. *ApJ*, **552**, L45.

Damour, T. & Deruelle, N., 1985. *Ann. Inst. H. Poincaré (Physique Théorique)*, **43**, 107.

Damour, T. & Deruelle, N., 1986. *Ann. Inst. H. Poincaré (Physique Théorique)*, **44**, 263.

Damour, T. & Esposito-Farèse, G., 1998. *Phys. Rev. D*, **58(042001)**, 1.

Damour, T. & Esposito-Farèse, G., 1992. *Classical and Quantum Gravity*, **9**, 2093.

Damour, T. & Esposito-Farese, G., 1996. *Phys. Rev. D*, **54**, 1474.

Damour, T. & Ruffini, R., 1974. *C. R. Acad. Sc. Paris, Serie A*, **279**, 971.

Damour, T. & Schäfer, G., 1988. *Nuovo Cim.*, **101**, 127.

Damour, T. & Schäfer, G., 1991. *Phys. Rev. Lett.*, **66**, 2549.

Damour, T. & Taylor, J. H., 1991. *ApJ*, **366**, 501.

Damour, T. & Taylor, J. H., 1992. *Phys. Rev. D*, **45**, 1840.

Danzmann, K. & et al., 1995. In: *First Edoardo Amaldi Conference on Gravitational Wave Experiments*, p. 100, eds Coccia, E., Pizzella, G. & Ronga, F.

Danzmann, K., 2000. *Advances in Space Research*, **25**, 1129.

Daugherty, J. K. & Harding, A. K., 1983. *ApJ*, **273**, 761.

Daugherty, J. K. & Harding, A. K., 1986. *ApJ*, **309**, 362.

Davies, J. G., Lyne, A. G., Smith, F. G., Izvekova, V. A., Kuzmin, A. D. & Shitov, Y. P., 1984. *MNRAS*, **211**, 57.

Deich, W. T. S., Cordes, J. M., Hankins, T. H. & Rankin, J. M., 1986. *ApJ*, **300**, 540.

Dennett-Thorpe, J. & de Bruyn, A. G., 2002. *Nature*, **415**, 57.

Deshpande, A. A. & Ramachandran, R., 1998. *MNRAS*, **300**, 577.

Deshpande, A. A. & Rankin, J. M., 1999. *ApJ*, **524**, 1008.

Deshpande, A. A. & Rankin, J. M., 2001. *MNRAS*, **322**, 438.

Deutsch, A. J., 1955. *Annales d'Astrophysique*, **18**, 1.

Dicke, R. H., 1946. *Rev. Sci. Instrum.*, **17**, 268.

Doroshenko, O. & Kopeikin, S. M., 1995. *MNRAS*, **274**, 1029.

Doroshenko, O., Löhmer, O., Kramer, M., Jessner, A., Wielebinski, R., Lyne, A. G. & Lange, C., 2001. *A&A*, **379**, 579.

Drake, F. D. & Craft, H. D., 1968. *Nature*, **220**, 231.

Duncan, R. C. & Thompson, C., 1992. *ApJ*, **392**, L9.

Edwards, R. T. & Bailes, M., 2001. *ApJ*, **547**, L37.

Edwards, R. T. & Stappers, B. W., 2002. *A&A*, **393**, 733.

Edwards, R. T., Bailes, M., van Straten, W. & Britton, M. C., 2001. *MNRAS*, **326**, 358.

Ekers, R. D., van Gorkom, J. H., Schwarz, U. J. & Goss, W. M., 1983. *A&A*, **122**, 143.

Erber, T., 1966. *Reviews of Modern Physics*, **38**, 626.

Everett, J. E. & Weisberg, J. M., 2001. *ApJ*, **553**, 341.

Faulkner, A. J., 2004. *PhD thesis*, University of Manchester.

Faulkner, A. J., Kramer, M., Lyne, A. G., Manchester, R. N., Stairs, I. H., Hobbs, G. B., Possenti, A., Lorimer, D. R., McLaughlin, M. A., D'Amico, N., Camilo, F. & Burgay, M., 2004. *ApJ*, in press.

Fich, M., Blitz, L. & Stark, A. A., 1989. *ApJ*, **342**, 272.

Fichtel, C. E., Hartman, R. C., Kniffen, D. A., Thompson, D. J., Bignami, G. F., Ögelman, H., Özel, M. E. & Tümer, T., 1975. *ApJ*, **198**, 163.

Fiedler, R. L., Dennison, B., Johnston, K. J. & Hewish, A., 1987. *Nature*, **326**, 675.

Fiedler, R., Dennison, K., Johnston, J., Waltman, E. & Simon, R., 1994. *ApJ*, **430**, 581.

Figer, D. F., McLean, I. S. & Morris, M., 1999. *ApJ*, **514**, 202.

Fomalont, E. B., Goss, W. M., Beasley, A. J. & Chatterjee, S., 1999. *AJ*, **117**, 3025.

Foster, R. S. & Backer, D. C., 1990. *ApJ*, **361**, 300.

Foster, R. S. & Cordes, J. M., 1990. *ApJ*, **364**, 123.

Foster, R. S., Ray, P. S., Lundgren, S. C., Backer, D. C., Dexter, M. R. & Zepka, A., 1996. In: *Pulsars: Problems and Progress, IAU Colloquium 160*, p. 25, eds Johnston, S., Walker, M. A. & Bailes, M., Astronomical Society of the Pacific, San Francisco.

Foster, R. S., Fairhead, L. & Backer, D. C., 1991. *ApJ*, **378**, 687.

Fowler, L. A., Wright, G. A. E. & Morris, D., 1981. *A&A*, **93**, 54.

Frail, D. A. & Weisberg, J. M., 1990. *AJ*, **100**, 743.

Freire, P. C., Camilo, F., Kramer, M., Lorimer, D. R., Lyne, A. G., Manchester, R. N. & D'Amico, N., 2003. *MNRAS*, **340**, 1359.

Freire, P. C., Gupta, Y., Ransom, S. M. & Ishwara-Chandra, C. H., 2004. *ApJ*, **606**, L53.

Freire, P. C., Kramer, M. & Lyne, A. G., 2001. *MNRAS*, **322**, 885.

Gangadhara, R. T. & Gupta, Y., 2001. *ApJ*, **555**, 31.

Gil, J. A., Gronkowski, P. & Rudnicki, W., 1984. *A&A*, **132**, 312.

Gil, J. A., Jessner, A. & Kramer, M., 1993a. *A&A*, **271**, L13.

Gil, J. A., Kijak, J. & Seiradakis, J. H., 1993b. *A&A*, **272**, 268.

Gil, J. & Mitra, D., 2001. *ApJ*, **550**, 383.

Gil, J. A. & Sendyk, M., 2000. *ApJ*, **541**, 351.

Glendenning, N. K., 1992. Physics Review D, **46**, 4161.

Glendenning, N. K., 2000. *Compact stars : nuclear physics, particle physics, and general relativity*, Springer, New York.

Gold, T., 1968. *Nature*, **218**, 731.

Goldreich, P. & Julian, W. H., 1969. *ApJ*, **157**, 869.

Gómez, G. C., Benjamin, R. A. & Cox, D. P., 2001. *AJ*, **122**, 908.

Gothoskar, P. & Gupta, Y., 2000. *ApJ*, **531**, 345.

Gotthelf, E. V., Vasisht, G., Boylan-Kolchin, M. & Torii, K., 2000. *ApJ*, **542**, L37.

Gould, D. M., 1994. *PhD thesis*, The University of Manchester.

Gould, D. M. & Lyne, A. G., 1998. *MNRAS*, **301**, 235.

Guelin, M., Guibert, J., Hutchmeier, W. & Weliachew, L., 1969. *Nature*, **221**, 249.

Gunn, J. E. & Ostriker, J. P., 1970. *ApJ*, **160**, 979.

Gupta, Y., 1995. *ApJ*, **451**, 717.

Gupta, Y., 2000. In: *Sources and Scintillations: refraction and scattering in radio astronomy, IAU Colloquium 182*, p. 25, eds Strom, R., Peng, B., Walker, M. & Nan, R., Kluwer Academic Publishers, Netherlands.

Gupta, Y., Bhat, N. D. R. & Rao, A. P., 1999. *ApJ*, **520**, 173.

Gupta, Y., Rickett, B. J. & Lyne, A. G., 1994. *MNRAS*, **269**, 1035.

Hagen, J. & Farley, D., 1973. *Rad. Sci.*, **8**, 775.

Halpern, J. P. & Holt, S. S., 1992. *Nature*, **357**, 222.

Halpern, J. P., Camilo, F., Gotthelf, E. V., Helfand, D. J., Kramer, M., Lyne, A. G., Leighly, K. M. & Eracleous, M., 2001. *ApJ*, **552**, L125.

Han, J. L. & Manchester, R. N., 2001. *MNRAS*, **320**, L35.

Han, J. L., 2004. In: *The Magnetized Interstellar Medium*, p 3, eds Uyaniker, B., Reich, W. & Wielebinski, R.

Hankins, T. H., 1972. *ApJ*, **177**, L11.

Hankins, T. H. & Rickett, B. J., 1975. In: *Methods in Computational Physics Volume 14 – Radio Astronomy*, p. 55, Academic Press, New York.

Hankins, T. H., Kern, J. S., Weatherall, J. C. & Eilek, J. A., 2003. *Nature*, **422**, 141.

Harding, A. K., Muslimov, A. G. & Zhang, B., 2002. *ApJ*, **576**, 366.

Hartman, R. C., Bertsch, D. L., Bloom, S. D., Chen, A. W., Deines-Jones, P., Esposito, J. A., Fichtel, C. E., Friedlander, D. P., Hunter, S. D., McDonald, L. M., Sreekumar, P., Thompson, D. J., Jones, B. B., Lin, Y. C., Michelson, P. F., Nolan, P. L., Tompkins, W. F., Kanbach, G., Mayer-Hasselwander, H. A., Mücke, A., Pohl, M., Reimer, O., Kniffen, D. A., Schneid, E. J., von Montigny, C., Mukherjee, R. & Dingus, B. L., 1999. *ApJS*, **123**, 79.

Hartmann, D. H., 1995. *ApJ*, **447**, 646.

Haslam, C. G. T., Stoffel, H., Salter, C. J. & Wilson, W. E., 1982. *A&AS*, **47**, 1.

Heger, A., Woosley, S. E., Langer, N. & Spruit, H. C., 2003. In: *Stellar Rotation, Proc. of IAU Symposium S215*, in press, eds Maeder, A. & Eenens, P., PASP, San Francisco. (astro-ph/0301374).

Heiles, C., 2002. In: *Single-Dish Radio Astronomy: Techniques & Applications*, p. 131, eds Stanimirovic, S., Altschuler, D. R., Goldsmith, P. F. & Salter, C. J., Astronomical Society of the Pacific, San Francisco.

Heiles, C., Perillat, P., Nolan, M., Lorimer, D. R., Rhat, R., Ghosh, T., Lewis, B. M., O'Neil, K., Salter, C. & Stanimirovic, S., 2001. *PASP*, **113**, 1274.

Helfand, D. J., Manchester, R. N. & Taylor, J. H., 1975. *ApJ*, **198**, 661.

Hewish, A., Bell, S. J., Pilkington, J. D. H., Scott, P. F. & Collins, R. A., 1968. *Nature*, **217**, 709.

Hibschman, J. A. & Arons, J., 2001a. *ApJ*, **546**, 382.

Hibschman, J. A. & Arons, J., 2001b. *ApJ*, **560**, 871.

Hill, A. S., Stinebring, D. R., Barnor, H. A., Berwick, D. E. & Webber, A. B., 2003. *ApJ*, **599**, 457.

Hobbs, G., Lyne, A. G., Kramer, M., Martin, C. E. & Jordan, C., 2004. *MNRAS*, in press.

Holloway, N. J., 1973. *Nature*, **246**, 6.

Hotan, A., van Straten, W. & Manchester, R. N., 2004. *Publ. Astr. Soc. Aust.*, in press.

Hoyle, F., Narlikar, J. & Wheeler, J. A., 1964. *Nature*, **203**, 914.

Hulse, R. A. & Taylor, J. H., 1975. *ApJ*, **195**, L51.

Istomin, Y. N., 1991. *Sov. Astron. Lett.*, **17**, 301.

Izvekova, V. A., Kuz'min, A. D., Malofeev, V. M. & Shitov, Y. P., 1981. *Astrophys. Space Sci.*, **78**, 45.

Jackson, J. D., 1962. *Classical Electrodynamics*, Wiley.
Jackson, M. S., Halpern, J. P., Gotthelf, E. V. & Mattox, J. R., 2002. *ApJ*, **578**, 935.
Jacoby, B. A. & Anderson, S. B., 2001, in Encycopedia of Astronomy and Astrophysics, Bristol: Institute of Physics, p. 795.
Jacoby, B. A., Bailes, M., van Kerkwijk, M. H., Ord, S., Hotan, A., Kulkarni, S. R. & Anderson, S. B., 2003. *ApJ*, **599**, L99.
Jaffe, A. H. & Backer, D. C., 2003. *ApJ*, **583**, 616.
Jauncey, D. L., Kedziora-Chudczer, L., Lovell, J. E. J., Macquart, J., Nicolson, G. D., Perley, R. A., Reynolds, J. E., Tzioumis, A. K., Wieringa, M. H. & Bignall, H. E., 2001. *Ap&SS*, **278**, 87.
Jenet, F. A., Cook, W. R., Prince, T. A. & Unwin, S. C., 1997. *PASP*, **109**, 707.
Johnston, H. M. & Kulkarni, S. R., 1991. *ApJ*, **368**, 504.
Johnston, S., 2002. *Publ. Astr. Soc. Aust.*, **19**, 277.
Johnston, S. & Galloway, D., 1999. *MNRAS*, **306**, L50.
Johnston, S., Manchester, R. N., Lyne, A. G., Bailes, M., Kaspi, V. M., Qiao, G. & D'Amico, N., 1992. *ApJ*, **387**, L37.
Johnston, S., Nicastro, L. & Koribalski, B., 1998. *MNRAS*, **297**, 108.
Johnston, S. & Romani, R. W., 2002. *MNRAS*, **332**, 109.
Johnston, S. & Romani, R. W., 2003. *ApJ*, **590**, L95.
Johnston, S., van Straten, W., Kramer, M. & Bailes, M., 2001. *ApJ*, **549**, L101.
Jones, P. B., 1980. *ApJ*, **236**, 661.
Joshi, B. C., Lyne, A. G., Kramer, M., Lorimer, D. R., Jordan, C., Holloway, A., Ikin, T. & Stairs, I. H., 2003. In: *Radio Pulsars*, p. 321, eds Bailes, M., Nice, D. J. & Thorsett, S. E., Astronomical Society of the Pacific, San Francisco.
Joshi, K. J. & Rasio, F. A., 1997. *ApJ*, **479**, 948.
Jouteux, S., Ramachandran, R., Stappers, B. W., Jonker, P. G. & van der Klis, M., 2002. *A&A*, **384**, 532.
Kanbach, G., 2002. In: *WE-Heraeus Seminar on Neutron Stars, Pulsars and Supernova Remnants. MPE Report 278.*, p. 91, eds Becker, W., Lesch, H. & Truemper, J., Max-Plank-Institut für extraterrestrische Physik.
Kapoor, R. C. & Shukre, C. S., 1998. *ApJ*, **501**, 228.
Karastergiou, A., von Hoensbroech, A., Kramer, M., Lorimer, D., Lyne, A., Doroshenko, O., Jessner, A., Jordan, A. & Wielebinski, R., 2001. *A&A*, **379**, 270.
Karastergiou, A., Kramer, M., Johnston, S., Lyne, A. G., Bhat, N. D. R. & Gupta, Y., 2002. *A&A*, **391**, 247.
Karastergiou, A., Johnston, S. & Kramer, M., 2003. *A&A*, **404**, 325.
Kardeshev, N. S., Nikolaev, Y. N., Novikov, A. Y., Popov, M. V., Soglasnov, V. A., Kuz'min, A. D., Smirnova, T. V., Seiber, W. & Wielebinski, R., 1986. *A&A*, **163**, 114.
Kargaltsev, O., Pavlov, G. G. & Romani, R. W., 2004. *ApJ*, **602**, 327.
Kaspi, V. M., 2004. In: *Young Neutron Stars and Their Environments, IAU Symposium 218*, p. 231, eds Camilo, F. & Gaensler, B. M., Astronomical Society of the Pacific, San Francisco.
Kaspi, V. M., Bailes, M., Manchester, R. N., Stappers, B. W. & Bell, J. F., 1996. *Nature*, **381**, 584.

Kaspi, V. M., Bailes, M., Manchester, R. N., Stappers, B. W., Sandhu, J. S., Navarro, J. & D'Amico, N., 1997. *ApJ*, **485**, 820.

Kaspi, V. M. & Helfand, D. J., 2002. In: *Neutron Stars in Supernova Remnants*, p. 3, eds Slane, P. O. & Gaensler, B. M., Astronomical Society of the Pacific, San Francisco.

Kaspi, V. M., Lackey, J. R., Mattox, J., Manchester, R. N., Bailes, M. & Pace, R., 2000a. *ApJ*, **528**, 445.

Kaspi, V. M., Lyne, A. G., Manchester, R. N., Crawford, F., Camilo, F., Bell, J. F., D'Amico, N., Stairs, I. H., McKay, N. P. F., Morris, D. J. & Possenti, A., 2000b. *ApJ*, **543**, 321.

Kaspi, V. M., Ransom, S. M., Backer, D. C., Ramachandran, R., Demorest, P., Arons, J. & Spitkovskty, A., 2004. *ApJ*, submitted, (astro-ph/0401614).

Kaspi, V. M., Taylor, J. H. & Ryba, M., 1994. *ApJ*, **428**, 713.

Kassim, N. E. & Frail, D. A., 1996. *MNRAS*, **283**, L51.

Kassim, N. E. & Lazio, T. J. W., 1999. *ApJ*, **527**, L101.

Kazbegi, A. Z., Machabeli, G. Z. & Melikidze, G. I., 1991. *MNRAS*, **253**, 377.

Kerr, F. J., Bowers, P. F., Jackson, P. D. & Kerr, M., 1986. *A&AS*, **66**, 373.

Kijak, J. & Gil, J., 1998. *MNRAS*, **299**, 855.

Kijak, J. & Gil, J., 2002. *A&A*, **392**, 189.

Kijak, J. & Gil, J., 2003. *A&A*, **397**, 969.

Klein, B., 2004. *PhD thesis*, University of Bonn.

Komesaroff, M. M., 1970. *Nature*, **225**, 612.

Konacki, M., Maciejewski, A. J. & Wolszczan, A., 1999. *ApJ*, **513**, 471.

Konacki, M., Maciejewski, A. J. & Wolszczan, A., 2000. *ApJ*, **544**, 921.

Konacki, M. & Wolszczan, A., 2003. *ApJ*, **591**, L147.

Kopeikin, S. M., 1995. *ApJ*, **439**, L5.

Kopeikin, S. M., 1996. *ApJ*, **467**, L93.

Kopeikin, S. M, 2003. In: *Radio Pulsars (ASP Conf. Ser.)*, p. 111, eds Bailes, M., Nice, D. J. & Thorsett, S. E., Astronomical Society of the Pacific, San Francisco.

Kopeikin, S. M. & Schäfer, G., 1999. *Phys. Rev. D*, **60**, 124002.

Koribalski, B. S., Johnston, S., Weisberg, J. & Wilson, W., 1995. *ApJ*, **441**, 756.

Kramer, M., 1998. *ApJ*, **509**, 856.

Kramer, M., Bell, J. F., Manchester, R. N., Lyne, A. G., Camilo, F., Stairs, I. H., D'Amico, N., Kaspi, V. M., Hobbs, G., Morris, D. J., Crawford, F., Possenti, A., Joshi, B. C., McLaughlin, M. A., Lorimer, D. R. & Faulkner, A. J., 2003a. *MNRAS*, **342**, 1299.

Kramer, M., Karastergiou, A., Gupta, Y., Johnston, S., Bhat, N. D. R. & Lyne, A. G., 2003b. *A&A*, **407**, 655.

Kramer, M., Lyne, A. G., Hobbs, G., Lohmer, O., Carr, P., Jordan, C. & Wolszczan, A., 2003c. *ApJ*, **593**, 31.

Kramer, M., Löhmer, O. & Karastergiou, A., 2003d. In: *Radio Pulsars (ASP Conf. Ser.)*, p. 99, eds Bailes, M., Nice, D. J. & Thorsett, S. E., Astronomical Society of the Pacific, San Francisco.

Kramer, M., Doroshenko, O. & Xilouris, K. M., 1999a. In: *Pulsar Timing, General Relativity, and the Internal Structure of Neutron Stars*, p. 47, eds Arzoumanian, Z., van der Hooft, F. & van den Heuvel, E. P. J., North Holland, Amsterdam.

Kramer, M., Lange, C., Lorimer, D. R., Backer, D. C., Xilouris, K. M., Jessner, A. & Wielebinski, R., 1999b. *ApJ*, **526**, 957.

Kramer, M., Xilouris, K. M., Camilo, F., Nice, D., Lange, C., Backer, D. C. & Doroshenko, O., 1999c. *ApJ*, **520**, 324.

Kramer, M., Jessner, A., Müller, P. & Wielebinski, R., 1996a. In: *Pulsars: Problems and Progress, IAU Colloquium 160*, p. 13, eds Johnston, S., Walker, M. A. & Bailes, M., Astronomical Society of the Pacific, San Francisco.

Kramer, M., Xilouris, K. M., Jessner, A. & Wielebinski, R.; Timofeev, M., 1996b. *A&A*, **306**, 867.

Kramer, M., Johnston, S. & van Straten, W., 2002. *MNRAS*, **334**, 523.

Kramer, M., Klein, B., Lorimer, D. R., Müller, P., Jessner, A. & Wielebinski, R., 2000. In: *Pulsar Astronomy - 2000 and Beyond, IAU Colloquium 177*, p. 37, eds Kramer, M., Wex, N. & Wielebinski, R., Astronomical Society of the Pacific, San Francisco.

Kramer, M., Wielebinski, R., Jessner, A., Gil, J. A. & Seiradakis, J. H., 1994. *A&AS*, **107**, 515.

Kramer, M., Xilouris, K. M., Jessner, A., Lorimer, D. R., Wielebinski, R. & Lyne, A. G., 1997. *A&A*, **322**, 846.

Kramer, M., Xilouris, K. M., Lorimer, D. R., Doroshenko, O., Jessner, A., Wielebinski, R., Wolszczan, A. & Camilo, F., 1998. *ApJ*, **501**, 270.

Kramer, M. & Xilouris, K. M., 2000. In: *Pulsar Astronomy - 2000 and Beyond, IAU Colloquium 177*, p. 229, eds Kramer, M., Wex, N. & Wielebinski, R., Astronomical Society of the Pacific, San Francisco.

Krause-Polstorff, J. & Michel, F. C., 1985. *MNRAS*, **213**, 43P.

Kuiper, L., Hermsen, W., Verbunt, F., Thompson, D. J., Stairs, I. H., Lyne, A. G., Strickman, M. S. & Cusumano, G., 2000. *A&A*, **359**, 615.

Kulkarni, S. R. & Heiles, C., 1980. *AJ*, **85**, 1413.

Kuz'min, A. D. & Izvekova, V. A., 1993. *MNRAS*, **260**, 724.

Kuzmin, A. D. & Losovskii, B. Y., 1997. *Astronomy Letters*, **23**, 283.

Kuzmin, A. D., Hamilton, P. A., Shitov, Y. P., McCulloch, P. M., McConnell, D. & Pugatchev, V. D., 2003. *MNRAS*, **344**, 1187.

Lai, D., Bildsten, L. & Kaspi, V. M., 1995. *ApJ*, **452**, 819.

Lambert, H. C. & Rickett, B. J., 1999. *ApJ*, **517**, 299.

Lambert, H. C. & Rickett, B. J., 2000. *ApJ*, **531**, 883.

Lange, C., Camilo, F., Wex, N., Kramer, M., Backer, D., Lyne, A. & Doroshenko, O., 2001. *MNRAS*, **326**, 274.

Lange, C., Kramer, M., Wielebinski, R. & Jessner, A., 1998. *A&A*, **332**, 111.

Large, M. I., Vaughan, A. E. & Mills, B. Y., 1968. *Nature*, **220**, 340.

Lattimer, J. M. & Prakash, M., 2001. *ApJ*, **550**, 426.

Lattimer, J. M., Prakash, M., Masak, D. & Yahil, A., 1990. *ApJ*, **355**, 241.

Lazio, T. J. W. & Cordes, J. M., 1998a. *ApJS*, **118**, 201.

Lazio, T. J. W. & Cordes, J. M., 1998b. *ApJ*, **505**, 715.

Leahy, D. A., Darbro, W., Elsner, R. F., Weisskopf, M. C., Sutherland, P. G., Kahn, S. & Grindlay, J. E., 1983. *ApJ*, **266**, 160.

Lee, L. C. & Jokipii, J. R., 1975. *ApJ*, **201**, 532.

Lesch, H., Jessner, A., Kramer, M. & Kunzl, T., 1998. *A&A*, **332**, L21.

Lewin, W. H. G., van Paradijs, J. & Taam, R. E., 1993. *Space Science Reviews*, **62**, 223.

Link, B., Epstein, R. I. & Lattimer, J. M., 1999. *Phys. Rev. Lett.*, **83**, 3362.

Löhmer, O., Kramer, M., Mitra, D., Lorimer, D. R. & Lyne, A. G., 2001. *ApJ*, **562**, L157.

Lommen, A. N., 2002. In: *WE-Heraeus Seminar on Neutron Stars, Pulsars and Supernova Remnants. MPE Report 278.*, p. 114, eds Becker, W., Lesch, H. & Truemper, J., Max-Plank-Institut für extraterrestrische Physik.

Lommen, A. N. & Backer, D. C., 2001. *ApJ*, **562**, 297.

Lorimer, D. R., 2001. Arecibo Technical Memo No. 2001–01.

Lorimer, D. R., Bailes, M., Dewey, R. J. & Harrison, P. A., 1993. *MNRAS*, **263**, 403.

Lorimer, D. R. & Freire, P. F., 2004. In: *Binary Pulsars*, in press, eds Rasio, F. & Stairs, I. H. Astronomical Society of the Pacific, San Francisco.

Lorimer, D. R., Jessner, A., Seiradakis, J. H., Lyne, A. G., D'Amico, N., Athanasopoulos, A. Xilouris, K. M., Kramer, M. & Wielebinski, R., 1998. *A&AS*, **128**, 541.

Lorimer, D. R., Kramer, M., Müller, P., Wex, N., Jessner, A., Lange, C. & Wielebinski, R., 2000. *A&A*, **358**, 169.

Lorimer, D. R., McLaughlin, M. A., Arzoumanian, Z. A., Xilouris, K., Cordes, J. M., Lommen, A. N., Fruchter, A. S., Chandler, A. M. & Backer, D. C., 2004a. *MNRAS*, **347**, L21.

Lorimer, D. R., Xilouris, K. M., Fruchter, A. S., Stairs, I. H., Camilo, F., McLaughlin, M. A., Vazquez, A. M., Eder, J. E., Roberts, M. S. E., Hessels, J. W. T., Ransom, S. M. & Kaspi, V. M., 2004b. *MNRAS*, in press.

Lyne, A. G., 1984. *Nature*, **310**, 300.

Lyne, A. G., 1992. In: *X-ray Binaries and Recycled Pulsars*, p. 79, eds van den Heuvel, E. P. J. & Rappaport, S. A., Kluwer, Dordrecht.

Lyne, A. G., 1999. In: *Pulsar Timing, General Relativity, and the Internal Structure of Neutron Stars*, p. 141, eds Arzoumanian, Z., van der Hooft, F. & van den Heuvel, E. P. J., North Holland, Amsterdam.

Lyne, A. G., 2003. In: *Radio Pulsars (ASP Conf. Ser.*, p. 11, eds Bailes, M., Nice, D. J. & Thorsett, S. E., Astronomical Society of the Pacific, San Francisco.

Lyne, A. G. & Bailes, M., 1990. *MNRAS*, **246**, 15.

Lyne, A. G. & Lorimer, D. R., 1994. *Nature*, **369**, 127.

Lyne, A. G. & Manchester, R. N., 1988. *MNRAS*, **234**, 477.

Lyne, A. G. & Rickett, B. J., 1968. *Nature*, **219**, 1339.

Lyne, A. G. & Smith, F. G., 1982. *Nature*, **298**, 825.

Lyne, A. G. & Smith, F. G., 2005. *Pulsar Astronomy, 3rd ed.*, Cambridge University Press, Cambridge.

Lyne, A. G., Biggs, J. D., Brinklow, A., Ashworth, M. & McKenna, J., 1988. *Nature*, **332**, 45.

Lyne, A. G., Brinklow, A., Middleditch, J., Kulkarni, S. R., Backer, D. C. & Clifton, T. R., 1987. *Nature*, **328**, 399.

Lyne, A. G., Burgay, M., Kramer, M., Possenti, A., Manchester, R. N., Camilo, F., McLaughlin, M., Lorimer, D. R., Joshi, B. C., Reynolds, J. E. & Freire, P. C. C., 2004. *Science*, **303**, 1153.

Lyne, A. G., Manchester, R. N., Lorimer, D. R., Bailes, M., D'Amico, N., Tauris, T. M., Johnston, S., Bell, J. F. & Nicastro, L., 1998. *MNRAS*, **295**, 743.

Lyne, A. G., Manchester, R. N. & Taylor, J. H., 1985. *MNRAS*, **213**, 613.

Lyne, A. G., Mankelow, S. H., Bell, J. F. & Manchester, R. N., 2000. *MNRAS*, **316**, 491.

Lyne, A. G., Pritchard, R. S. & Graham-Smith, F., 2001. *MNRAS*, **321**, 67.

Lyne, A. G., Pritchard, R. S. & Smith, F. G., 1993. *MNRAS*, **265**, 1003.

Malhotra, R., Black, D., Eck, A. & Jackson, A., 1992. *Nature*, **356**, 583.

Malofeev, V., 1996. In: *Pulsars: Problems and Progress, IAU Colloquium 160*, p. 271, eds Johnston, S., Walker, M. A. & Bailes, M., Astronomical Society of the Pacific, San Francisco.

Malofeev, V. M. & Malov, O. I., 1997. *Nature*, **389**, 697.

Malofeev, V. M., Shishov, V. I., Sieber, W., A., J., Kramer, M. & Wielebinski, R., 1996. *A&A*, **308**, 180.

Manchester, R. N., 1971. *ApJS*, **23**, 283.

Manchester, R. N., 2001. In: *Young Supernova Remnants*, p. 305, eds Holt, S. S. & Hwang, U., American Inst. Physics, New York.

Manchester, R. N. & Lyne, A. G., 1977. *MNRAS*, **181**, 761.

Manchester, R. N. & Taylor, J. H., 1972. *Astrophys. Lett.*, **10**, 67.

Manchester, R. N. & Taylor, J. H., 1977. *Pulsars*, Freeman, San Francisco.

Manchester, R. N., Lyne, A. G., Camilo, F., Bell, J. F., Kaspi, V. M., D'Amico, N., McKay, N. P. F., Crawford, F., Stairs, I. H., Possenti, A., Morris, D. J. & Sheppard, D. C., 2001. *MNRAS*, **328**, 17.

Manchester, R. N., Taylor, J. H. & Huguenin, G. R., 1975. *ApJ*, **196**, 83.

Marcy, G. W. & Butler, R. P., 2000. *PASP*, **112**, 137.

Maron, O., Kijak, J., Kramer, M. & Wielebinski, R., 2000. *A&AS*, **147**, 195.

Marshall, F. E., Gotthelf, E. V., Zhang, W., Middleditch, J. & Wang, Q. D., 1998. *ApJ*, **499**, L179.

Matsakis, D. N., Taylor, J. H. & Eubanks, T. M., 1997. *A&A*, **326**, 924.

McGary, R. S., Brisken, W. F., Fruchter, A. S., Goss, W. M. & Thorsett, S. E., 2001. *AJ*, **121**, 1192.

McKinnon, M., 1997. *ApJ*, **475**, 763.

McKinnon, M. & Stinebring, D., 1998. *ApJ*, **502**, 883.

McLaughlin, M. A. & Cordes, J. M., 2000. *ApJ*, **538**, 818.

McLaughlin, M. A. & Cordes, J. M., 2003. *ApJ*, **596**, 982.

McLaughlin, M. A. & Cordes, J. M., 2004. In: *4th AGILE Science Workshop on X-ray and Gamma-ray Astrophysics of Galactic Sources*, in press, ed. Pellizzoni, A. (astro-ph/0310748).

McLaughlin, M. A., Mattox, J. R., Cordes, J. M. & Thompson, D. J., 1996. *ApJ*, **473**, 763.

McLaughlin, M. A., Cordes, J. M., Hankins, T. H. & Moffett, D. A., 1999. *ApJ*, **512**, 929.

McLaughlin, M. A., Stairs, I. H., Kaspi, V. M., Lorimer, D. R., Kramer, M., Lyne, A. G., Manchester, R. N., Camilo, F., Hobbs, G., Possenti, A., D'Amico, N. & Faulkner, A. J., 2003. *ApJ*, **591**, L135.

Melrose, D. B., 1981. In: *Pulsars, IAU Symposium No. 95*, p. 133, eds Sieber, W. & Wielebinski, R., Reidel, Dordrecht.

Melrose, D. B., 1989. *Solar Physics*, **120**, 369.

Melrose, D. B., 1992. *Philos. Trans. Roy. Soc. London A*, **341**, 105.

Melrose, D. B., 2000. In: *Pulsar Astronomy - 2000 and Beyond, IAU Colloquium 177*, p. 721, eds Kramer, M., Wex, N. & Wielebinski, R., Astronomical Society of the Pacific, San Francisco.

Melrose, D. B., 2004. In: *Young Neutron Stars and Their Environments*,

IAU Symposium 218, p. 349, eds Camilo, F. & Gaensler, B. M., Astronomical Society of the Pacific, San Francisco.

Mestel, L., 1971. *Nature*, **233**, 149.

Mestel, L., & Pryce, M. H. L., 1992. *MNRAS*, **254**, 355.

Mezger, P. G., Zylka, R., Philipp, S. & Launhardt, R., 1999. *A&A*, **348**, 457.

Michel, F. C., 1991. *Theory of Neutron Star Magnetospheres*, University of Chicago Press, Chicago.

Michel, F. C., 1992. In: *The Magnetospheric Structure and Emission Mechanisms of Radio Pulsars, IAU Colloquium 128*, p. 236, eds Hankins, T. H., Rankin, J. M. & Gil, J. A., Pedagogical University Press, Zielona Góra, Poland.

Michel, F. C. & Li, H., 1999. *Phys. Rep.*, **318**, 227.

Middleditch, J. & Kristian, J., 1984. *ApJ*, **279**, 157.

Middleditch, J. & Priedhorsky, W. C., 1986. *ApJ*, **306**, 230.

Migliazzo, J. M., Gaensler, B. M., Backer, D. C., Stappers, B. W., van der Swaluw, E. & Strom, R. G., 2002. *ApJ*, **567**, L141.

Mignani, R. P., Caraveo, P. A. & Bignami, G. F., 1997. *ApJ*, **474**, L51.

Mignani, R. P., de Luca, A. & Caraveo, P. A., 2000. *ApJ*, **543**, 318.

Mignani, R. P., de Luca, A. & Caraveo, P. A., 2004. In: *Young Neutron Stars and Their Environments, IAU Symposium 218*, p. 391, eds Camilo, F. & Gaensler, B. M., Astronomical Society of the Pacific, San Francisco.

Mignani, R. P., de Luca, A., Caraveo, P. A. & Becker, W., 2002. *ApJ*, **580**, L147.

Mitra, D. & Li, X. H., 2004. *A&A*, **421**, 215.

Mitra, D. & Ramachandran, R., 2001. *A&A*, **370**, 586.

Mitra, D. & Rankin, J. M., 2002. *ApJ*, **577**, 322.

Mitra, D., Wielebinski, R., Kramer, M. & Jessner, A., 2003. *A&A*, **398**, 993.

Moffett, D. A. and Hankins, T. H., 1996. *ApJ*, **468**, 779.

Molkov, S. V., Cherepashchuk, A. M., Lutovinov, A. A., Revnivtsev, M. G., Postnov, K. A. & Sunyaev, R. A., 2004. *Astronomy Letters*, in press, (astro-ph/0402416).

Morris, D., Kramer, M., Thum, C. & et al., 1997. *A&A*, **322**, L17.

Morris, D. J., Hobbs, G., Lyne, A. G., Stairs, I. H., Camilo, F., Manchester, R. N., Possenti, A., Bell, J. F., Kaspi, V. M., Amico, N. D., McKay, N. P. F., Crawford, F. & Kramer, M., 2002. *MNRAS*, **335**, 275.

Mueller, H., 1948. *Opt. Soc. America*, **338**, 661.

Muno, M. P., Baganoff, F. K., Bautz, M. W., Brandt, W. N., Broos, P. S., Feigelson, E. D., Garmire, G. P., Morris, M. R., Ricker, G. R. & Townsley, L. K., 2003. *ApJ*, **589**, 225.

Murray, S. S., Slane, P. O., Seward, F. D., Ransom, S. M. & Gaensler, B. M., 2002. *ApJ*, **568**, 226.

Muslimov, A. G. & Harding, A. K., 1997. *ApJ*, **485**, 735.

Muslimov, A. G. & Tsygan, A. I., 1992. *MNRAS*, **255**, 61.

Naghizadeh-Khouei, J. & Clarke, D., 1993. *A&A*, **274**, 968.

Narayan, R. 1992. *Philos. Trans. Roy. Soc. London A*, **341**, 151.

Narayan, R. & Vivekanand, M., 1982. *A&A*, **113**, L3.

Nicastro, L. & Johnston, S., 1995. *MNRAS*, **273**, 122.

Nicastro, L., Nigro, F., D'Amico, N., Lumiella, V. & Johnston, S., 2001. *A&A*, **368**, 1055.

Nice, D. J., 1999. *ApJ*, , 927.

Nice, D. J. & Taylor, J. H., 1995. *ApJ*, **441**, 429.

Nice, D. J., Fruchter, A. S. & Taylor, J. H., 1995. *ApJ*, **449**, 156.
Nordtvedt, K., 1968a. *Phys. Rev.*, **169**, 1014.
Nordtvedt, K., 1968b. *Phys. Rev.*, **170**, 1186.
Oppenheimer, J. R. & Volkoff, G., 1939. *Phys. Rev.*, **55**, 374.
Ord, S. M., 2002. *PhD thesis*, University of Manchester.
Ord, S. M., Bailes, M. & van Straten, W., 2002a. *ApJ*, **574**, L75.
Ord, S. M., Bailes, M. & van Straten, W., 2002b. *MNRAS*, **337**, 409.
Oster, L. & Sieber, W., 1976. *ApJ*, **203**, 233.
Pacini, F., 1967. *Nature*, **216**, 567.
Pacini, F., 1968. *Nature*, **219**, 145.
Pacini, F., 1971. *ApJ*, **163**, L17.
Pavlov, G. G. & Shibanov, I. A., 1978. *Astronomicheskii Zhurnal*, **55**, 373.
Pavlov, G. G., Stringfellow, G. S. & Cordova, F. A., 1996. *ApJ*, **467**, 370.
Pavlov, G. G., Zavlin, V. E. & Sanwal, D., 2002. In: *WE-Heraeus Seminar on Neutron Stars, Pulsars and Supernova Remnants. MPE Report 278.*, p. 273, eds Becker, W., Lesch, H. & Truemper, J., Max-Plank-Institut für extraterrestrische Physik.
Peters, P. C., 1964. *Phys. Rev.*, **136**, 1224.
Petrova, S. A., 2001. *A&A*, **378**, 883.
Pfahl, E. & Loeb, A., 2004. *ApJ*, in press, (astro-ph/0309744).
Phillips, J. A., 1992. *ApJ*, **385**, 282.
Phillips, J. A., Thorsett, S. E. & Kulkarni, S. R., 1993. *Planets Around Pulsars*, Astronomical Society of the Pacific Conference Series, ASP, San Francisco
Phillips, J. A. & Wolszczan, A., 1991. *ApJ*, **382**, L27.
Phinney, E. S., 1992. *Philos. Trans. Roy. Soc. London A*, **341**, 39.
Phinney, E. S. & Kulkarni, S. R., 1994. *Ann. Rev. Astr. Ap.*, **32**, 591.
Pines, D. & Alpar, M. A., 1985. *Nature*, **316**, 27.
Pons, J. A., Walter, F. M., Lattimer, J. M., Prakash, M., Neuhäuser, R. & An, P., 2002. *ApJ*, **564**, 981.
Possenti, A., Cerutti, R., Colpi, M. & Mereghetti, S., 2002. *A&A*, **387**, 993.
Possenti, A., D'Amico, N., Manchester, R. N., Camilo, F., Lyne, A. G., Sarkissian, J. & Corongiu, A., 2003. *ApJ*, **599**, 475.
Press, W. H., Teukolsky, S. A., Vetterling, W. T. & Flannery, B. P., 1992. *Numerical Recipes: The Art of Scientific Computing*, 2nd edition, Cambridge University Press, Cambridge.
Prince, T. A., Anderson, S. B., Kulkarni, S. R. & Wolszczan, W., 1991. *ApJ*, **374**, L41.
Radhakrishnan, V. & Cooke, D. J., 1969. *Astrophys. Lett.*, **3**, 225.
Radhakrishnan, V. & Shukre, C. S., 1985. In: *Supernovae, Their Progenitors and Remnants*, p. 155, eds Srinivasan, G. & Radhakrishnan, V., Indian Academy of Sciences, Bangalore.
Ramachandran, R., Deshpande, A. A. & Indrani, C., 1998. *A&A*, **339**, 787.
Ramachandran, R., Mitra, D., Deshpande, A. A., McConnell, D. M. & Ables, J. G., 1997. *MNRAS*, **290**, 260.
Ramaty, R. & Lingenfelter, R. E., 1983. *Space Science Reviews*, **36**, 305.
Rand, R. J. & Lyne, A. G., 1994. *MNRAS*, **268**, 497.
Rankin, J. M., 1983a. *ApJ*, **274**, 333.
Rankin, J. M., 1983b. *ApJ*, **274**, 359.
Rankin, J. M., 1986. *ApJ*, **301**, 901.
Rankin, J. M., 1993. *ApJ*, **405**, 285.

Rankin, J. M. & Rathnasree, N., 1995. *ApJ*, **452**, 814.

Ransom, S. M., 2001. *PhD thesis*, Harvard University.

Ransom, S. M., Cordes, J. M. & Eikenberry, S. S., 2003a. *ApJ*, **589**, 911.

Ransom, S. M., Hessels, J. W. T., Stairs, I. H., Kaspi, V. M., Backer, D. C., Greenhill, L. J. & Lorimer, D. R., 2003b. In: *Radio Pulsars (ASP Conf. Ser.*, p. 371, eds Bailes, M., Nice, D. J. & Thorsett, S. E., Astronomical Society of the Pacific, San Francisco.

Ransom, S. M., Eikenberry, S. S. & Middleditch, J., 2002. *AJ*, **124**, 1788.

Ransom, S. M., Greenhill, L. J., Herrnstein, J. R., Manchester, R. N., Camilo, F., Eikenberry, S. S. & Lyne, A. G., 2001. *ApJ*, **546**, L25.

Ransom, S. M., Kaspi, V. M., Demorest, P., Ramachandran, R., Backer, D. C., Pfahl, E., Ghigo, F., Arons, J. & Kaplan, D., 2004. *ApJ*, **609**, L71.

Rasio, F. A., 1994. *ApJ*, **427**, L107.

Rasio, F. A., Nicholson, P. D., Shapiro, S. L. & Teukolsy, S. A., 1992. *Nature*, **355**, 325.

Rickett, B. J., 1969. *Nature*, **221**, 158.

Rickett, B. J., 1977. *Ann. Rev. Astr. Ap.*, **15**, 479.

Rickett, B. J., 1990. *Ann. Rev. Astr. Ap.*, **28**, 561.

Rickett, B. J., 1996. In: *Pulsars: Problems and Progress, IAU Colloquium 160*, p. 439, eds Johnston, S., Walker, M. A. & Bailes, M., Astronomical Society of the Pacific, San Francisco.

Rickett, B. J., Coles, W. A. & Bourgois, G., 1984. *A&A*, **134**, 390.

Rickett, B. J., Hankins, T. H. & Cordes, J. M., 1975. *ApJ*, **201**, 425.

Rickett, B. J., Quirrenbach, A., Wegner, R., Krichbaum, T. P. & Witzel, A., 1995. *A&A*, **293**, 479.

Ritchings, R. T., 1976. *MNRAS*, **176**, 249.

Roberts, M. S. E., Hessels, J. W. T., Ransom, S. M., Kaspi, V. M., Freire, P. C. C., Crawford, F. & Lorimer, D. R., 2002. *ApJ*, **577**, L19.

Robertson, H. P., 1938. *Ann. Math.*, **38**, 101.

Robinson, B. J., Cooper, B. F. C., Gardiner, F. F., Wielebinski, R. & Landecker, T. L., 1968. *Nature*, **218**, 1143.

Rohlfs, K. & Wilson, T. L., 2000. *Tools of Radio Astronomy*, Springer-Verlag, Berlin.

Romani, R. W., 1987. *ApJ*, **313**, 718.

Romani, R. W., 1992. *ApJ*, **399**, 621.

Romani, R. W., 1996. *ApJ*, **470**, 469.

Romani, R. W., 2003. In: *Radio Pulsars*, 331, eds Bailes, M., Nice, D. J. & Thorsett, S. E., Astronomical Society of the Pacific, San Francisco.

Romani, R. W., Blandford, R. D. & Cordes, J. M., 1987. *Nature*, **328**, 324.

Romani, R. & Johnston, S., 2001. *ApJ*, **557**, L93.

Romani, R. W., Miller, A. J., Cabrera, B., Nam, S. W. & Martinis, J. M., 2001. *ApJ*, **563**, 221.

Romani, R. W., Narayan, R. & Blandford, R., 1986. *MNRAS*, **220**, 19.

Romani, R. W. & Yadigaroglu, I.-A., 1995. *ApJ*, **438**, 314.

Rowe, E. T., 1995. *A&A*, **296**, 275.

Roy, A. E., 1988. *Orbital Motion*, Institute of Physics Publishing, Bristol and Philadelphia.

Ruderman, M. A, 1976. *ApJ*, **203**, 213.

Ruderman, M. A. & Sutherland, P. G., 1975. *ApJ*, **196**, 51.

Ruderman, M. A., Shaham, J. & Tavani, M., 1989 *ApJ*, **336**, 507.

Ryle, M., 1952. *Royal Society of London Proceedings Series A*, **211**, 351.

Sallmen. S. & Backer, D. C., 1995. In: *Millisecond Pulsars: A Decade of Surprise*, p. 340, eds Fruchter, A. S., Tavani, M. & Backer, D. C., Astron. Soc. Pac. Conf. Ser. Vol. 72.

Scheuer, P. A. G., 1968. *Nature*, **218**, 920.

Seidelmann, P. K., Guinot, B. & Doggett, L. E., 1992. In: *Explanatory Supplement to the Astronomical Almanac*, p. 39, ed. Seidelmann, P. K., University Science Books, Mill Valley, California.

Seiradakis, J. H., 1992. *IAU Circ. No. 5532*.

Seiradakis, J. H., Gil, J. A., Graham, D. A., Jessner, A., Kramer, M., Malofeev, V. M., Sieber, W. & Wielebinski, R., 1995. *A&AS*, **111**, 205.

Shapiro, I. I., 1964. *Phys. Rev. Lett.*, **13**, 789.

Shapiro, S. L. & Teukolsky, S. A., 1983. *Black Holes, White Dwarfs and Neutron Stars. The Physics of Compact Objects*, Wiley-Interscience, New York.

Shearer, A. & Golden, A., 2001. *ApJ*, **547**, 967.

Shearer, A., Redfern, R. M., Gorman, G., Butler, R., Golden, A., O'Kane, P., Beskin, G. M., Neizvestny, S. I., Neustroev, V. V., Plokhotnichenko, V. L. & Cullum, M., 1997. *ApJ*, **487**, L181.

Shearer, A., Redfern, M., Pedersen, H., Rowold, T., O'Kane, P., Butler, R., O'Bryne, C. & Cullum, M., 1994. *ApJ*, **423**, L51.

Shearer, A., Stappers, B., O'Connor, P., Golden, A., Strom, R., Redfern, M. & Ryan, O., 2003. *Science*, **301**, 493.

Shemar, S. L. & Lyne, A. G., 1996. *MNRAS*, **282**, 677.

Shibanov, I. A., Zavlin, V. E., Pavlov, G. G. & Ventura, J., 1992. *A&A*, **266**, 313.

Shibazaki, N., Murakami, T., Shaham, J. & Nomoto, K., 1989. *Nature*, **342**, 656.

Shitov, Y. P., 1983. *Sov. Astron.*, **27**, 314.

Shitov, Y. P., Pugachev, V. D., 1998 *New Astronomy*, **3**, 101.

Shklovskii, I. S., 1970. *Sov. Astron.*, **13**, 562.

Shrauner, J. A., 1997. *PhD thesis*, Princeton University.

Sieber, W., 1973. *A&A*, **28**, 237.

Sieber, W., 1982. *A&A*, **113**, 311.

Sieber, W., 1997. *A&A*, **321**, 519.

Sieber, W. & Oster, L., 1975. *A&A*, **38**, 325.

Sieber, W. & Oster, L., 1977. *A&A*, **61**, 445.

Sigurdsson, S., Richer, H. B., Hansen, B. M., Stairs, I. H. & Thorsett, S. E., 2003. *Science*, **301**, 193.

Smarr, L. L. & Blandford, R., 1976. *ApJ*, **207**, 574.

Smirnova, T. V. & Shishov, V. I., 1989. *Sov. Astron. Lett.*, **15**, 191.

Smith, F. G., 1969. *Nature*, **223**, 934.

Smith, F. G., 1970. *MNRAS*, **149**, 1.

Smith, F. G., 1973. *MNRAS*, **161**, 9P.

Smith, F. G., 2003. *Reports on Progress in Physics*, **66**, 173.

Smith, F. G. & Thompson, J. H., 1988. *Optics*, John Wiley and Sons Ltd.

Spitkovsky, A., 2004. In: *Young Neutron Stars and Their Environments, IAU Symposium 218*, p. 357, eds Camilo, F. & Gaensler, B. M., Astronomical Society of the Pacific, San Francisco.

Spitkovsky, A. & Arons, J., 2002. In: *Neutron Stars in Supernova Remnants*,

p. 81, eds Slane, P. O. & Gaensler, B. M., Astronomical Society of the Pacific, San Francisco.

Shitov, Y. P. & Pugachev, V. D., 1997. *New Astronomy*, **3**, 101.

Staelin, D. H., 1969. *Proc. I.E.E.E.*, **57**, 724.

Staelin, D. H. & Reifenstein, III, E. C., 1968. *Science*, **162**, 1481.

Stairs, I. H., 2003. *Living Reviews in Relativity*, **6**, 5.

Stairs, I. H., 2004. *Science*, **304**, 547.

Stairs, I. H., Manchester, R. N., Lyne, A. G., Kaspi, V. M., Camilo, F., Bell, J. F., D'Amico, N., Kramer, M., Crawford, F., Morris, D. J., McKay, N. P. F., Lumsden, S. L., Tacconi-Garman, L. E., Cannon, R. D., Hambly, N. C. & Wood, P. R., 2001. *MNRAS*, **325**, 979.

Stairs, I. H., Splaver, E. M., Thorsett, S. E., Nice, D. J. & Taylor, J. H., 2000a. *MNRAS*, **314**, 459.

Stairs, I. H., Thorsett, S. E., Taylor, J. H. & Arzoumanian, Z., 2000b. In: *Pulsar Astronomy - 2000 and Beyond, IAU Colloquium 177*, p. 121, eds Kramer, M., Wex, N. & Wielebinski, R., Astronomical Society of the Pacific, San Francisco.

Stairs, I. H., Lyne, A. G. & Shemar, S. L., 2000c. *Nature*, **406**, 484.

Stairs, I. H., Thorsett, S. E., Taylor, J. H. & Wolszczan, A., 2002. *ApJ*, **581**, 501.

Standish, E. M., 1990. *A&A*, **233**, 252.

Stanimirovic, S., Altschuler, D. R., Goldsmith, P. F. & Salter, C. J. *Single-Dish Radio Astronomy: Techniques & Applications*, Astronomical Society of the Pacific, San Francisco, 2002.

Stephenson, F. R., Green, D. A., 2002. *Historical supernovae and their remnants*, Oxford: Clarendon Press, 2002, Oxford

Stinebring, D. R., Cordes, J. M., Rankin, J. M., Weisberg, J. M. & Boriakoff, V., 1984. *ApJS*, **55**, 247.

Stinebring, D. R., Kaspi, V. M., Nice, D. J., Ryba, M. F., Taylor, J. H., Thorsett, S. E. & Hankins, T. H., 1992. *Rev. Sci. Instrum.*, **63**, 3551.

Stinebring, D. R., McLaughlin, M. A., Cordes, J. M., Becker, K. M., Goodman, J. E. E., Kramer, M. A., Sheckard, J. L. & Smith, C. T., 2001. *ApJ*, **549**, L97.

Stokes, G., 1852. *Trans. Camb. Phil. Soc.*, **9**, 3/399.

Sturrock, P. A., 1971. *ApJ*, **164**, 529.

Taylor, J. H., 1974. *A&AS*, **15**, 367.

Taylor, J. H., 1987. In: *The Origin and Evolution of Neutron Stars, IAU Symposium No. 125*, 383, eds Helfand, D. J. & Huang, J.-H., Reidel, Dordrecht.

Taylor, J. H., 1992. *Philos. Trans. Roy. Soc. London A*, **341**, 117.

Taylor, J. H. & Cordes, J. M., 1993. *ApJ*, **411**, 674.

Taylor, J. H. & Huguenin, G. R., 1969. *Nature*, **221**, 816.

Taylor, J. H. & Weisberg, J. M., 1989. *ApJ*, **345**, 434.

Taylor, J. H., Manchester, R. N. & Huguenin, G. R., 1975. *ApJ*, **195**, 513.

Thompson, A. R., Moran, J. M. & Swenson, G. W., 1986. *Interferometry and Synthesis in Radio Astronomy*, John Wiley and Sons, New York.

Thompson, C., 2001. In: *The Rome 2000 Mini-workshop*, eds Feroci, M., Mereghetti, S. & Stella, L. (astro-ph/0110679).

Thompson, D., 2004. In: *Young Neutron Stars and Their Environments, IAU Symposium 218*, p. 399, eds Camilo, F. & Gaensler, B. M., Astronomical Society of the Pacific, San Francisco.

Thorsett, S. E., 1991. *ApJ*, **377**, 263.
Thorsett, S. E. & Chakrabarty, D., 1999. *ApJ*, **512**, 288.
Thorsett, S. E., Arzoumanian, Z. & Taylor, J. H., 1993. *ApJ*, **412**, L33.
Thorsett, S. E., Arzoumanian, Z., Camilo, F. & Lyne, A. G., 1999. *ApJ*, **523**, 763.
Torii, K., Tsunemi, H., Dotani, T. & Mitsuda, K., 1997. *ApJ*, **489**, L145.
Torres, D. F., Butt, Y. M. & Camilo, F., 2001. *ApJ*, **560**, L155.
Toscano, M., Bailes, M., Manchester, R. & Sandhu, J., 1998. *ApJ*, **506**, 863.
Toscano, M., Sandhu, J. S., Bailes, M., Manchester, R. N., Britton, M. C., Kulkarni, S. R., Anderson, S. B. & Stappers, B. W., 1999. *MNRAS*, **307**, 925.
Trümper, J., Pietsch, W., Reppin, C., Voges, W., Staubert, R. & Kendziorra, E., 1978. *ApJ*, **219**, L105.
Usov, V. V. & Melrose, D. B., 1995. *Aust. J. Phys.*, **48**, 571.
Usov, V. V. & Melrose, D. B., 1996. *ApJ*, **464**, 306.
van Kerkwijk, M. H., 1996. In: *Pulsars: Problems and Progress, IAU Colloquium 160*, p. 489, eds Johnston, S., Walker, M. A. & Bailes, M., Astronomical Society of the Pacific, San Francisco.
van Straten, W., 2003. *PhD thesis*, Swinburne University of Technology.
van Straten, W., Bailes, M., Britton, M., Kulkarni, S. R., Anderson, S. B., Manchester, R. N. & Sarkissian, J., 2001. *Nature*, **412**, 158.
van Vleck, J. H. & Middleton, D., 1966. *Proc. I.E.E.E*, **54**, 2.
Vivekanand, M. & Narayan, R., 1981. *J. Astrophys. Astr.*, **2**, 315.
Vivekanand, M., Ables, J. & McConnell, D., 1998. *ApJ*, **501**, 823.
Vivekanand, M., Narayan, R. & Radhakrishnan, V., 1982. *J. Astrophys. Astr.*, **3**, 237.
Voûte, J. L. L., Kouwenhoven, M. L. A., van Haren, P. C., Langerak, J. J., Stappers, B. W., Driesens, D., Ramachandran, R. & Beijaard, T. D., 2002. *A&A*, **385**, 733.
von Hoensbroech, A. & Xilouris, K. M., 1997. *A&AS*, **126**, 121.
von Hoensbroech, A., Kijak, J. & Krawczyk, A., 1998. *A&A*, **334**, 571.
Wallace, P. T., Peterson, B. A., Murdin, P. G., Danziger, I. J., Manchester, R. N., Lyne, A. G., Goss, W. M., Smith, F. G., Disney, M. J., Hartley, K. F., Jones, D. H. P. & Wellgate, G. W., 1977. *Nature*, **266**, 692.
Walker, M. A., 2001. *Ap&SS*, **278**, 149.
Wardle, J. & Kronberg, P., 1974. *ApJ*, **194**, 249.
Weaver, H. & Williams, D. R. W., 1973. *A&AS*, **8**, 1.
Weisberg, J. M., 1996. In: *Pulsars: Problems and Progress, IAU Colloquium 160*, p. 447, eds Johnston, S., Walker, M. A. & Bailes, M., Astronomical Society of the Pacific, San Francisco.
Weisberg, J. M. & Taylor, J. H., 2002. *ApJ*, **576**, 942.
Weisberg, J. M. & Taylor, J. H., 2003. In: *Radio Pulsars*, p. 93, eds Bailes, M., Nice, D. J. & Thorsett, S. E., Astronomical Society of the Pacific, San Francisco.
Weisberg, J. M. & Taylor, J. H., 2004. In: *Binary Pulsars*, in press, eds Rasio, F. & Stairs, I. H. Astronomical Society of the Pacific, San Francisco.
Weisberg, J. M., Boriakoff, V. & Rankin, J., 1979. *A&A*, **77**, 204.
Weisberg, J. M., Romani, R. W. & Taylor, J. H., 1989. *ApJ*, **347**, 1030.
Wex, N., 1995a. *PhD thesis*, Jena University, Jena, Germany.
Wex, N., 1995b. *Class. Quantum Grav.*, **12**, 983.

Wex, N. 1997a. Private communication (Implementation of radial velocities in TEMPO).

Wex, N., 1997b. *A&A*, **317**, 976.

Wex, N., 1998. *MNRAS*, **298**, 997.

Wex, N. 1999. Private communication (Implementation of Laplace-Lagrange parameters in TEMPO).

Wex, N., 2000. In: *Pulsar Astronomy - 2000 and Beyond, IAU Colloquium 177*, p. 113, eds Kramer, M., Wex, N. & Wielebinski, R., Astronomical Society of the Pacific, San Francisco.

Wex, N., 2001. In: *Gyros, Clocks, Interferometers: Testing Relativistic Gravity in Space, Notes in Physics, vol. 562*, p. 381, eds C. Lämmerzahl, C. E. & Hehl, F.

Wex, N. & Kopeikin, S., 1999. *ApJ*, **513**, 388.

Wex, N., Kalogera, V. & Kramer, M., 2000. *ApJ*, **528**, 401.

Wheaton, W. A., Doty, J. P., Primini, F. A., Cooke, B. A., Dobson, C. A., Goldman, A., Hecht, M., Howe, S. K., Hoffman, J. A. & Scheepmaker, A., 1979. *Nature*, **282**, 240.

Wijnands, R. & van der Klis, M., 1998. *Nature*, **394**, 344.

Will, C., 2001. *Living Reviews in Relativity*, **4**, 4.

Wolszczan, A., 1991. *Nature*, **350**, 688.

Wolszczan, A., 1994. *Science*, **264**, 538.

Wolszczan, A. & Cordes, J. M., 1987. *ApJ*, **320**, L35.

Wolszczan, A., Cordes, J. M. & Stinebring, D. R., 1984. In: *Millisecond Pulsars*, p. 63, eds Reynolds, S. P. & Stinebring, D. R., NRAO, Green Bank.

Wolszczan, A. & Frail, D. A., 1992. *Nature*, **355**, 145.

Xilouris, K. M., 1991. *A&A*, **248**, 323.

Xilouris, K. M., Kramer, M., Jessner, A., von Hoensbroech, A., Lorimer, D., Wielebinski, R., Wolszczan, A. & Camilo, F., 1998. *ApJ*, **501**, 286.

Xilouris, K. M., Kramer, M., Jessner, A., Wielebinski, R. & Timofeev, M., 1996. *A&A*, **309**, 481.

Young, M. D., Manchester, R. N. & Johnston, S., 1999. *Nature*, **400**, 848.

Zavlin, V. E. & Pavlov, G. G., 2002. In: *WE-Heraeus Seminar on Neutron Stars, Pulsars and Supernova Remnants. MPE Report 278.*, p. 263, eds Becker, W., Lesch, H. & Truemper, J., Max-Plank-Institut für extraterrestrische Physik.

Zavlin, V. E., Pavlov, G. G., Sanwal, D., Manchester, R. N., Trümper, J., Halpern, J. P. & Becker, W., 2002. *ApJ*, **569**, 894.

Zavlin, V. E., Pavlov, G. G., Sanwal, D. & Trümper, J., 2000. *ApJ*, **540**, L25.

Zhang, L. & Cheng, K. S., 2003. *A&A*, **398**, 639.

Zhang, B. & Harding, A. K., 2000. *ApJ*, **535**, L51.

Zhang, B. & Qiao, G. J., 1996. *A&A*, **310**, 135.

Index